1·13·95

Elementary Geometry

Elementary Geometry

John Roe

Jesus College, Oxford

Oxford   New York   Tokyo
OXFORD UNIVERSITY PRESS
1993

Oxford University Press, Walton Street, Oxford OX2 6DP

Oxford   New York   Toronto
Delhi   Bombay   Calcutta   Madras   Karachi
Kuala Lumpur   Singapore   Hong Kong   Tokyo
Nairobi   Dar es Salaam   Cape Town
Melbourne   Auckland   Madrid
and associated companies in
Berlin   Ibadan

Oxford is a trade mark of Oxford University Press

Published in the United States
by Oxford University Press Inc., New York

A catalogue record for this book is available from the British Library

Library of Congress Cataloging in Publication Data
Roe, John, 1959–
Elementary geometry | John Roe. – 1st ed.
1. Geometry.   I. Title.
QA453.R66   1993   516.2—dc20   92-41660
CIP

ISBN 0-19-853457-4
ISBN 0-19-853456-6 (pbk.)

Typeset by the author
using LaTeX
Printed in Great Britatin by
Biddles Ltd, Guildford and King's Lynn

Preface

This is a textbook for a course on geometry for students of mathematics and physics. It aims to consolidate the often somewhat diverse knowledge that students may have of geometry, and then to extend their knowledge in directions suggested by modern mathematical research. As background the reader needs to have studied or to be studying concurrently a first course in linear algebra going as far as determinants. Later in the book we use some ideas from calculus of several variables including Green's theorem; but a reader who does not have this background should be able to make it to Chapter 11 without too much trouble. The asides addressed to those who are familiar with topology or group theory can similarly be ignored by those who are not.

I take it that the reader will have studied some geometry already. This will almost certainly have been what is called *Euclidean* geometry, whether it was discussed in Euclid's own language of points, lines, and parallels, or in the more modern language of vectors and transformations. One of the most surprising discoveries in the history of mathematics was that there are alternatives to this structure that seems so natural and inevitable. Several such *non-Euclidean geometries* are briefly introduced in this book. Its heart, however, is Euclidean geometry, which forms our intuition and which has proved (depending on your point of view) so enormously fruitful as a mathematical system, or so astonishingly accurate as a physical theory.

The first four chapters of the book therefore give a rapid introduction to Euclidean geometry, in two stages. Chapters 1 and 2 discuss a stripped-down version of Euclidean geometry called *affine* geometry, which has parallel lines but not circles. Chapters 3 and 4 add extra structure to arrive at Euclidean geometry proper. Within each pair of chapters, the first takes a more classical approach, starting from fundamental properties of parallelism, measurement, and so on; the second takes a more algebraic approach, taking the properties of vectors as fundamental. Readers will have different backgrounds, and these chapters should make it clear that there are several ways to set up the same geometric system. For the rest of the book, however, we use only the algebraic approach.

In Chapter 5 coordinates are introduced, along with the fundamental idea of Descartes and Fermat that a curve may be described by an equation relating the coordinates. It is their work which first made it possible to solve geometric problems by algebraic methods. Chapter 6 discusses special topics in plane Euclidean geometry, some classical (circle theorems, construction problems), some less so (oriented angles, inversive geometry). Chapter 7 studies plane curves, especially conics, in more detail. Here we meet the important idea of simplifying the curve by choosing an appropriate coordinate system. We also discuss higher-degree curves and their relation to problems like Fermat's last theorem and modern developments in algebra.

In Chapter 8 we use vector methods to study solid geometry, especially the vector product which is special to dimension three. We use the vector product to investigate lines, planes, and rotations, and to give an account of Hamilton's *quaternions*, which historically are the source of vector algebra and are still important in connection with 'spin' in quantum mechanics. Chapter 9 discusses area and volume from an axiomatic point of view; we learn about the famous 'method of exhaustion' which is one of the triumphs of Greek mathematics and we relate it to the modern theory of integration. Chapter 10 discusses quadrics, the simplest surfaces in three dimensions. The geometry of quadrics is intimately related to the algebra of symmetric matrices, and we give a direct proof of the fundamental diagonalization theorem for such matrices based on Rayleigh's principle. Chapters 11 and 12 are about differential geometry, for curves in Chapter 11 and for surfaces in Chapter 12. Slightly more mathematical technique is needed in these last two chapters, but the rewards are great, culminating in the isoperimetric inequality and the Gauss–Bonnet theorem relating the geometry and the topology of a closed surface.

Each chapter ends with a set of exercises, offering the student a chance to check his or her understanding. Hints or solution are provided for most (though not all!) of the exercises, at the end of the book. I am grateful to Oxford University Press for permission to reproduce a number of Oxford examination questions among these exercises. I would also like to thank Sir Ernst Gombrich and John Stillwell for their permission to reproduce the quotations appearing on pages 46 and 94.

This book originated from the notes of lectures given to first-year students at Oxford. I would like to express my gratitude to all the students and colleagues who have commented on the text, and in particular to David Acheson, William Appleby, Jonathan Frank, Jerry Kaminker, Giles Keen, Frances Kirwan, Raj Mody, and John Stillwell. I am also grateful to all my teachers, and here there is one name that I would wish to single out. My father, Michael Roe, introduced me to geometry when I was about eight years old. This book is dedicated to him.

Contents

1
Axioms for geometry

1.1 What is geometry?

Cut out a large triangle from a big sheet of paper. Not a right-angled triangle or an isosceles triangle or any special kind; just a 'general' triangle is what is needed. Label its corners A, B, and C, with AC being the longest side. Arrange the triangle so that AC is horizontal.

Measure the sides AB and BC and find their midpoints P and Q. Now fold in the corners of the triangle as follows. First fold B over the horizontal line PQ so that it lies on the side AC. Next, fold A over a vertical line through P, and finally fold C over a vertical line through Q. If you have done your construction accurately, you should find that the three folded angles A, B, and C now fit together exactly to give a straight line (see Figure 1.1).

Congratulations! You have now proved that the sum of the angles of a triangle is 180 degrees. Or have you?

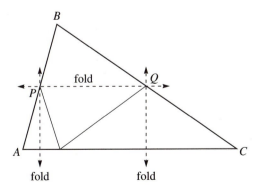

Fig. 1.1. A 'proof' that the angles of a triangle add up to 180 degrees.

There are two or three things that might worry you if you think hard about this supposed proof. First of all, if it works at all it only works for the one triangle that you cut out of the paper. It doesn't say anything about all the other triangles that you might have cut out instead. Of course, if you try it for three or four different triangles and it works for all of them, you might suspect that it will always work. But it's hard to be sure. Maybe the triangles that you tried were special in some way! And however many you did try, there would always be infinitely many more that you did not.

But, second, can you be so certain that the proof works even for one triangle? Surely you would have noticed if there was a gap of 10 degrees when you fitted the angles together. But would you have noticed a gap of 1 degree? of $\frac{1}{10}$ degree? of $\frac{1}{100}$ degree? The best you could say would be that the angles added up to 180 degrees within the limits of accuracy of your observation. Moreover, even if the angles seemed not to fit together properly, you would be unlikely to conclude that the theorem that the angle sum is 180 was wrong. Instead, you would look for some other cause of error: for instance, you might wonder whether you had really cut the sides straight. This is rather mysterious, since it suggests that you have some idea in your head of an 'ideal' triangle, to which the paper triangle is just an approximation. So how could approximate results about approximate paper triangles ever prove exact results about these 'ideal' triangles?

A third question might also worry you. How do you know that the angles of the triangle do not change when you fold the paper over? The whole proof depends on the assumption that the angles A, B, and C are the same when they are sitting at the three corners of the triangle as they are when they are folded to the centre. Of course this seems reasonable (as long as you don't tear the paper); but it wouldn't be so reasonable if the paper was replaced by a sheet of bubble gum. It seems an unavoidable conclusion that to our idea of an 'ideal' triangle we must add the idea of an 'ideal motion' or *congruence* of such a triangle, moving it from one place to another while preserving lengths and angles.

Even to formulate questions of this kind is a considerable achievement, and it is natural that when we first study geometry we do not worry too much about them. Certainly the ancient Egyptians, who as far as we know were the first people to 'do geometry', seem not to have been concerned by them. It is often said that, in Egypt, geometry met the need for an accurate surveying system whereby the ownership of land in the fertile valley of the Nile could be traced after boundary markers had been washed away by the annual floods, and this is perhaps confirmed by the literal meaning of the word 'geometry', which is 'earth measurement'. The historian Herodotus relates that in 1300 BC 'if a man lost any of his land by the annual overflow of the Nile, he had to report the loss to Pharaoh who would then send an overseer to measure the loss and make a proportionate abatement of the tax'.

If we want to make progress in geometry, though, we must eventually face questions like those raised above, and the first people to do this seem to have been the Greeks. They made two vital contributions to geometry. The first is that they made geometry *abstract*. For the Egyptians, a 'line' perhaps meant the line of a stretched cord, a 'point' a peg in the ground. Greek thinkers took seriously the fact that no physical peg or pencil mark can ever be more than an approximation to what we mean by a point (for it will always have some size) and similarly no physical cord can ever be more than an approximation to what we mean by a line (for it will never be completely straight). They introduced the idea of considering 'idealized' points and lines, inhabitants of some abstract space to which our physical space merely aspires. The laws of geometry apply with perfect accuracy only to this idealized world; there is always some approximation involved in applying them to messy physical reality. In the language of modern applied mathematics we might say that geometry provides a 'mathematical model' which can apply with greater or lesser accuracy to a given physical situation. The Greek philosopher Plato (*c.* 400 BC) put the emphasis rather differently. 'The objects of geometric knowledge are eternal', he wrote; 'geometry compels us to contemplate true reality rather than the realm of change'.

But how was one to obtain any information about this idealized world? The Greeks had an answer to this question, and that answer was their second great contribution: they made geometry *deductive*. The way to find out about geometric truth, they said, was to build up one's conclusions systematically on the basis of unquestionable premises or *axioms*. Starting from things which nobody could deny, and basic laws of thought, they would reason and prove their way towards previously unguessed knowledge. This whole process was codified by Euclid around 300 BC, in his book the *Elements*, the most successful scientific textbook ever written. Euclid, who was a teacher at what we might now call the University of Alexandria, brought together many of his own and his predecessors' results in this wonderful book. In it, on the basis of five 'postulates' and five 'common notions', he deduced step by step the whole of geometry as it was known in his day. Euclid's work was so successful that for hundreds of years it was regarded as the supreme example of human reasoning. Philosophers who wished to commend the rationality of their systems of thought used to say that they were built up *more geometrico* — in the (Euclidean) geometric manner.

Here is an outline of the way Euclid's deductive system worked. He began with two kinds of fundamental concept:

- *Undefined terms* are things like 'point', 'line' and so on. These are the basic building blocks of geometry. Obviously we cannot define everything[1] so we have to begin somewhere, and it is only sensible to

---

[1]Perhaps this is not so obvious; at any rate, it appears that Euclid himself tried to define

begin with the most basic and elementary terms.

- *Axioms* (also known as *postulates*) are the basic assumptions about the terms of geometry. For instance, 'Through any two points there is exactly one line'. A more famous example is the 'parallel axiom': Given any line $\mathcal{L}$ and any point P, there exists exactly one line through P parallel to $\mathcal{L}$.

From this starting point he went on to build up systematically:

- *Defined terms* such as 'isosceles triangle': an isosceles triangle is *defined* to be one that has two sides of equal length.

- *Theorems*[2] such as 'The base angles of an isosceles triangle are equal' or 'The perpendicular bisectors of the sides of a triangle meet at a point'. Each theorem has to be accompanied by a *proof*, which gives the reason why it is true in terms of previous definitions, theorems, and axioms.

A deductive system of this kind is also known as an *axiomatic system*. Euclid's work has had such influence that some would say mathematics is nothing more than the study of axiomatic systems.

The force of the axiomatic method is this: if you believe the axioms, you have to believe the theorems too. There is an amusing illustration of this principle in John Aubrey's *Brief Life* of the philosopher Thomas Hobbes (1588–1679):

He was 40 yeares old before he looked on Geometry; which happened accidentally. Being in a Gentleman's Library, Euclid's Elements lay open, and 'twas[3] the *47 El. libri 1*. He read the Proposition. *By G—*, sayd he (he would now and then sweare an emphaticall Oath by way of emphasis) *this is impossible!* So he reads the Demonstration of it, which referred him back to such a Proposition; which proposition he read. That referred him back to another, which he also read. *Et sic deinceps* [and so on] that at last he was demonstratively convinced of that trueth. This made him in love with Geometry.

---

everything. The first definition in the *Elements* says, 'A point is that which has no parts'. But what is a *part*? Euclid doesn't say, and since he makes no further mention of the concept in this context, he leaves the term 'point' practically undefined.

[2] Also known as propositions, corollaries, or lemmas, among other things. There is a slight distinction among these terms: *theorems* are really first-class results, *propositions* are more run-of-the-mill, *corollaries* are subsidiary results easily deduced from a theorem, and *lemmas* are auxiliary results used in the proof of a major proposition or theorem. Which name you use for any particular result is largely a matter of taste.

[3] This was Euclid's proof of the theorem of Pythagoras.

Here is the deductive method working to perfection; Hobbes believes the axioms, and so (eventually) he has to believe the theorems too. But why should he believe the axioms? I suppose that the old geometers would have answered that you 'just see' that they are true. But eventually this simple point of view had to be modified, and some modern mathematicians would say that the axioms are no more than 'rules of the game'; you can change them if you want, but then you are playing a different game. What do you think?

In this book we will follow the Greek deductive method. However, the axioms we use will be slightly different from Euclid's. Our axioms are designed to build on our knowledge of algebra, which is a subject of which the Greeks knew very little. By using this knowledge, we can make more rapid progress than Euclid did: he more or less had to invent some parts of algebra as he went along.

Hobbes, incidentally, would not have approved. Aubrey goes on:

> He would often complain that Algebra (though of great use) was too much admired, and so followed after, that it made men not contemplate and consider so much the nature and power of Lines, which was a great hinderance to the Groweth of Geometrie; for that though algebra did rarely well and quickly, and easily in right lines, yet 'twould not *bite* in *solid* Geometry.

The algebra that Hobbes might have been looking for which will 'bite' in solid geometry is of course the algebra of vectors. We will develop vector algebra in this book.

There are two ways of looking at vector algebra. One way is to see its rules as convenient summaries of geometrical facts which have been proved by Euclidean methods. The other is to regard its rules as fundamental axioms in themselves, and derive the rest of geometry from them. Whichever approach you favour, you will be able to use this book. In the remainder of this chapter we will explore the first approach, deriving the rules of vector algebra from Euclidean-style geometric axioms. In Chapter 2 and in most of the rest of the book, our study will be based on vectors. Thus, if you are happy to take the rules of vector algebra as fundamental, you can omit the rest of this chapter and go straight to the beginning of Chapter 2.

1.2 The axioms of incidence

We will introduce seven undefined terms in all as we develop geometry: *point, line*[4], *plane, ruler, distance, area,* and *volume.* We will also use the word *space* to mean the set of all points. Everything else will be defined in terms

---

[4]This means a *straight* line, but we will not repeat the word 'straight' every time.

of these. In this section we will deal with the axioms which just involve the first three of these terms (point, line, and plane), and which do not involve any quantitative measurement.

Classical geometry provides a rich vocabulary to describe the various possible relationships between points, lines, and planes. For example, a point *P* may *lie on* the line $\mathcal{L}$, or the line $\mathcal{L}$ may *pass through* or *meet* or *contain* the point *P*. These all mean the same thing, which in the notation of set theory is written $P \in \mathcal{L}$. Similarly a line may *lie in* a plane, two lines may *meet at* a point, and so on. It helps to have all these different terms available, though from a severely abstract point of view they are only picturesque ways of saying whether two sets intersect or not. A catch-all term for these relationships is 'incidence relations', and the axioms in this section, which deal with the properties of incidence relations, are often known as the *axioms of incidence*. For each axiom, I suggest that you draw some sketches and convince yourself that the axiom is true in your intuitive picture of geometry. This process should not be too hard in the case of the first of our axioms:

AXIOM I (LINE AXIOM): Through any two distinct points there is exactly one line.

If *P* and *Q* are points, this allows us to introduce the notation $\overleftrightarrow{PQ}$ for the unique line containing both *P* and *Q*. The two-sided arrow is meant to remind you that the line extends as far as you like past *P* and *Q* in both directions. Later on, we will distinguish $\overleftrightarrow{PQ}$ from PQ, which will refer just to the segment of $\overleftrightarrow{PQ}$ that is between *P* and *Q*.

Even from this single axiom we can derive a consequence which is worth noticing:

(1.2.1)   **Proposition:** *Two distinct lines can meet in at most one point.*

Proof: If they met at two points, say P_1 and P_2, then there would be at least two lines through P_1 and P_2, contrary to the line axiom. □

Of course, it is perfectly possible for two lines not to meet at all. In intuitive terms they can either be *parallel* (in the same direction but never meeting) or *skew* (in different directions but not in the same plane). We will have to discuss parallel lines carefully later on.

Before we introduce any further axioms it may be helpful to explain a modern way of thinking about axiomatic systems in general. This way of thinking emphasizes the fact that the basic terms of the axiomatic system are *undefined*. They could mean anything! Until we know what they mean, there is no point in asking whether or not the axioms are *true*. Suppose that alien beings have landed on Earth by flying saucer, and their leader tells you that

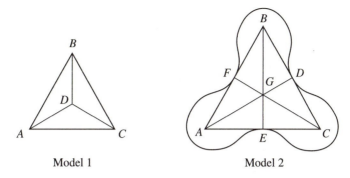

Fig. 1.2. Models for the line axiom.

Through any two distinct blurgs there is exactly one phogon.

Unless you know what a blurg and a phogon are, you will have no way of telling whether or not this statement is true. In the same way, there may be many different *interpretations* of the undefined terms 'point' and 'line' in an axiomatic system for geometry. Some interpretations will make all the axioms true and some will not. An interpretation which makes all the axioms true is called a *model* for the axiomatic system; because the theorems are all deduced logically from the axioms, they will be true in a model as well.

(1.2.2)   **Example:** Here are two different models for the single axiom that we have introduced so far. They are much smaller than the usual model we have in mind for geometry; in fact, Model 1 has just four points and Model 2 has seven. Because they have only finitely many points, we can describe them just by listing all the points and lines in a table, as follows:

	Points	Lines
Model 1	A, B, C, D	AB, AC, AD, BC, BD, CD
Model 2	A, B, C, D, E, F, G	$AFB, BDC, CEA, AGD,$ BGE, CGF, DEF

The models may be easier to visualize if you draw a diagram like Figure 1.2; but remember that only the labelled points are really part of the model. The rest of the diagram is just to help you see what is going on. It is easy to check that in both models the line axiom is satisfied.

Such models are called *finite geometries*, and their study is an active area of mathematics in its own right, with applications to the design of scientific experiments and error-correcting codes for computers. We will not pursue it any further in this book, however.

We see, therefore, that an axiomatic system may have many different models. Is it possible for an axiomatic system to have *no* models? Certainly it is. Suppose that the alien leader now tells you

> There are two distinct phogons through the same two blurgs.

You know that it is lying to you. Its statements are *inconsistent* with each other and so there is no possible model that will make both of them true. In general we can define an axiomatic system to be *consistent* if it has a model. The existence of a model guarantees that no contradiction can be deduced from the axioms.

Is geometry consistent?   That is now a natural question to ask. It is perfectly reasonable to answer, 'Yes, because I can see a model with the inner eye of mathematical intuition'. That would be a version of the Greek idea; the points and lines of Plato's ideal world form the model which guarantees consistency. Another answer might be to try to construct a model for geometry out of some other mathematical objects. In Chapter 2 we will construct such a model in terms of the real number system. But of course this just pushes the consistency question one stage further back. In the 1930s the logician Gödel proved that a watertight guarantee of consistency can never be given. It seems that mathematicians, too, walk by faith.

Setting aside these very general issues, let us introduce some more geometrical axioms. We say that three or more points are *collinear* if there is a line through all of them. In terms of this concept we can state the analogue of the line axiom for planes:

AXIOM 2 (PLANE AXIOM): Through any three noncollinear points there is exactly one plane.

Part of our intuition about geometry is that lines are long, planes are wide, and space is not flat. This is expressed mathematically by the next axiom:

AXIOM 3 (DIMENSION AXIOM): Any line contains at least two distinct points; and any plane contains at least two distinct lines. Moreover, there are at least two distinct planes.

We call this the 'dimension axiom' because what it says is essentially that lines are one-dimensional, planes are two-dimensional, and space is at least three-dimensional. Finally, we need an axiom about how lines and planes meet one another:

AXIOM 4 (LINE–PLANE INTERSECTION AXIOM): If two distinct points of a line lie in some plane $\mathcal{P}$, then the whole line lies in $\mathcal{P}$.

(1.2.3) **Proposition:** *Given a line $\mathcal{L}$, and a point $P \notin \mathcal{L}$, there is a unique plane containing both $\mathcal{L}$ and P.*

Proof: By the dimension axiom, there are at least two distinct points Q and R on $\mathcal{L}$. The three points P, Q, and R are not collinear, since the unique line $\mathcal{L}$ through Q and R does not pass through P. So by the plane axiom, there is a unique plane $\mathcal{P}$ containing P, Q, and R. Then $\mathcal{P}$ contains two points of $\mathcal{L}$, namely Q and R, so it contains the whole line $\mathcal{L}$ by the line–plane intersection axiom. $\square$

We have not yet arrived at the most famous and controversial axiom of Euclidean geometry. To state it, we must first make a definition:

(1.2.4) **Definition:** *Two lines are* coplanar *if there is a plane that contains both of them. They are* parallel *if they are coplanar and do not meet.*

We will also say that a line is parallel to itself. This is just a matter of convention, in the same way that it is a matter of convention whether you regard 0 as a natural number or not. The convention of saying that a line is parallel to itself makes some things easier later on.

AXIOM 5 (PARALLEL AXIOM): Let $\mathcal{L}$ be a line and P be a point. Then there is one and only one line that passes through P and is parallel to $\mathcal{L}$.

The form of the parallel axiom given above is also sometimes called *Playfair's axiom.* Euclid in fact stated the parallel axiom in a different, although equivalent, form.

The parallel axiom was a source of controversy for over two thousand years. It was not that people disbelieved it. But they felt that it was not sufficiently obvious to qualify as a fundamental and incontrovertible truth, which is what axioms were supposed to be. In particular, one difference between the parallel axiom and the other axioms is that the former involves the idea of *infinity* in a rather important way. One has to talk about arbitrarily long lines to make sense of the parallel axiom; if you know that two lines aren't parallel, you still have no idea how far you may have to trace along them before they actually meet. For the other axioms one can get away with thinking about line *segments* (the finite pieces of line between two points)

rather than the infinitely long lines themselves. Now Greek art tended to cel-
ebrate the compactness and orderliness of the finite, rather than the openness
and transcendental possibilities of the infinite — think, for example, of the
Parthenon as compared to a Gothic cathedral — and it has been suggested
that the way the parallel axiom involves ideas of the infinite made the Greeks
doubtful about accepting it as basic. If this was the case, they showed mathematical wisdom, since many errors in mathematics arise from the unthinking
generalization to the infinite of what is known to be true for the finite.

Be that as it may, a succession of geometers undertook to do without the
parallel axiom. They tried to do this in several ways. Some attempted to
prove the offending axiom as a theorem on the basis of Euclid's other axioms,
or on the basis of these axioms augmented by some extra ones supposed to
be more apparent than the parallel axiom itself. Others proposed to work by
contradiction: starting from the assumption that the parallel axiom is false,
they would deduce successively more bizarre results in the hope of arriving
at an outright contradiction. These investigations were pursued by Greek,
Arabic, and Western European mathematicians over many years[5], and their
final solution in the nineteenth century is one of the great surprises in the
history of mathematics.

The solution was that one *cannot* do without the parallel axiom. Suppose
that we followed the second approach suggested above, replacing the parallel
axiom by some contrary axiom, for instance that through any point not on a line
$\mathcal{L}$ there are *at least two* lines parallel to $\mathcal{L}$. We will then arrive at a succession
of theorems, bizarre indeed to the Euclidean mind, such as that the sum of the
angles of any triangle is less than 180°, or that similar triangles are congruent.
But *we will never reach a contradiction*. In fact, as was discovered at the
beginning of the nineteenth century by Lobatchewsky, Bolyai, and Gauss,
what we are doing is proving the theorems of an *alternative geometry* —
one just as valid as Euclid's. Which geometry applies to our physical space
is a matter of physics (Lobatchewsky suggested that one could check it by
astronomical observations); which geometry we study mathematically is up
to us, since one is no more and no less consistent than the other.

The discovery of this new geometry had profound philosophical implica-
tions. Remember that Euclid's work was viewed as the paradigm of deductive
reasoning, and Euclidean geometry as an example of incontrovertible truth.
Now, it seemed, there was an alternative to this 'incontrovertible truth'. How
could such a thing be? Was reason not sufficient to arrive at truth, even in
so rational a subject as mathematics? The modern answer would be that
reason has to have something to reason about, and that of course if you start
from different axioms you will reach different conclusions; but this answer

---

[5]The book by Gray, [19], contains a very readable account of the history of the attempts made
on the parallel axiom.

presupposes a way of thinking which itself is partly a consequence of the discovery of non-Euclidean geometry.

For the rest of this book we will accept the parallel axiom, as Euclid did, and go on to work out its consequences. Euclidean geometry is indeed now only one among many possible geometries; but it is one of the most important.

(1.2.5)   **The transitivity of parallelism:** Parallelism is a relation between straight lines. This relation is *reflexive* (every line is parallel to itself) and *symmetric* (if $\mathcal{L}_1$ is parallel to $\mathcal{L}_2$, then $\mathcal{L}_2$ is parallel to $\mathcal{L}_1$). Our intuition suggests that the relation is also *transitive*: if $\mathcal{L}_1$ is parallel to $\mathcal{L}_2$ and $\mathcal{L}_2$ is parallel to $\mathcal{L}_3$, then $\mathcal{L}_1$ is parallel to $\mathcal{L}_3$. The parallel axiom allows us to prove this result, at least in the plane.

(1.2.6)   **Proposition:** *Let $\mathcal{L}_1$, $\mathcal{L}_2$, and $\mathcal{L}_3$ be three coplanar lines. If $\mathcal{L}_1$ is parallel to $\mathcal{L}_2$ and $\mathcal{L}_2$ is parallel to $\mathcal{L}_3$, then $\mathcal{L}_1$ is parallel to $\mathcal{L}_3$.*

Proof: Suppose $\mathcal{L}_1$ and $\mathcal{L}_3$ are *not* parallel. Since they are coplanar, the only way they can fail to be parallel is if they are distinct and meet at some point, P say. But then there are two lines through P parallel to $\mathcal{L}_2$ (namely $\mathcal{L}_1$ and $\mathcal{L}_3$), and this contradicts the parallel axiom. $\square$

It seems reasonable to suppose that this result also holds for parallel lines not necessarily in the same plane. To show this, though, we need to introduce one further axiom — the last of the axioms of incidence.

AXIOM 6 (PLANE–PLANE INTERSECTION AXIOM): If two distinct planes meet, then their intersection is a line.

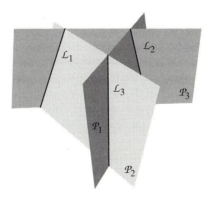

Fig. 1.3. Intersection of three planes.

(1.2.7) **Lemma:** *Suppose three distinct planes are given, each of which intersects the other two. Then their lines of intersection either all meet at a point, or are all parallel to one another.*

Proof: (See Figure 1.3.) Let $\mathcal{P}_1$, $\mathcal{P}_2$, and $\mathcal{P}_3$ be the three planes, and let $\mathcal{L}_1 = \mathcal{P}_2 \cap \mathcal{P}_3$, $\mathcal{L}_2 = \mathcal{P}_3 \cap \mathcal{P}_1$, and $\mathcal{L}_3 = \mathcal{P}_1 \cap \mathcal{P}_2$ be their lines of intersection.

The lines $\mathcal{L}_1$ and $\mathcal{L}_2$ are in the same plane $\mathcal{P}_3$, so they either meet or are parallel.

If they *do* meet, say at a point O, then O belongs to the plane $\mathcal{P}_2$ (since it lies on $\mathcal{L}_1$), and similarly it belongs to the plane $\mathcal{P}_1$ (since it lies on $\mathcal{L}_2$). Therefore, O lies on the line $\mathcal{L}_3$ of intersection of these two planes; so the three lines meet at O.

If they *do not* meet but are parallel, then line $\mathcal{L}_3$ cannot meet $\mathcal{L}_1$; for if it did, then their point of intersection would have to be common to all three lines, by the previous case. Since $\mathcal{L}_1$ and $\mathcal{L}_3$ are coplanar and do not meet, they are parallel. The same argument shows that $\mathcal{L}_2$ and $\mathcal{L}_3$ are parallel. $\square$

(1.2.8) **Proposition:** *Let $\mathcal{L}_1$, $\mathcal{L}_2$, and $\mathcal{L}_3$ be any* three lines. *If $\mathcal{L}_1$ is parallel to $\mathcal{L}_2$ and $\mathcal{L}_2$ is parallel to $\mathcal{L}_3$, then $\mathcal{L}_1$ is parallel to $\mathcal{L}_3$.*

Proof: If the three lines are coplanar, then proposition 1.2.6 applies. Otherwise, there is a point P on $\mathcal{L}_3$ that does not lie in the plane $\mathcal{P}_3$ of $\mathcal{L}_1$ and $\mathcal{L}_2$. Let $\mathcal{P}_1$ be the unique plane containing P and $\mathcal{L}_1$ and let $\mathcal{P}_2$ be the unique plane containing P and $\mathcal{L}_2$. Let $\mathcal{L}$ be the line of intersection of $\mathcal{P}_1$ and $\mathcal{P}_2$.

Since $\mathcal{L}_1$ and $\mathcal{L}_2$ are parallel, lemma 1.2.7 shows that the line $\mathcal{L}$ is parallel to each of them. Since it passes through P and is parallel to $\mathcal{L}_2$, it must be equal to the line $\mathcal{L}_3$ which also has this property, by the parallel axiom. Thus $\mathcal{L}_3 = \mathcal{L}$ is parallel to $\mathcal{L}_1$ also. $\square$

1.3 Measurement axioms

Euclid's system did not give a high profile to actual *numbers*. For instance, Euclid's statement of the Pythagorean theorem ultimately amounts to the assertion that the square drawn on the hypotenuse can be dissected into pieces which can then be reassembled to form the two squares on the other two sides. There is no mention of numbers representing the lengths or areas. In fact, *algebra* in contrast to geometry was not worked out in Greece, and its development in the Western world had to await contact with Islamic thought in the Middle Ages.

As usual, there was a reason for the Greek attitude. They took the view that the only things that could properly be called *numbers* were the whole numbers $1, 2, 3, \ldots$ and the fractions or *rational numbers* $\frac{m}{n}$ derived from them. Quantities like the (ratios of) lengths of lines, which they called *proportions*, were considered to be of quite a different nature. They knew,

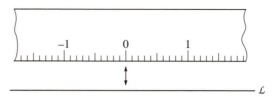

Fig. 1.4. A ruler on a line

in fact, that the proportion of the diagonal of a square to its side could not be represented as a fraction $\frac{m}{n}$. We would express this by saying '$\sqrt{2}$ is irrational', but they did not use this kind of language.

In this book, though, it will be assumed that you are familiar with the *real number system*: the number system which includes positive and negative, rational and irrational numbers, but not imaginary numbers like i (the square root of minus 1). It is possible to derive the properties of real numbers from those of other simpler mathematical objects, such as the whole numbers that the Greeks used, and such derivations ultimately go back to the discussion of the theory of proportion in Book V of Euclid; but nowadays we think of them as belonging to analysis rather than to geometry.

Using our knowledge of the real number system, we can state some further axioms. These say that every line in our geometry looks like the standard 'number line' **R**; or to put that a bit more precisely, there is a one-to-one correspondence between the line and **R**, which we can think of as giving a numerical 'coordinate' to each point on the line. To specify such a correspondence, our intuition suggests that we must choose an *origin* and a *scale*: in other words, we must select one point A on the line to have coordinate 0 and another point B to have coordinate 1. Once this has been done, the coordinates of all other points are uniquely determined.

In order to make this intuition precise, we will introduce a new undefined term: a *ruler* on a line $\mathcal{L}$. A ruler on $\mathcal{L}$ is a one-to-one correspondence $\mathcal{L} \to \mathbf{R}$ which should be thought of as giving a 'uniform scale' along $\mathcal{L}$ (see Figure 1.4). The real number corresponding to a point X will be called the *coordinate* of X (relative to the given ruler).

> AXIOM 7 (RULER AXIOM): Let A and B be distinct points, and let $\mathcal{L}$ be the line through A and B. Then there is a ruler on $\mathcal{L}$, in terms of which A corresponds to 0 and B corresponds to 1.

A ruler in terms of which A has coordinate 0 and B has coordinate 1 will be called a ruler *based on* the points A and B.

What would happen if we had selected a different origin and scale? This question brings out the important distinction between the point X itself, which is a geometric object having nothing to do with any choice of ruler, and its coordinate, which is a real number depending not only on X but also on the choice of ruler or 'coordinate system'. As long as we stick to one coordinate system this distinction is blurred; but in more advanced geometry it is often necessary to 'transform the coordinates'. Our present interest is in the one-dimensional case of rulers on a line; but the same principles apply in higher dimensions, and we will return to the subject in Chapter 5.

In this book we will usually represent quantities which depend on a choice of 'coordinate system' by *small* letters. Thus the coordinate of a point X, relative to some ruler, is a real number x. If we are also considering a second ruler, x' will denote the coordinate of X relative to this new ruler. The two rulers are related by a change of origin and a change of scale: we expect, therefore, that the relationship will be of the form

(1.3.1) $$x' = U(x) = lx + m, \qquad (l \neq 0).$$

For example, one could think of the line as a temperature scale, with the first ruler given by centigrade measurement and the second by Fahrenheit measurement; then the *coordinate transformation U* is given by

$$U(x) = \frac{9}{5}x + 32.$$

The next axiom makes explicit our assumption about the relationship between two rulers on the same line.

> AXIOM 8 (RULER COMPARISON AXIOM): Two different rulers on the same line are related by a transformation of the form 1.3.1.

Rulers allow us to talk about the *ratios* of distances between points on a line. We cannot yet talk about distances themselves, because we have not assumed that there is any absolute standard of distance; but we can talk about one pair of points on a given line being twice as far apart (or half as far apart or whatever) as another. More formally, we can make the following definition:

(1.3.2)   **Definition:** *Let A, B, C, and D be points on a line $\mathcal{L}$. We define the ratio[6] AB : CD to be*

$$AB : CD = (a - b) : (c - d)$$

where a, b, c, and d are the coordinates of A, B, C, and D relative to some ruler on $\mathcal{L}$.

---

[6]A *ratio* $x : y$ is just given by a couple of real numbers, not both of which are zero; and two ratios $x : y$ and $x' : y'$ are *equal* if $xy' = yx'$. A ratio is just like a fraction, except that either one of x and y is allowed to be zero.

There is, however, a possible problem hidden in this innocuous-looking definition. We have defined the ratio $AB : CD$ in terms of some particular choice of ruler on the line $\mathcal{L}$. How do we know that if we had chosen some other ruler, we would have got the same result? Only if our concept of ratio does not depend on the choice of ruler can it be said to be truly geometrical. Similar questions arise in mathematics whenever the definition of some concept involves an arbitrary choice. One has to check that the concept is *well-defined*.

Of course, this kind of check is what we have the ruler comparison axiom for. If we work with a second ruler, it tells us that we have

$$a' - b' : c' - d' = U(a) - U(b) : U(c) - U(d)$$
$$= (la + m) - (lb + m) : (lc + m) - (ld + m) = a - b : c - d.$$

So the ratio does not depend on the choice of ruler, and this makes our definition a good one.

(1.3.3)   **Example:** Distance ratios along a line $\mathcal{L}$ satisfy various algebraic relations which are easily proved in terms of coordinates. For instance, if $AB : CD = x : y$ and $AB : DE = x : z$, then $AB : CE = x : y + z$. To give a formal proof of this, introduce a ruler on $\mathcal{L}$. The equation $AB : CD = x : y$ is equivalent to $y(a - b) = x(c - d)$, and the equation $AB : DE = x : z$ is equivalent to $z(a - b) = x(d - e)$. These equations can easily be combined to give

$$(y + z)(a - b) = x(c - e)$$

which is equivalent to $AB : CE = x : y + z$. We will not bother to work out all possible relations of this kind now; later on we will refer to this sort of calculation as 'using the properties of ratios'.

If A and C are thought of as fixed points, the ratio $AB : BC$ determines the point B. This is part of the content of the next proposition:

(1.3.4)   **Proposition:** *Let A and C be fixed points on a line $\mathcal{L}$. Then for every possible ratio $x : y$ except $1 : -1$, there is exactly one point B on $\mathcal{L}$ for which $AB : BC = x : y$. There is no point B for which $AB : BC = 1 : -1$.*

Proof: Choose a ruler relative to which A has coordinate 0 and C has coordinate 1. For a point B with coordinate λ, the ratio $AB : BC$ equals $\lambda : 1 - \lambda$. If we put this equal to $x : y$, we get $y\lambda = x(1 - \lambda)$, which has the unique solution $\lambda = x/(x+y)$ provided that $x + y \neq 0$, i.e. $x : y \neq 1 : -1$. $\square$

Using the properties of ratios, one sees that if $AB : BC = x : y$, then $AB : AC = x : x + y$ and $AC : BC = x + y : y$. Any of these ratios therefore suffices to determine the point B. The ratio $AB : AC$ cannot equal $1 : 0$, and the ratio $AC : BC$ cannot equal $0 : 1$.

An enthusiast for ratios might regard it as a blemish that there is one ratio which does not correspond to any point on the line. One way of remedying this might be to regard the ratio $AB : BC = 1 : -1$ as defining B to be some new kind of 'ideal point' or 'point at infinity' on the line $\mathcal{L}$. It turns out that this can be done, and has many interesting properties; for instance, two parallel lines 'meet at a point at infinity'. This leads to the subject of *projective geometry*, which will be discussed later (Section 2.4).

Distance ratios allow us to define what we mean by a *line segment*:

(1.3.5)   **Definition:** *Let A and C be two distinct points. The* line segment *AC is the set of all those points B on $\overline{AC}$ such that $AB : BC = x : y$, where $x, y \geq 0$.*

If we work with the ruler where A has coordinate 0 and C has coordinate 1, then B has coordinate $x/(x + y)$. If x and y are both non-negative this ranges between 0 and 1; so another way of defining AC is as the set of all those points on $\overline{AC}$ that have coordinates between 0 and 1 in this particular ruler. One refers to such points B as being *between* A and C. If A and C are the same, then we define AC to be the set consisting just of the single point A.

(1.3.6)   **Definition:** *Let AC be a line segment. The* midpoint *of AC is defined to be the unique point B on AC such that $AB : BC = 1 : 1$. (If A and C are the same, then we let $B = A$.)*

Notice that the midpoint of AC is the same as the midpoint of CA. This is necessary for our definition to make sense, since you cannot tell just from the set of points AC which end is A and which end is C.

Rulers allow us to speak of the ratios of distances along a particular line. But they still leave all the lines in our geometry rather isolated from one another; we do not yet have any way to compare distance ratios between two different lines. Our final axiom in this section, the similarity axiom, makes such a comparison possible.

AXIOM 9 (SIMILARITY AXIOM): Let $\mathcal{L}_1$ and $\mathcal{L}_2$ be two parallel lines in a plane, and let $\mathcal{M}$ and $\mathcal{N}$ be two other lines, meeting the parallel lines at points M_1, M_2, N_1, and N_2, and meeting each other at X. Then the ratios $XM_1 : XM_2$ and $XN_1 : XN_2$ are the same.

In the language of classical geometry, XM_1N_1 and XM_2N_2 are *similar triangles* (see Figure 1.5); they have the same shape, even though they do not have the same size. One triangle is a 'scaled' version of the other and so the scale factor between pairs of corresponding sides is the same.

There is a converse to the similarity axiom: if the ratios are the same, the lines are parallel. The proof uses the similarity axiom together with the parallel axiom:

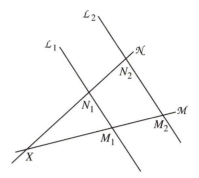

Fig. 1.5. The similarity axiom.

(1.3.7) **Proposition:** *Let M and N be two lines meeting at a point X. Let M_1 and M_2 be points on M, N_1 and N_2 points on N, none of which are equal to X, with $XM_1 : XM_2 = XN_1 : XN_2$. Then the lines $\overline{M_1N_1}$ and $\overline{M_2N_2}$ are parallel.*

Proof: Consider a line L through M_2 parallel to $\overline{M_1N_1}$. Where does it meet N? By the similarity axiom, it meets N at a point N for which $XN_1 : XN = XM_1 : XM_2$. But the point N_2 has this same property, so $N = N_2$. Therefore $\overline{M_2N_2}$ is equal to L and so is parallel to $\overline{M_1N_1}$. □

Remark: There is more to this proof than meets the eye. How do we know that L meets N at all? From the line–plane intersection axiom we know that the whole figure lies in a plane; and we know that L is not parallel to N because $\overline{M_1N_1}$ is not parallel to N and parallelism is transitive. We also used the parallel axiom and the line axiom in the proof; can you see where?

(1.3.8) **Theorem:** (THALES' THEOREM) *Let L_1, L_2, and L_3 be three lines in some plane P, and let M and N be lines in P, not parallel to any of the lines L_i. Let M and N intersect the lines L_i at the points M_i and N_i respectively, for $i = 1, 2, 3$. Then*

- *If the lines L_1, L_2, and L_3 are all parallel, then the ratios $M_1M_2 : M_2M_3$ and $N_1N_2 : N_2N_3$ are equal.*

- *Conversely, if the lines L_1 and L_2 are parallel, and the ratios $M_1M_2 : M_2M_3$ and $N_1N_2 : N_2N_3$ are equal, then the third line L_3 is parallel to L_1 and L_2.*

Proof: Suppose that the three lines are parallel. Let K be the line $\overline{M_3N_1}$. K is not parallel to L_1 (since it meets it but is not equal to it) so it is not parallel to L_2 either (transitivity) and so K must meet L_2, say at a point K. Then apply the similarity axiom twice:

$$M_1M_2 : M_2M_3 = N_1K : KM_3 = N_1N_2 : N_2N_3.$$

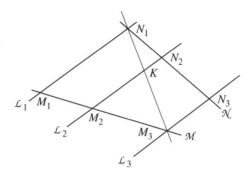

Fig. 1.6. Thales' theorem.

The converse is proved in the same way, using the converse (1.3.7) of the similarity axiom. □

A *transversal* to a system of parallel lines in a plane is a line which is not parallel to them and therefore cuts all of them. Thales' theorem may be restated in the following form: a system of three parallel lines cuts off equal ratios on any transversal (see Figure 1.6).

Thales himself lived around 600 BC and is the earliest Greek geometer whose name has come down to us. He was a merchant, a scientist who made observations on electricity (from amber) and magnetism (from the lodestone), and a traveller, as well as a mathematician. He devised a method of measuring the distances of ships at sea, which presumably must have involved the use of some version of the similarity axiom. On one of his visits to Egypt he is reputed to have measured the height of the Great Pyramid by observing the length of its shadow and comparing that to the length of the shadow of an object of known height. There is no evidence that Thales actually enunciated the theorem which bears his name, but stories like these make it clear that he understood it.

1.4 Parallelograms

(1.4.1)   **Definition:** *We say that four points A, B, C, and D form a* parallelogram *if the midpoint of AC is equal to the midpoint of BD.*

The four points A, B, C, D are called the *vertices* of the parallelogram. We allow the possibility that two or more of them may coincide; but if they do not, then the lines $\overleftrightarrow{AB}$, $\overleftrightarrow{BC}$, $\overleftrightarrow{CD}$, and $\overleftrightarrow{DA}$ are called the *sides* of the parallelogram. The two line segments AC and BD are called the *diagonals* of the parallelogram, and the common midpoint of the two diagonals is called the *midpoint* of the parallelogram.

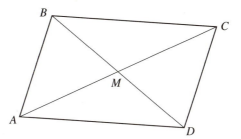

Fig. 1.7. A parallelogram.

If *ABCD* is a parallelogram, then cyclic reorderings like *BCDA* and reversals like *DCBA* are also parallelograms, but other reorderings like *ACBD* need not be.

(1.4.2)   **Proposition:** *Let A, B, and C be any three points. Then there is a unique fourth point D such that ABCD is a parallelogram. Moreover, D is in the plane of A, B, and C.*

Proof: From *A* and *C* we can find the midpoint *M* of *AC*; and then *D* is determined as the unique point on the line $\overrightarrow{BM}$ for which *BM* : *MD* = 1 : 1. (If it happens that *B* = *M*, then we must take *D* = *M* also.)

Now let $\mathcal{P}$ be a plane containing *A*, *B*, and *C*, say; such a plane exists (by the plane axiom). Since $\mathcal{P}$ contains *A* and *C*, it contains the whole line $\overleftrightarrow{AC}$ (by the line–plane intersection axiom) and therefore contains *M*. Now $\mathcal{P}$ contains two points, namely *B* and *M*, on the line $\overleftrightarrow{BD}$, and so $\mathcal{P}$ contains the whole of that line, by the same argument as before. So $\mathcal{P}$ contains *D* as well as *A*, *B*, and *C*. □

Usually, one defines a parallelogram as a quadrilateral (four-sided figure) such that both pairs of opposite sides are parallel. We did not give this definition because we also want to consider 'squashed flat' parallelograms where *A*, *B*, *C*, and *D* all lie in a line. However, if we leave out this exceptional case, our definition is equivalent to the usual one:

(1.4.3)   **Proposition:** *Let A, B, C, and D be four points. Then*

(i) *If ABCD is a parallelogram, then $\overleftrightarrow{AB}$ is parallel to $\overleftrightarrow{CD}$ and $\overleftrightarrow{AD}$ is parallel to $\overleftrightarrow{BC}$.*

(ii) *Conversely, if $\overleftrightarrow{AB}$ is parallel to $\overleftrightarrow{CD}$ and $\overleftrightarrow{AD}$ is parallel to $\overleftrightarrow{BC}$, and A, B, C, and D are not collinear, then ABCD is a parallelogram.*

Proof: Let us start by supposing that *ABCD* is a parallelogram, and let *M* be its midpoint. The two lines $\overleftrightarrow{AC}$ and $\overleftrightarrow{BD}$ meet at *M*, and the ratios *AM* : *MC* and $BM : BD$ are equal (both being $1 : 1$), so by the converse of the similarity axiom (1.3.7), the lines $\overrightarrow{AB}$ and $\overrightarrow{CD}$ are parallel. Similarly, the lines $\overrightarrow{AD}$ and $\overrightarrow{BC}$ are parallel. (This argument might fail if $\overrightarrow{AC} = \overrightarrow{BD}$, but in that case all four points are collinear anyway, and so *all* the sides are parallel.)

Now for the second part, assume that $\overrightarrow{AB}$ is parallel to $\overrightarrow{CD}$ and $\overrightarrow{AD}$ is parallel to $\overrightarrow{BC}$, and we must prove that $ABCD$ is a parallelogram. We have seen in 1.4.2 that given the three points A, B, and C there is a unique fourth point D' such that $ABCD'$ is a parallelogram; and by what we have just proved, $\overrightarrow{AD'}$ is parallel to $\overrightarrow{BC}$ and $\overrightarrow{CD'}$ is parallel to $\overrightarrow{AB}$. Now apply the parallel axiom: the lines $\overrightarrow{AD}$ and $\overrightarrow{AD'}$ are both parallel to $\overrightarrow{BC}$ and pass through A, so they must be the same: $\overrightarrow{AD'} = \overrightarrow{AD}$. Similarly, $\overrightarrow{CD'} = \overrightarrow{CD}$. Therefore

$$D = \overrightarrow{AD} \cap \overrightarrow{CD} = \overrightarrow{AD'} \cap \overrightarrow{CD'} = D';$$

so $ABCD = ABCD'$ is a parallelogram.

Where did we use the assumption that A, B, C, and D are not collinear? We needed to know that D is the *unique* point of intersection of the lines $\overrightarrow{AD}$ and $\overrightarrow{CD}$, and this will be true provided that these lines are not the same. If they were the same, then A, C, and D would be collinear, and then B would be collinear with these three because $\overrightarrow{AB}$ is parallel to $\overrightarrow{CD}$. It is this possibility that is ruled out by our assumption. □

The requirement in the converse part of this proposition that the four points not be collinear is a nuisance. It means that we will need to give special arguments in our proofs to handle the possibility that certain sets of points are collinear. An example occurs in the proof of the next result, which says that the result of 'stretching' a parallelogram uniformly in one direction is still a parallelogram.

(1.4.4)   **Theorem:** (STRETCH THEOREM) *Let $ABB'A'$ be a parallelogram (with distinct vertices). Let C be a point on $\overrightarrow{AB}$ and C' a point on $\overrightarrow{A'B'}$ such that $AB : AC = A'B' : A'C'$. Then $ACC'A'$ is a parallelogram also.*

Proof: If A, B, B', and A' are not collinear, then it is enough to prove that $\overrightarrow{CC'}$ is parallel to $\overrightarrow{AA'}$, and this follows from the second part of Thales' theorem (1.3.8). If they are collinear, then all six points mentioned in the theorem lie on the same line. Introduce a ruler on this line and let small letters denote coordinates relative to this ruler. Because $ABB'A'$ is a parallelogram, $(a + b')/2 = (a' + b)/2$, that is $b - a = b' - a'$. By the definition of equality for ratios, $(b - a)(c' - a') = (b' - a')(c - a)$. Therefore, $c - a = c' - a'$, and so $ACC'A'$ is a parallelogram in this case too. □

The most important fact that we shall need about parallelograms is

(1.4.5)   **Theorem:** (DESARGUES' THEOREM) *Let A, B, C, A', B', and C' be six points. Suppose that $ABB'A'$ is a parallelogram and that $BCC'B'$ is a parallelogram. Then $ACC'A'$ is a parallelogram too.*

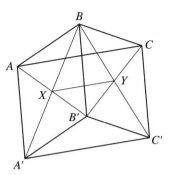

Fig. 1.8. Desargues' theorem.

Remark: This result is in fact a very *special* case of Desargues' theorem; a more general version is given in the next chapter (2.3.5). Notice that the two parallelograms $ABB'A'$ and $BCC'B'$ need not be in the same plane.

Proof: (See Figure 1.8.) We will deal first of all with the general case, where the points A, C, A', C' are not collinear. We have proved in 1.4.3 that a nonflat quadrilateral is a parallelogram if and only if each side is parallel to the opposite one. Thus to show that $ACC'A'$ is a parallelogram we need only prove that $\overleftrightarrow{AA'}$ is parallel to $\overleftrightarrow{CC'}$ and that $\overleftrightarrow{AC}$ is parallel to $\overleftrightarrow{A'C'}$.

The first of these is easy. $\overleftrightarrow{AA'}$ is parallel to $\overleftrightarrow{BB'}$, and $\overleftrightarrow{BB'}$ is parallel to $\overleftrightarrow{CC'}$, so by the transitivity of parallelism (1.2.8), $\overleftrightarrow{AA'}$ is parallel to $\overleftrightarrow{CC'}$.

The second is trickier. Let X and Y be the midpoints of the parallelograms $ABB'A'$ and $BCC'B'$ respectively. The lines $\overleftrightarrow{AX}$ and $\overleftrightarrow{CY}$ meet at B', and $AX : XB' = 1 : 1 = CY : YB'$, so by the converse to the similarity axiom (1.3.7), the lines $\overleftrightarrow{AC}$ and $\overleftrightarrow{XY}$ are parallel. By a similar argument using the lines $\overleftrightarrow{A'X}$ and $\overleftrightarrow{C'Y}$ (which meet at A'), the lines $\overleftrightarrow{XY}$ and $\overleftrightarrow{A'C'}$ are parallel. By the transitivity of parallelism, then, $\overleftrightarrow{AC}$ is parallel to $\overleftrightarrow{A'C'}$.

This completes the proof in the general case, but there are now some annoying special cases to sort out. First of all, suppose that all the points $A, \ldots, C'$ are collinear, and let them have coordinates $a, \ldots, c'$ relative to some ruler on the line through them. Then since $ABB'A'$ and $BCC'B'$ are parallelograms,

$$a + b' = b + a', \quad b + c' = c + b',$$

and it is easy to deduce that $a + c' = c + a'$, so that $ACC'A'$ is a parallelogram. The only remaining special case is that in which A, C, C', A' are collinear but B, B' are not collinear with them. In this case we may argue as follows. We can construct a unique point A'' such that $ACC'A''$ is a parallelogram (by 1.4.2). By applying the case of the theorem that we have proved to the parallelograms

$ACC'A''$ and $CBB'C'$, we deduce that $ABB'A''$ is a parallelogram. But $ABB'A'$ is a parallelogram too, so by the uniqueness statement of 1.4.2, $A' = A''$. Therefore $ACC'A' = ACC'A''$ is a parallelogram. This completes the proof. $\square$

1.5   **Vectors**

We're now going to use our results on parallelograms to develop the properties of *vectors*. In doing this, we are taking leave of Euclid and the Greeks and jumping forward in time to the mid-nineteenth century. There were several reasons why the ancient geometers did not arrive at the vector concept. In the first place, they did not have the algebraic notation and symbolism which we use, and which make vector methods so powerful. Secondly, the Greek picture of geometry was quite static; the idea of 'moving a figure around' was used as little as possible. Some say that the Greeks were worried by Zeno's 'proofs' that motion is logically impossible. But the natural way to understand vectors is in terms of just such 'motions'. The famous American designer R. Buckminster Fuller said he liked vectors because they embody 'the dynamic qualities of represented experiences'.

What is a vector? The school textbooks usually define a vector as 'a quantity having magnitude and direction', such as the *velocity vector* of an object moving through space. It is helpful to represent a vector as an 'arrow' attached to a point of space. But one is not supposed to think of the vector as being firmly rooted just at one point. For instance, one wants to be able to *add* vectors, and the recipe for doing that is to pick up one vector and move it around without changing its length or direction until its tail lies on the head of the other one.

It is better, then, to think of a vector as an *instruction to move* ('Proceed one mile east-south-east') rather than as an arrow pointing from one fixed point to another. The instruction makes sense wherever you are, even if it may be rather difficult to carry out, whereas the arrow is not much use unless you are already at its origin. The 'instruction' idea makes vector addition simple; to add two vectors, you just carry out one instruction after the other. Not every instruction to move is a vector; for instance, 'Move three miles in the direction of Oxford' is not. For an instruction to be a vector, it must specify movement through the same distance and in the same direction for every point.

This idea of an 'instruction' is expressed mathematically as a *function* or *mapping* (the two words are synonymous). Remember that a *mapping* f from space to itself just means any kind of rule that associates a new point $f(P)$ to each point P. To single out those mappings which move points the same distance in the same direction, we will use our results on parallelograms.

(1.5.1)   **Definition:** *A vector is a mapping* **v** *(from space to itself) which associates to each point A a new point* **v**(*A*), *having the property that for any two points A and A′, the midpoint of A***v**(*A′*) *is equal to the midpoint of A′***v**(*A*).

Thus, if **v** is a vector and *A* and *A′* are any two points, then *A*, *A′*, **v**(*A′*), and **v**(*A*) form a parallelogram. The set of all vectors associated to space S will be denoted by $\vec{S}$.

(1.5.2)   **Example:** There is a special vector called **0**, the zero vector. Considered as an instruction, **0** says 'Don't move! Stay exactly where you are!'. More mathematically, **0**(*A*) = *A* for all points *A*.

(1.5.3)   **Proposition:** *Let X and Y be points. Then there is exactly one vector* **v** *such that* **v**(*X*) = *Y*.

Proof: How shall we define **v**? To define the mapping **v**, we must define what **v**(*A*) is for any point *A*. Since **v**(*X*) must equal *Y*, we know that *A*, *X*, *Y*, and **v**(*A*) must form a parallelogram. But as we said before (1.4.2), there is exactly one point *D* such that *AXYD* is a parallelogram, so we may *define* **v**(*A*) to be this point.

The only thing to be sure about is that **v** really is a vector. To see this we must check that for any two points *A* and *A′*, *AA′***v**(*A′*)**v**(*A*) is a parallelogram. But **v**(*A*)*AXY* and *XY***v**(*A′*)*A′* are parallelograms; so Desargues' theorem (1.4.5) tells us that *AA′***v**(*A′*)**v**(*A*) is a parallelogram too. □

The unique vector which maps *X* to *Y*, whose existence is assured by this proposition, will be denoted $\overrightarrow{XY}$. It is clear that $\overrightarrow{XX} = \mathbf{0}$ for any point *X*.

Vectors can be added, as we said before, by joining the head of one to the tail of another. In terms of mappings, this is just the operation of *composition*, applying one mapping to the result of another.

(1.5.4)   **Proposition:** *Let* **u** *and* **v** *be vectors, and define a new mapping* **w** *to be the composite of* **u** *and* **v**, *in other words*

$$\mathbf{w}(A) = \mathbf{u}(\mathbf{v}(A)).$$

Then **w** *is a vector. It is characterized by the fact that for any point A, the points A,* **u**(*A*), **w**(*A*), *and* **v**(*A*) *form a parallelogram.*

Proof: To show that **w** is a vector, it is necessary to check that for any points *A* and *A′*, the points *A*, **w**(*A*), **w**(*A′*), and *A′* form a parallelogram. But since **v** is a vector, *A***v**(*A*)**v**(*A′*)*A′* is a parallelogram; and since **u** is a vector, **v**(*A*)**w**(*A*)**w**(*A′*)**v**(*A′*) is a parallelogram. So Desargues' theorem (1.4.5) gives the result we want.

For any point *A*, let *A′* = **v**(*A*). Then the four points *A*, **u**(*A*), **w**(*A*), and **v**(*A*) are respectively *A*, **u**(*A*), **u**(*A′*), and *A′*, so they form a parallelogram because **u** is a vector. This characterizes **w**(*A*) because a parallelogram is determined by three of its points (1.4.2). □

The vector **w** defined in this proposition is denoted by **u** + **v**. The last part of the proposition is sometimes called the *parallelogram law* of vector addition.

Vector addition has various natural properties:

(1.5.5) **Proposition:**

(i) *For any vector* **u**, **u** + **0** = **0** + **u** = **u**.

(ii) *For any vector* **v** *there is a unique vector* −**v** *such that* **v** + (−**v**) = **0**.

(iii) *For any two vectors* **u** *and* **v**, **u** + **v** = **v** + **u**.

(iv) *For any three vectors* **u**, **v**, *and* **w**, (**u** + **v**) + **w** = **u** + (**v** + **w**).

Remark: In the language of abstract algebra, this result tells us that vectors form an *abelian group* under the operation of vector addition.

Proof:

(i) This is obvious.

(ii) Suppose $\mathbf{v}(X) = Y$; let $-\mathbf{v} = \overrightarrow{YX}$. Then **v** + (−**v**) maps Y to Y, so it must be the unique vector which does this, which is the zero vector. Any other vector **u** such that **v** + **u** = **0** must map Y to X, so it must be equal to −**v**.

(iii) By proposition 1.5.4, **u** + **v** is characterized by the fact that for any point A, the points A, $\mathbf{u}(A)$, $(\mathbf{u} + \mathbf{v})(A)$, and $\mathbf{v}(A)$ form a parallelogram. Similarly, **v** + **u** is characterized by the fact that for any point A the points A, $\mathbf{v}(A)$, $(\mathbf{v} + \mathbf{u})(A)$, and $\mathbf{u}(A)$ form a parallelogram. But these two conditions characterize the same point, so $(\mathbf{v} + \mathbf{u})(A) = (\mathbf{u} + \mathbf{v})(A)$ for any A.

(iv) This is true because of the 'associativity of function composition'. Both sides of the equation, applied to a point A, mean the same thing, namely $\mathbf{u}(\mathbf{v}(\mathbf{w}(A)))$. □

As well as being added to one another, vectors can also be multiplied by real numbers (often called *scalars* in this context — they *scale* the magnitude of a vector, while leaving its direction unchanged). If **v** is a vector and λ is a scalar, then $\lambda\mathbf{v}$ is the vector that sends a point A in the same direction as $\mathbf{v}(A)$ but λ times as far. More formally, the mapping $\lambda\mathbf{v}$ is defined as follows: if $\mathbf{v}(X) = Y$, then $(\lambda\mathbf{v})(X) = Z$, where Z is the unique point on $\overrightarrow{XY}$ for which $XY : XZ = 1 : \lambda$.

It is necessary to check that $\lambda\mathbf{v}$ is a vector, which means that for any two points A and A', the four points A, A', $(\lambda\mathbf{v})(A')$, and $(\lambda\mathbf{v})(A)$ should form a parallelogram. Since we already know that A, A', $\mathbf{v}(A')$, and $\mathbf{v}(A)$ form a parallelogram, this is an immediate consequence of the stretch theorem (1.4.4).

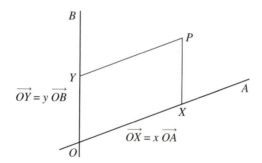

Fig. 1.9. The proof of proposition 1.5.7.

(1.5.6)   **Example:** Reversing the definition, we can see that if $\mathcal{L}$ is any line and O and A are two distinct points on it, then any other point $P \in L$ has $\overrightarrow{OP} = p\overrightarrow{OA}$. Moreover, the assignment $P \mapsto p$ is a ruler on $\mathcal{L}$; in fact, it is the ruler based on O and A.

This can be thought of as a simple kind of 'vector equation' for a line. The next proposition gives an analogous 'vector equation' for a plane:

(1.5.7)   **Proposition:** *Let O, A, and B be three noncollinear points in a plane $\mathcal{P}$. Then for any point $P \in \mathcal{P}$ there are scalars x and y such that*

(1.5.8)
$$\overrightarrow{OP} = x\overrightarrow{OA} + y\overrightarrow{OB},$$

and any point P for which equation 1.5.8 holds lies in $\mathcal{P}$.

Proof: (See Figure 1.9.) Geometrically, equation 1.5.8 says that there is a parallelogram $OXPY$, where $X \in \overrightarrow{OA}$ and $Y \in \overrightarrow{OB}$. If there is such a parallelogram, then $P \in \mathcal{P}$, because the four vertices of a parallelogram are coplanar. Conversely, if $P \in \mathcal{P}$ then we can construct such a parallelogram; let X be the point of intersection of $\overrightarrow{OA}$ with the line through P parallel to $\overrightarrow{OB}$, and let Y be the point of intersection of $\overrightarrow{OB}$ with the line through P parallel to $\overrightarrow{OA}$. (There is an easy special case when P belongs to one of the lines $\overrightarrow{OA}$ or $\overrightarrow{OB}$.) □

(1.5.9)   **Proposition:** *For any vectors $\mathbf{u}$ and $\mathbf{v}$, and scalars λ and μ, the following identities hold.*

(i) *Associative law:* $(\lambda\mu)\mathbf{v} = \lambda(\mu\mathbf{v})$

(ii) *Distributive laws:* $(\lambda + \mu)\mathbf{v} = \lambda\mathbf{v} + \mu\mathbf{v}, \quad \lambda(\mathbf{u} + \mathbf{v}) = \lambda\mathbf{u} + \lambda\mathbf{v}$

(iii) *Special scalar multiples:* $0\mathbf{v} = \mathbf{0}, \ 1\mathbf{v} = \mathbf{v}, \ (-1)\mathbf{v} = -\mathbf{v}.$

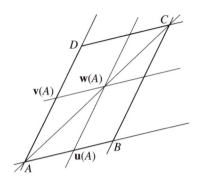

Fig. 1.10. Distributivity of scalar multiplication over addition.

Proof: The only difficult one is the second distributive law. Its proof is once again based on the similarity axiom.

Let $\mathbf{w} = \mathbf{u} + \mathbf{v}$. According to proposition 1.5.4, $\mathbf{w}$ is characterized by the property that for any point A, $A\mathbf{u}(A)\mathbf{w}(A)\mathbf{v}(A)$ is a parallelogram. If we can prove that the four points A, $B = (\lambda\mathbf{u})(A)$, $C = (\lambda\mathbf{w})(A)$, and $D = (\lambda\mathbf{v})(A)$ form a parallelogram as well, then by 1.5.4 we will have shown that $\lambda\mathbf{w} = \lambda\mathbf{u} + \lambda\mathbf{v}$, which is what we want.

As usual, we assume for the general case that the points A, $\mathbf{u}(A)$, $\mathbf{w}(A)$, and $\mathbf{v}(A)$ aren't collinear (see Figure 1.10). Notice that the lines $\overleftrightarrow{CD}$ and $\overleftrightarrow{\mathbf{v}(A)\mathbf{w}(A)}$ cut off equal ratios on the transversals $\overleftrightarrow{AC}$ and $\overleftrightarrow{AD}$, so they are parallel by the converse to the similarity axiom (1.3.7). But $\overleftrightarrow{AB}$ and $\overleftrightarrow{\mathbf{v}(A)\mathbf{w}(A)}$ are parallel since $A\mathbf{u}(A)\mathbf{w}(A)\mathbf{v}(A)$ is a parallelogram, so $\overleftrightarrow{AB}$ is parallel to $\overleftrightarrow{CD}$ by the transitivity of parallelism. By a similar argument, $\overleftrightarrow{AD}$ is parallel to $\overleftrightarrow{BC}$; so $ABCD$ is a parallelogram.

In the special case in which all the points are collinear, we calculate using a ruler in the same kind of way as in previous proofs. □

Any system consisting of a field of scalars and a collection of objects called vectors equipped with addition and scalar multiplication operations satisfying certain standard rules is called a *vector space*. What we have proved so far can be summed up by saying that geometrical vectors form a vector space (over the field of real numbers).

There are many other mathematical processes which also lead to vector spaces. The subject of *linear algebra* studies vector spaces abstractly — using only the rules governing addition and multiplication, and without regard to how the 'vectors' arise. Thus the general results of linear algebra are useful to geometry (and we will use them in the rest of the book), while the concrete vectors of geometry provide a source of ideas for studying the more abstract subject of linear algebra.

1.6 Exercises

1. $\mathcal{L}_1$ and $\mathcal{L}_2$ are lines, and $\mathcal{P}$ is a plane. $\mathcal{L}_1$ does not lie in $\mathcal{P}$, but meets it in a point P. $\mathcal{L}_2$ lies in $\mathcal{P}$, but does not contain P. Does $\mathcal{L}_1$ meet $\mathcal{L}_2$? Which of the axioms of incidence are used in solving this question?

2. Three distinct points A, B, and C lie in a plane $\mathcal{P}_1$, and also in another plane $\mathcal{P}_2$. Must the two planes be the same? If so, prove it from the axioms of incidence; if not, say why.

3. A three-legged table never wobbles, but a four-legged one may do so. Why? Why do people make four-legged tables, then?

4. Investigate how many of the axioms of incidence are satisfied by the two finite geometries described in 1.2.2.

5. A finite geometry is defined as follows. There are eight points, $P_0, \dots, P_7$. A *line* is a set of two points. A *plane* is a set of four points $\{P_a, P_b, P_c, P_d\}$ for which the *Nim-sum* $a \oplus b \oplus c \oplus d$ is zero. (The Nim-sum of two numbers is defined by writing them out in binary notation and then adding them in binary but ignoring all carries.) Show that this finite geometry obeys all the axioms of incidence.

6. Let A, B, C, and D be four points in space, not necessarily coplanar. Let W, X, Y, and Z be the midpoints of the line segments AB, BC, CD, and DA respectively. Prove that $WXYZ$ is a parallelogram. (To avoid special cases you may assume that W, X, Y, and Z are not collinear.)

7. Check that the proof of Desargues' theorem (1.4.5) remains valid when A, B', and C are collinear.

8. $ABCD$ is a parallelogram (not squashed flat) and W is the midpoint of side BC. The lines $\overrightarrow{AW}$ and $\overrightarrow{BD}$ meet at X. Show that $DX : XB = 2 : 1$.

9. Let $AA'B'B$ be a parallelogram. Let O be a point on $\overrightarrow{AA'}$, and let $\overrightarrow{OB'}$ meet $\overrightarrow{AB}$ at P. Prove that $AP : PB = OA : AA'$.

10. $\mathcal{L}$ and $\mathcal{L}'$ are two nonparallel lines in an affine plane. A, B, and C are three points on $\mathcal{L}$, and A', B', and C' are three points on $\mathcal{L}'$. If $\overrightarrow{AB'}$ is parallel to $\overrightarrow{BA'}$ and $\overrightarrow{BC'}$ is parallel to $\overrightarrow{CB'}$, show that $\overrightarrow{AC'}$ is parallel to $\overrightarrow{CA'}$.

    (This is a version of *Pappus' theorem*. Does it remain true if the two lines are parallel?)

11. Let A, B, and C be three noncollinear points, and let $\mathcal{L}$ be a line meeting $\overrightarrow{BC}$ at X, $\overrightarrow{CA}$ at Y, and $\overrightarrow{AB}$ at Z. Let $BX : XC = \alpha_1 : \alpha_2$, $CY : YA = \beta_1 : \beta_2$, and $AZ : ZB = \gamma_1 : \gamma_2$. Prove that $\alpha_1 \beta_1 \gamma_1 + \alpha_2 \beta_2 \gamma_2 = 0$ (*Menelaus' theorem*).

12. As in the preceding question, let A, B, and C be three noncollinear points and let X, Y, and Z be points on $\overrightarrow{BC}$, $\overrightarrow{CA}$, and $\overrightarrow{AB}$ dividing them in the ratios $\alpha_1 : \alpha_2$, $\beta_1 : \beta_2$, and $\gamma_1 : \gamma_2$ respectively. Suppose that the lines $\overrightarrow{AX}$, $\overrightarrow{BY}$, and $\overrightarrow{CZ}$ all meet at P. Prove that $\alpha_1 \beta_1 \gamma_1 - \alpha_2 \beta_2 \gamma_2 = 0$ (*Ceva's theorem*).

2

Vector geometry

2.1 Affine spaces

In Chapter 1 we explained how Euclid built up the whole of geometry from a small number of fundamental assumptions or *axioms*. Following Euclid's method, we stated nine axioms and from them developed the theory of *vectors*. We defined a vector to be a certain kind of mapping from space S to itself: one having the property that for any two points X and Y, the quadrilateral $XY\mathbf{v}(Y)\mathbf{v}(X)$ is a parallelogram.

There is an alternative approach to the same geometrical results. Instead of defining vectors by Euclidean constructions one can reverse the process and define the Euclidean concepts (line, plane, ruler, and so on) in vector terms. Either approach leads to the same geometry in the end. The vector one is more modern and direct, but its algebraic assumptions are not 'intuitively obvious' in the way that the Euclidean-style axioms are. This chapter will develop geometry from the vector point of view.

The algebraic assumptions in question are the axioms for a *vector space*. Such a space is simply a set V of objects called *vectors*, equipped with two binary operations: *addition* (one vector can be added to another) and *scalar multiplication* (a vector can be multiplied by a 'scalar' or real number). The operations are assumed to obey the laws set out in Figure 2.1; these laws play the same rôle in the vector approach to geometry as the nine Euclidean-style axioms did in the approach of Chapter 1. To be strictly accurate, we should refer to a *real* vector space, because we have assumed that the scalars are real numbers; but we will not be using any other kind of vector space in this book and so we will omit the word 'real'.

The geometrical space S that we want to study is closely related to, but not the same as, the set V of vectors. The relationship between them is described in mathematical language by saying that V *acts* or *operates* on S; this simply means that for any point $X \in S$ and any vector $\mathbf{v} \in V$ there is given a new

$$
\begin{aligned}
\mathbf{u} + \mathbf{0} &= \mathbf{0} + \mathbf{u} = \mathbf{u}. \\
\mathbf{u} + \mathbf{v} &= \mathbf{v} + \mathbf{u}. \\
\mathbf{u} + (-\mathbf{u}) &= (-\mathbf{u}) + \mathbf{u} = \mathbf{0}. \\
\mathbf{u} + (\mathbf{v} + \mathbf{w}) &= (\mathbf{u} + \mathbf{v}) + \mathbf{w}. \\
(\lambda + \mu)\mathbf{u} &= \lambda\mathbf{u} + \mu\mathbf{u}. \\
\lambda(\mathbf{u} + \mathbf{v}) &= \lambda\mathbf{u} + \lambda\mathbf{v}. \\
\lambda(\mu\mathbf{u}) &= (\lambda\mu)\mathbf{u}. \\
0\mathbf{u} = \mathbf{0}, \quad &1\mathbf{u} = \mathbf{u}.
\end{aligned}
$$

Fig. 2.1. The axioms for a vector space.

point $\mathbf{v}(X)$, the result of translating X by the vector $\mathbf{v}$. Informally, we can say that each vector gives an 'instruction to move' to the points of S. This operation must have the following two properties:

(i) For any two vectors $\mathbf{u}$ and $\mathbf{v}$, and any $X \in S$,

$$(\mathbf{u} + \mathbf{v})(X) = \mathbf{u}(\mathbf{v}(X)).$$

(ii) For any two points X and Y in S, there is exactly one vector $\mathbf{v} \in V$ such that $\mathbf{v}(X) = Y$. (This unique vector is denoted $\overrightarrow{XY}$.)

Like the laws of a vector space, these two properties are fundamental assumptions that we will work with in this chapter. We embody them in the next definition.

(2.1.1)   **Definition:** *An* affine space *consists of a set S, a vector space V, and an operation of V on S which satisfies the two laws (i) and (ii) above.*

We will usually refer to an affine space by the letter S denoting its set of points. The associated vector space V will usually be denoted by $\vec{S}$; it is called the *space of vectors of* the affine space S.

The results of Chapter 1 may be summed up by saying that any model S for the nine geometrical axioms of that chapter is in fact an affine space. In fact, the vector space axioms were proved in 1.5.5 and 1.5.9, and the two laws for an affine space were proved in 1.5.4 and 1.5.3. To justify our earlier assertion that the Euclidean and the vector approaches to geometry are equivalent we need to reverse this process and to show that any affine space (of the correct dimension) is a model for the axioms of Chapter 1. This means that we must *define* the terms line, plane, ruler, and so on in any affine space, and show that our definitions obey the axioms.

A by-product of our investigation will be a proof of the *relative consistency* of the axioms of Chapter 1. Remember that an axiomatic system is *consistent* if no contradiction can ever be deduced from its axioms. It is said to be consistent *relative* to another axiomatic system if any contradiction in the first system would imply a contradiction in the second. We will find that our system is consistent relative to the theory of the real numbers. Since our axioms explicitly mention the real numbers, they would not make any sense if the theory of the real numbers itself were inconsistent; so relative consistency is the best that we can hope for.

(2.1.2)   **Review of linear algebra:** We will need to use the fundamental concepts of *linear dependence* and *independence* from the theory of vector spaces. Here is a brief reminder; for more information, consult one of the standard textbooks on linear algebra such as Halmos [20], Lipschutz [29], or Morris [32]. Let $\mathbf{v}_1, \ldots, \mathbf{v}_k$ be a set of vectors in a vector space V. It is called *linearly dependent* if there are scalars $\lambda_1, \ldots, \lambda_k$, not all zero, such that the *linear combination*

$$\sum_{i=1}^{k} \lambda_i \mathbf{v}_i = \lambda_1 \mathbf{v}_1 + \cdots + \lambda_k \mathbf{v}_k$$

is equal to the zero vector $\mathbf{0}$; and it is called *linearly independent* otherwise. Linearly independent sets of vectors have the useful property that one can 'equate coefficients'; if $\mathbf{v}_1, \ldots, \mathbf{v}_k$ is a linearly independent set and

$$\lambda_1 \mathbf{v}_1 + \cdots + \lambda_k \mathbf{v}_k = \mu_1 \mathbf{v}_1 + \cdots + \mu_k \mathbf{v}_k,$$

then $\sum(\lambda_i - \mu_i)\mathbf{v}_i = \mathbf{0}$ and so by linear independence $\lambda_i = \mu_i$ for all i. For linearly dependent sets of vectors, this conclusion does not follow.

A linearly independent set $\mathbf{v}_1, \ldots, \mathbf{v}_n$ is called a *basis* for V if it has the additional property that any vector $\mathbf{v} \in V$ can be written as a linear combination $\sum \lambda_i \mathbf{v}_i$ of $\mathbf{v}_1, \ldots, \mathbf{v}_n$. Because of linear independence, the coefficients $\lambda_1, \ldots, \lambda_n$ are uniquely determined by $\mathbf{v}$; they are called the *components* of $\mathbf{v}$ relative to the given basis. There are many possible bases for a given V, but it is a theorem of linear algebra that any two bases have the same number n of elements. This number is called the *dimension* of the vector space V.

A *subspace*[1] of a vector space V is a nonempty subset of V that is closed under the operations of addition and scalar multiplication, and thereby becomes a vector space in its own right. Any subspace of V has a dimension which is no greater than the dimension of V. Subspaces of dimensions 1 and 2 will be particularly important. A subspace of dimension 1 consists of all multiples of some nonzero vector $\mathbf{u}$; a subspace of dimension 2 consists of all linear

---

[1]Sometimes called a *vector subspace*, to distinguish it from the affine subspaces that we are about to introduce.

combinations $x\mathbf{u} + y\mathbf{v}$ of two nonzero vectors $\mathbf{u}$ and $\mathbf{v}$, neither of which is a multiple of the other.

(2.1.3) **Definition:** *Let S be an affine space, with associated vector space $\vec{S}$. Let U be a vector subspace of $\vec{S}$, and let X be a point of S. The subset*

$$U[X] = \{\mathbf{u}(X) : \mathbf{u} \in U\}$$

of S is called the affine subspace *of S through the point X parallel to the vector subspace U. It is called a* line *if U has dimension 1, and a* plane *if U has dimension 2.*

The results 1.5.6 and 1.5.7 assure us that these 'new' affine definitions of line and plane agree with the classical concepts studied in Chapter 1. Notice that U can be reconstructed from $U[X]$ as the set of all vectors $\overrightarrow{Y_1 Y_2}$ with $Y_1, Y_2 \in U[X]$. It follows that if $U_1[X_1] \subseteq U_2[X_2]$, then $U_1 \subseteq U_2$.

A *ruler* on an affine line $\mathcal{L} = U[X]$ is a one-to-one correspondence between $\mathcal{L}$ and the real numbers $\mathbf{R}$, which is supposed to give a 'uniform scale' along the line. Such a ruler can be defined in affine terms as a map sending a point $P \in U[X]$ to the real number p such that $\overrightarrow{XP} = p\mathbf{u}$, where $\mathbf{u}$ is a fixed nonzero vector in U. Again, 1.5.6 shows that this is consistent with Chapter 1.

Two affine lines $U_1[X_1]$ and $U_2[X_2]$ are said to be *parallel* if they are both parallel to the same one-dimensional vector subspace, in other words if $U_1 = U_2$. We need to check that this definition also agrees with the classical one, according to which two lines are parallel if they are coplanar (*i.e.* in the same plane) and nonintersecting. This is a little harder:

(2.1.4) **Proposition:** *Two affine lines are parallel if and only if they are coplanar and nonintersecting.*

Proof: Let $\mathcal{L}_1$ and $\mathcal{L}_2$ be two lines, written as $U_1[X_1]$ and $U_2[X_2]$, where U_1 and U_2 are one-dimensional vector subspaces. Let $\mathbf{u}_1$ and $\mathbf{u}_2$ be nonzero

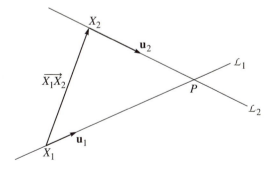

Fig. 2.2. Criterion for intersection of two lines.

vectors in U_1 and U_2. The lines will intersect if and only if there is a point P such that $\overline{X_1P}$ is a multiple of $\mathbf{u}_1$ and $\overline{X_2P}$ is a multiple of $\mathbf{u}_2$; that is, if and only if $\overline{X_1X_2}$ can be written as a linear combination of $\mathbf{u}_1$ and $\mathbf{u}_2$ (see Figure 2.2).

Suppose first that we know that the two lines are coplanar and nonintersecting. Let $\mathbf{v}$ be the vector $\overline{X_1X_2}$. Because $\mathcal{L}_1$ and $\mathcal{L}_2$ are coplanar, the three vectors $\mathbf{u}_1$, $\mathbf{u}_2$, and $\mathbf{v}$ belong to some two-dimensional vector subspace W (the vector space associated to the plane containing the two lines). Now three vectors in a two-dimensional space cannot be independent; there must be some linear relationship

$$\alpha_1\mathbf{u}_1 + \alpha_2\mathbf{u}_2 + \beta\mathbf{v} = \mathbf{0}$$

between them. However, the coefficient β must be zero; otherwise we could divide by it and so express $\mathbf{v}$ as a linear combination of $\mathbf{u}_1$ and $\mathbf{u}_2$, contrary to the hypothesis that the lines don't intersect. So $\mathbf{u}_1$ is a multiple of $\mathbf{u}_2$, and therefore $U_1 = U_2$ and the lines are parallel.

Conversely, suppose that $U_1 = U_2$. If $\overline{X_1X_2} \in U_1$ then the lines are the same. If not, then $\overline{X_1X_2}$ and $\mathbf{u}_1$ are two linearly independent vectors, spanning a two-dimensional vector space W say; and both lines are contained in the plane $W[X_1]$. Moreover, since $\mathbf{u}_2$ is a multiple of $\mathbf{u}_1$ and $\overline{X_1X_2}$ is not, $\overline{X_1X_2}$ cannot be expressed as a linear combination of $\mathbf{u}_1$ and $\mathbf{u}_2$; so the lines do not meet. $\square$

If $\mathcal{L} = U[X]$ and $\mathbf{v} \in U$, we will say that $\mathbf{v}$ is *parallel to* $\mathcal{L}$.

(2.1.5)   **Theorem:** *Let $\mathcal{S}$ be an affine space, and define the concepts 'line', 'plane', and 'ruler' in $\mathcal{S}$ as above. Then $\mathcal{S}$ is a model for all the axioms of Chapter 1, except possibly for the dimension axiom and the plane–plane intersection axiom. In order for $\mathcal{S}$ to be a model for these two axioms also, it is necessary and sufficient that $\dim \mathcal{S} = 3$.*

Proof: We just need to go through all the axioms and check that they are satisfied:

Line axiom: Through any two points there is a unique line. The unique line through distinct points X and Y is $U[X]$, where U is the one-dimensional subspace generated by $\overline{XY}$.

Plane axiom: Through any three noncollinear points there is a unique plane. Similarly, the unique plane through noncollinear points X, Y, and Z is $W[X]$, where W is the two-dimensional subspace generated by $\overline{XY}$ and $\overline{XZ}$.

Dimension axiom: Each line contains at least two points, each plane contains at least two lines, and there are at least two planes. It is clear that any line contains at least two (in fact, infinitely many) distinct points and each plane contains at least two (in fact, infinitely many) distinct lines. The statement that there are at least two distinct planes will be true if and only if the space V of vectors is three-dimensional or more.

Line–plane intersection axiom: If two points of a line lie in some plane, then the whole line lies in that plane. If two points X and Y of a line $U[X]$ lie in a plane $W[X]$, then $\overrightarrow{XY} \in W$. Since U consists of all multiples of $\overrightarrow{XY}$, $U \subseteq W$ and so $U[X] \subseteq W[X]$.

Parallel axiom: Given a line $\mathcal{L}$ and a point P, there is a unique line through P parallel to $\mathcal{L}$. It follows from our discussion of parallelism (2.1.4) that the unique line through Y parallel to the line $U[X]$ is $U[Y]$.

Plane–plane intersection axiom: If two planes intersect, their intersection is a line. If the dimension of V is four or more, then this axiom is not satisfied. For there are then four linearly independent vectors $\mathbf{v}_1$, $\mathbf{v}_2$, $\mathbf{v}_3$, and $\mathbf{v}_4$; if W_1 is the set of all linear combinations of $\mathbf{v}_1$ and $\mathbf{v}_2$, and W_2 is the set of all linear combinations of $\mathbf{v}_3$ and $\mathbf{v}_4$, then W_1 and W_2 are two two-dimensional subspaces whose intersection is just the zero vector. Therefore, $W_1[X]$ and $W_2[X]$ are two planes whose intersection is the single point X.

If the dimension of V is three or less, however, this axiom is satisfied. For in this case, the intersection of two distinct two-dimensional subspaces must be one-dimensional[2]. Therefore, two distinct planes $W_1[X]$ and $W_2[X]$ which meet at the point X must intersect in the line $(W_1 \cap W_2)[X]$.

Ruler axiom: Given two distinct points A and B, there is a ruler on the line $\overrightarrow{AB}$ that associates A with 0 and B with 1. Let A and B be distinct points on a line $\mathcal{L}$. The line may be written as $U[A]$, where U is the set of all multiples of the vector $\overrightarrow{AB}$. The corresponding ruler

$$P \in \mathcal{L} \mapsto p \in \mathbf{R} : \overrightarrow{AP} = p\overrightarrow{AB}$$

associates A with 0 and B with 1.

Ruler comparison axiom: Two rulers on the same line are related by a coordinate transformation of the form $x' = lx + m$, where x and x' are the coordinates of a point relative to the two rulers and $l \neq 0$ and m are constants. Let X be a point on some line $\mathcal{L}$. Let A, B and A', B' be two pairs of distinct points on $\mathcal{L}$, and let us compare the coordinates of X relative to the two rulers based on A, B and on A', B'. We have

$$\overrightarrow{AX} = x\overrightarrow{AB}, \quad \overrightarrow{A'X} = x'\overrightarrow{A'B'}.$$

But there are constants l and m, with $l \neq 0$, such that

$$\overrightarrow{AB} = l\overrightarrow{A'B'}, \quad \overrightarrow{A'A} = m\overrightarrow{A'B'}.$$

---

[2]This follows from the linear algebra formula that relates the dimensions of the sum and the intersection of two subspaces. Write

$$\dim(W_1 \cap W_2) = \dim W_1 + \dim W_2 - \dim(W_1 + W_2)$$

and notice that $\dim W_1 = \dim W_2 = 2$, whereas $\dim(W_1 + W_2) \leq 3$.

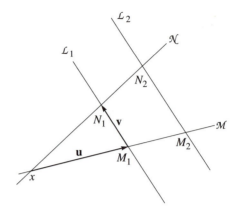

Fig. 2.3. Proof of the similarity axiom.

So

$$x'\overrightarrow{A'B'} = \overrightarrow{A'X} = \overrightarrow{A'A} + \overrightarrow{AX} = (lx + m)\overrightarrow{A'B'}.$$

It follows that $x' = lx + m$, as required.

Similarity axiom: Suppose that $\mathcal{L}_1$ and $\mathcal{L}_2$ are two parallel lines in a plane, and that $\mathcal{M}$ and $\mathcal{N}$ are two other lines meeting the parallel lines at points M_1, M_2, N_1, and N_2, and meeting one another at X (Figure 2.3). Then $XM_1 : XM_2 = XN_1 : XN_2$.

Let $\overrightarrow{XM_1} = \mathbf{u}$ and $\overrightarrow{M_1N_1} = \mathbf{v}$. These two vectors are linearly independent, and their sum $\mathbf{u} + \mathbf{v}$ is equal to $\overrightarrow{XN_1}$. Now, because XM_1M_2 are collinear, there is a scalar λ such that $\overrightarrow{XM_2} = \lambda\mathbf{u}$; because M_1N_1 and M_2N_2 are parallel, there is a scalar μ such that $\overrightarrow{M_2N_2} = \mu\mathbf{v}$; and because XN_1N_2 are collinear, there is a scalar ν such that $\overrightarrow{XN_2} = \nu(\mathbf{u} + \mathbf{v})$. But $\overrightarrow{XM_2} + \overrightarrow{M_2N_2} = \overrightarrow{XN_2}$, and so

$$(\lambda - \nu)\mathbf{u} + (\mu - \nu)\mathbf{v} = \mathbf{0}.$$

It follows from linear independence that $\lambda = \mu = \nu$. Hence

$$XM_1 : XM_2 = 1 : \lambda = 1 : \nu = XN_1 : XN_2. \qquad \square$$

As we mentioned before, this gives a proof of the consistency of the axioms of Chapter 1, relative to the consistency of the real numbers. If the theory of the real numbers is consistent, then so is the theory of three-dimensional vector spaces, and so therefore is the theory of three-dimensional affine spaces. But any model for the theory of three-dimensional affine spaces is, as we have just seen, a model for the axioms of Chapter 1.

2.2 Position vectors

Let S be an affine space. Then S is homogeneous; no point of S is singled out by some special geometric property. However, suppose that we *choose* some fixed point O in space to be an *origin*. The choice is arbitrary, in the same way that the choice of Greenwich to define the meridian of $0°$ longitude was arbitrary, but once we have made the choice we see that any other point $P \in S$ can be described by means of the vector $\overrightarrow{OP}$.

(2.2.1)   **Definition:** *The vector $\overrightarrow{OP}$ is called the* position vector *of the point P relative to the given origin O.*

Each point has a unique position vector, and each position vector describes a unique point; the choice of origin has allowed us to set up a *one-to-one correspondence* between points and vectors. Despite the existence of such a one-to-one correspondence, it is convenient to retain the distinction between points and vectors, because of the element of arbitrariness involved in a choice of origin.

The next result relates position vectors to distance ratios along a line. If X is a point on the line $\overline{AB}$, then the vectors $\overrightarrow{AX}$ and $\overrightarrow{XB}$ are linearly dependent, and so there are constants α and β such that $\alpha\overrightarrow{AX} = \beta\overrightarrow{XB}$. We define the ratio $AX : XB$ to be $\beta : \alpha$. It is easy to check that this definition is equivalent to the one given in the previous chapter in terms of a ruler on the line.

(2.2.2)   **Proposition:** (RATIO THEOREM) *Suppose that an origin O has been fixed. Let A and B be points, having position vectors $\overrightarrow{OA} = \mathbf{a}$ and $\overrightarrow{OB} = \mathbf{b}$ with respect to the origin. Then the point X on $\overline{AB}$ such that $AX : XB = \beta : \alpha$ has position vector*

$$\overrightarrow{OX} = \frac{\alpha\mathbf{a} + \beta\mathbf{b}}{\alpha + \beta}$$

with respect to the same origin.

Proof: Notice that $\overrightarrow{AB} = \mathbf{b} - \mathbf{a}$ (Figure 2.4). Since $AX : XB = \beta : \alpha$, $AX : AB = \beta : \beta + \alpha$, and so $\overrightarrow{AX} = \dfrac{\beta}{\beta + \alpha}(\mathbf{b} - \mathbf{a})$. Hence

$$\overrightarrow{OX} = \overrightarrow{OA} + \overrightarrow{AX} = \mathbf{a} + \frac{\beta}{\beta + \alpha}(\mathbf{b} - \mathbf{a}) = \frac{\alpha\mathbf{a} + \beta\mathbf{b}}{\alpha + \beta}$$

as required. $\square$

In particular, the *midpoint* of the line segment AB (the point M for which $AM : MB = 1 : 1$) has position vector $(\mathbf{a} + \mathbf{b})/2$. An equivalent way of expressing the ratio theorem is that *a general point on the line $\overline{AB}$ through two distinct points A and B has position vector* $\mathbf{r} = \lambda\mathbf{a} + (1 - \lambda)\mathbf{b}$, *where* $\lambda \in \mathbf{R}$. This is because the ratio $\beta : \alpha$ can always be expressed as $\lambda : 1 - \lambda$, where $\lambda = \beta/(\alpha + \beta)$.

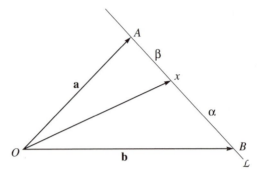

Fig. 2.4. The ratio theorem.

(2.2.3)   **Corollary:** *Let A, B, and C be points of an affine space S, having position vectors* **a**, **b**, *and* **c** *with respect to some origin. Then A, B, and C are collinear if and only if there exist constants* α, β, *and* γ, *not all zero, such that*

$$\alpha\mathbf{a} + \beta\mathbf{b} + \gamma\mathbf{c} = \mathbf{0}, \quad \alpha + \beta + \gamma = 0.$$

Proof: If the points are all the same, the result is obvious. Otherwise, two of them, say A and B, are different. Suppose that $C \in \overrightarrow{AB}$ with $AC : CB = \beta : \alpha$; then by the ratio theorem,

$$\mathbf{c} = \frac{\alpha\mathbf{a} + \beta\mathbf{b}}{\alpha + \beta},$$

and so the equations

$$\alpha\mathbf{a} + \beta\mathbf{b} + \gamma\mathbf{c} = \mathbf{0}, \quad \alpha + \beta + \gamma = 0,$$

are satisfied with $\gamma = -(\alpha + \beta)$. The argument can be reversed. $\square$

One can prove a similar result for planes:

(2.2.4)   **Proposition:** *Let A, B, C, and D be points of an affine space, with position vectors* **a**, **b**, **c**, *and* **d** *relative to some origin. Then A, B, C, and D are coplanar if and only if there exist constants* α, β, γ, *and* δ, *not all zero, such that*

$$\alpha\mathbf{a} + \beta\mathbf{b} + \gamma\mathbf{c} + \delta\mathbf{d} = \mathbf{0}, \quad \alpha + \beta + \gamma + \delta = 0.$$

Proof: Any plane through A is of the form $W[A]$, where W is a two-dimensional vector subspace. Therefore, the four points are coplanar if and only if the three vectors

$$\overrightarrow{AB} = \mathbf{b} - \mathbf{a}, \quad \overrightarrow{AC} = \mathbf{c} - \mathbf{a}, \quad \overrightarrow{AD} = \mathbf{d} - \mathbf{a}$$

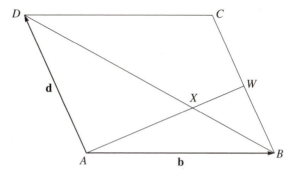

Fig. 2.5. Vector approach to Exercise 1.6.8.

all belong to a two-dimensional vector space W. This will happen if and only if they are linearly dependent, meaning that there are constants β, γ, and δ, not all zero, such that

$$\beta(\mathbf{b} - \mathbf{a}) + \gamma(\mathbf{c} - \mathbf{a}) + \delta(\mathbf{d} - \mathbf{a}) = \mathbf{0}.$$

This gives what we want, with $\alpha = -(\beta + \gamma + \delta)$. $\square$

(2.2.5)   **Example:** We will use vector methods to solve Exercise 1.6.8 from the previous chapter (Figure 2.5): if $ABCD$ is a nonflat parallelogram, W is the midpoint of BC, and X is the intersection of AW and DB, prove that $DX : XB = 2 : 1$. To say that $ABCD$ is a parallelogram is simply to say that $\overline{BC} = \overline{AD}$; you can find some equivalent definitions in Exercise 2.5.5.

We will take position vectors relative to A as origin, and we will write all our vectors in terms of $\overrightarrow{AB} = \mathbf{b}$ and $\overrightarrow{AD} = \mathbf{d}$. Because W is the midpoint of BC, its position vector $\mathbf{w}$ is equal to $\mathbf{b} + \frac{1}{2}\mathbf{d}$. Since X lies on the line AW, its position vector $\mathbf{x}$ can be written as $\mathbf{x} = \lambda\mathbf{w} = \lambda(\mathbf{b} + \frac{1}{2}\mathbf{d})$ for some scalar λ. On the other hand, since X lies on the line BD, one can also write $\mathbf{x} = \mu\mathbf{d} + (1 - \mu)\mathbf{b}$, by the ratio theorem. Comparing the coefficients of $\mathbf{b}$ and $\mathbf{d}$ in these two representations of $\mathbf{x}$, we obtain the equations

$$\lambda = 1 - \mu, \quad \mu = \tfrac{1}{2}\lambda.$$

It is easy to solve these two simultaneous equations to get $\lambda = \frac{2}{3}, \mu = \frac{1}{3}$. Thus $DX : XB = 1 - \mu : \mu = 2 : 1$.

Where did we use the hypothesis that the parallelogram is not flat? When we compared the coefficients of $\mathbf{b}$ and $\mathbf{d}$, we implicitly assumed that these two vectors are linearly independent. This will be true provided A, B, and D are not collinear, which is to say the parallelogram is not flat.

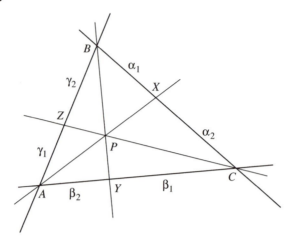

Fig. 2.6. Ceva's theorem.

2.3 Some theorems in affine geometry

In this section we will prove some famous theorems about affine spaces. They are interesting in themselves, but their proofs also illustrate how vector methods can be used to prove other results in geometry. The first, known as Ceva's theorem, gives a criterion in terms of ratios for three lines through the vertices of a proper triangle to be concurrent. A *proper triangle* is just made up of three noncollinear points A, B, and C, and lines are said to be *concurrent* if they all meet at a point.

(2.3.1) **Theorem:** (CEVA'S THEOREM) *Let ABC be a proper triangle, and let X, Y, and Z be points on $\overrightarrow{BC}$, $\overrightarrow{CA}$, and $\overrightarrow{AB}$ respectively. Suppose that*

$$BX : XC = \alpha_1 : \alpha_2, \quad CY : YA = \beta_1 : \beta_2, \quad AZ : ZB = \gamma_1 : \gamma_2.$$

Then the following conditions are equivalent:

- *The lines $\overrightarrow{AX}$, $\overrightarrow{BY}$, and $\overrightarrow{CZ}$ are concurrent or parallel.*
- $\alpha_1 \beta_1 \gamma_1 = \alpha_2 \beta_2 \gamma_2.$

Proof: (See Figure 2.6.) We will represent each point by its position vector relative to the origin C. The position vector of A will be denoted by **a**, the position vector of B will be denoted by **b**, and so on. Then by the ratio theorem

$$\mathbf{x} = \frac{\alpha_2 \mathbf{b}}{\alpha_1 + \alpha_2}, \quad \mathbf{y} = \frac{\beta_1 \mathbf{a}}{\beta_1 + \beta_2}.$$

Let us try to find the position vector of the point P which is the intersection of $\overrightarrow{AX}$ and $\overrightarrow{BY}$. By the ratio theorem once again,

$$\mathbf{p} = \lambda\mathbf{x} + (1-\lambda)\mathbf{a}, \quad \mathbf{p} = \mu\mathbf{y} + (1-\mu)\mathbf{b}$$

where $AP : PX = \lambda : 1 - \lambda$ and $BP : PY = \mu : 1 - \mu$. Substituting for $\mathbf{x}$ and $\mathbf{y}$, we get

$$(1-\lambda)\mathbf{a} + \frac{\alpha_2\lambda}{\alpha_1 + \alpha_2}\mathbf{b} = \frac{\beta_1\mu}{\beta_1 + \beta_2}\mathbf{a} + (1-\mu)\mathbf{b}.$$

Now $\mathbf{a}$ and $\mathbf{b}$ are linearly independent vectors (because the triangle ABC is proper); we may therefore equate the coefficients of $\mathbf{a}$ and $\mathbf{b}$. This gives us two simultaneous equations:

(2.3.2)

$$\frac{\beta_1\mu}{\beta_1 + \beta_2} = 1 - \lambda, \quad \frac{\alpha_2\lambda}{\alpha_1 + \alpha_2} = 1 - \mu.$$

These equations can be solved provided that $\alpha_1\beta_1 + \alpha_1\beta_2 + \alpha_2\beta_2 \neq 0$, and the solution is

$$\lambda = \frac{\beta_2(\alpha_1 + \alpha_2)}{\alpha_1\beta_1 + \alpha_1\beta_2 + \alpha_2\beta_2}, \quad \mu = \frac{\alpha_1(\beta_1 + \beta_2)}{\alpha_1\beta_1 + \alpha_1\beta_2 + \alpha_2\beta_2}.$$

Therefore

$$\mathbf{p} = \frac{\alpha_1\beta_1\mathbf{a} + \alpha_2\beta_2\mathbf{b}}{\alpha_1\beta_1 + \alpha_1\beta_2 + \alpha_2\beta_2}.$$

Any point on the line $\overrightarrow{CP}$ has a position vector which is a multiple of $\mathbf{p}$, and any point on the line $\overrightarrow{AB}$ has a position vector of the form $\xi\mathbf{a} + \eta\mathbf{b}$ with $\xi + \eta = 1$. Therefore, $\overrightarrow{CP}$ intersects $\overrightarrow{AB}$ at the point with position vector

$$\frac{\alpha_1\beta_1\mathbf{a} + \alpha_2\beta_2\mathbf{b}}{\alpha_1\beta_1 + \alpha_2\beta_2},$$

which divides AB in the ratio $\alpha_2\beta_2 : \alpha_1\beta_1$. If this ratio is to be equal to $\gamma_1 : \gamma_2$ we must have $\alpha_1\beta_1\gamma_1 = \alpha_2\beta_2\gamma_2$.

We still have to deal with the special case $\alpha_1\beta_1 + \alpha_1\beta_2 + \alpha_2\beta_2 = 0$, for which the equations 2.3.2 have no solution. This implies that the lines $\overrightarrow{AX}$ and $\overrightarrow{BY}$ are parallel. They are both parallel to the vector

$$\overrightarrow{AX} = \frac{\alpha_2\mathbf{b} - (\alpha_1 + \alpha_2)\mathbf{a}}{\alpha_1 + \alpha_2} = -\frac{\alpha_1\beta_1\mathbf{a} + \alpha_2\beta_2\mathbf{b}}{\alpha_1\beta_1}.$$

(To obtain the second of these expressions we multiplied numerator and denominator by $\alpha_1\beta_1/(\alpha_1 + \alpha_2)$ and used the identity $\alpha_1\beta_1 + \alpha_1\beta_2 + \alpha_2\beta_2 = 0$.) The line through C parallel to this vector will meet $\overrightarrow{AB}$ at the point with position vector

$$\frac{\alpha_1\beta_1\mathbf{a} + \alpha_2\beta_2\mathbf{b}}{\alpha_1\beta_1 + \alpha_2\beta_2},$$

and the argument can be completed as before. $\square$

It is interesting to compare this argument with the more classical one given in Exercise 1.6.12. Which is more elegant? Which would be easier to discover?

(2.3.3)   **Example:** A special case of Ceva's theorem is the well-known fact that the three *medians* of any triangle intersect in a point. A median is the line joining a vertex to the midpoint of the opposite side, so here we can take $\alpha_1 = \alpha_2 = \beta_1 = \beta_2 = \gamma_1 = \gamma_2 = 1$. Ceva's theorem then applies to show that the medians are concurrent.

Remark: The calculations that we have done in proving Ceva's theorem allow us to calculate all the ratios appearing in the figure. For example, from the formulae for the position vectors of P and Z, we find that

$$CP : PZ = \alpha_1\beta_1 + \alpha_2\beta_2 : \alpha_1\beta_2.$$

This leads to an interesting result about *harmonic division*. Let $\mathcal{L}$ be a line and A, B be points on $\mathcal{L}$. Two further points X and Y on $\mathcal{L}$ are said to *divide A and B harmonically* if the ratios $AX : XB$ and $AY : YB$ are the same except for sign; in other words, if there are α and β so that

$$AX : XB = \alpha : \beta, \quad AY : YB = -\alpha : \beta.$$

One also sometimes says that A, B, X, and Y form a *harmonic set*. In terms of coordinates relative to some ruler on $\mathcal{L}$, the condition for harmonic division is $(a - x)(y - b) = (x - b)(y - a)$.

The concept of harmonic division was studied by Pappus (about 320 AD), perhaps the last of the great Greek geometers. He proved the following remarkable theorem:

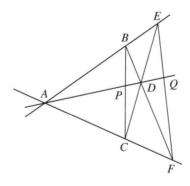

Fig. 2.7. Pappus' harmonic-division theorem.

(2.3.4)   **Proposition:** (Pappus' Harmonic-Division Theorem) *Let A, B, C, and D be four points in a plane; let $\overline{AB}$ and $\overline{CD}$ intersect at E, let $\overline{BD}$ and $\overline{AC}$ intersect at F. Let $\overline{AD}$ intersect $\overline{BC}$ at P and $\overline{EF}$ at Q. Then the points P and Q divide A and D harmonically.*

This theorem is remarkable because it defines a *metric* notion — harmonic division — in terms simply of the incidence relations between points and lines. From Figure 2.7, it is easy to see that given any three of the points A, P, D, and Q, you can construct the fourth one which makes them into a harmonic set using a straightedge alone. So the concept of harmonic division will make sense even in projective geometry, where distance ratios are not defined.

Proof: We will use the calculation of the ratios made in the remark following Ceva's theorem. Let $AB : BE = 1 : \lambda$ and $AC : CF = 1 : \mu$. Then, applying the remark to triangle AEF, we find that

$$AD : DQ = \lambda + \mu : \lambda\mu,$$

and so, using the properties of ratios,

$$AQ : QD = \lambda + \mu + \lambda\mu : -\lambda\mu.$$

On the other hand, applying the remark to triangle ABC (and watching out for minus signs!) we get

$$AD : DP = -(1 + \lambda)\mu - (1 + \mu)\lambda : \lambda\mu = -\lambda - \mu - 2\lambda\mu : \lambda\mu.$$

Therefore, using the properties of ratios again,

$$AP : PD = \lambda + \mu + \lambda\mu : \lambda\mu$$

and so P, Q divide A, D harmonically, as required. □

The last result that we shall discuss in this section is the general case of *Desargues' theorem*; a special case played a major part in the construction of vectors in Chapter 1. Girard Desargues (1593–1661) was an architect and military engineer from Lyons in France. His theorem is concerned with the perspective properties of triangles.

Let ABC and $A'B'C'$ be proper triangles (in the same or different planes). They are said to be *perspective from a point O* if the lines $\overline{AA'}$, $\overline{BB'}$, and $\overline{CC'}$ are concurrent at O (see Figure 2.8). This is perhaps the most natural concept of perspective; the triangles are projected onto one another through the point O. However, a triangle has three *sides* as well as three *vertices*, and this leads to another concept of perspective obtained by reversing the rôles of lines and points. The triangles ABC and $A'B'C'$ are said to be *perspective from a line $\mathcal{L}$* if the points of intersection of the pairs of corresponding sides (that is, $\overline{AB} \cap \overline{A'B'}$, $\overline{BC} \cap \overline{B'C'}$, and $\overline{CA} \cap \overline{C'A'}$) all lie on $\mathcal{L}$ (see Figure 2.9). Desargues' theorem states that these two apparently different concepts of perspective are in fact equivalent.

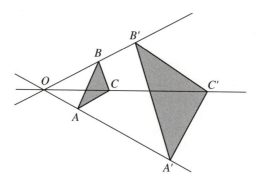

Fig. 2.8. Triangles perspective from a point.

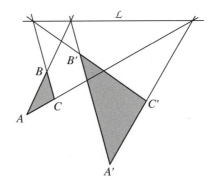

Fig. 2.9. Triangles perspective from a line.

(2.3.5)   **Theorem:** (DESARGUES' THEOREM) *If two proper triangles are perspective from a point, and if the pairs of corresponding sides of the two triangles intersect, then the three points of intersection of the pairs of corresponding sides are collinear, and so the triangles are perspective from a line.*

Proof: Let $P = \overline{BC} \cap \overline{B'C'}$, $Q = \overline{CA} \cap \overline{C'A'}$, $R = \overline{AB} \cap \overline{A'B'}$. Take position vectors relative to the origin O which is the point of intersection of $\overline{AA'}$, $\overline{BB'}$, and $\overline{CC'}$; and denote the position vector of each point by the corresponding boldface letter.

Because OAA', OBB', and OCC' are collinear,

$$\mathbf{a}' = \lambda\mathbf{a}, \quad \mathbf{b}' = \mu\mathbf{b}, \quad \mathbf{c}' = \nu\mathbf{c}$$

for some scalars λ, μ, and ν. Moreover, by the ratio theorem there are scalars ξ and η such that

$$\mathbf{r} = \xi\mathbf{a} + (1 - \xi)\mathbf{b} = \eta\mathbf{a}' + (1 - \eta)\mathbf{b}'.$$

Substitute for $\mathbf{a}'$ and $\mathbf{b}'$ in terms of $\mathbf{a}$ and $\mathbf{b}$, and compare coefficients[3] of $\mathbf{a}$ and $\mathbf{b}$, to obtain

$$-\xi + \mu\eta = \mu - 1, \quad \xi - \lambda\eta = 0.$$

Solving these equations, one obtains $\xi = -\dfrac{\lambda(\mu - 1)}{\lambda - \mu}$ and

$$\mathbf{r} = \frac{\mu(\lambda - 1)\mathbf{b} - \lambda(\mu - 1)\mathbf{a}}{\lambda - \mu}.$$

Similarly

$$\mathbf{p} = \frac{\nu(\mu - 1)\mathbf{c} - \mu(\nu - 1)\mathbf{b}}{\mu - \nu}$$

and

$$\mathbf{q} = \frac{\lambda(\nu - 1)\mathbf{a} - \nu(\lambda - 1)\mathbf{c}}{\nu - \lambda}.$$

Therefore

$$(\lambda - \mu)(\nu - 1)\mathbf{r} + (\mu - \nu)(\lambda - 1)\mathbf{p} + (\nu - \lambda)(\mu - 1)\mathbf{q} = \mathbf{0}.$$

Since also

$$(\lambda - \mu)(\nu - 1) + (\mu - \nu)(\lambda - 1) + (\nu - \lambda)(\mu - 1) = 0$$

the three points P, Q, and R are collinear, by the criterion (2.2.3) for collinearity. $\square$

The converse to Desargues' theorem — if two triangles are perspective from a line, then they are perspective from a point — is also true, and can be proved by similar methods. Of course, in stating Desargues' theorem we have assumed that the points of intersection P, Q, R actually exist. What if some of the lines were parallel? This would happen if any two of the three constants λ, μ, and ν used in the proof above were equal. And how is this version of Desargues' theorem related to the special case that we used in Chapter 1? To find a full answer to these questions one must study projective geometry.

2.4 A short introduction to projective geometry

Once upon a time, a mathematician was giving a course of lectures on geometry, and using a slide projector for illustrations. Unfortunately, the screen

---

[3]This is legitimate provided that the two vectors are linearly independent. They can fail to be linearly independent if and only if O, A, B, A', and B' are all collinear, in which case the point R is not defined.

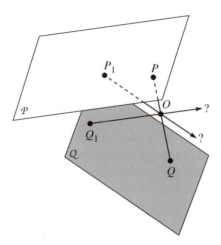

Fig. 2.10. Projection from one plane to another.

that he was using was not vertical. The effect was depressing. Circles came out elliptical, equilateral triangles just looked like any old triangles, and even parallel lines looked as though they would meet if extended far enough. But when he came to Desargues' theorem, he thought that his problems were over. Even though the projector distorted both distances and angles, it kept straight lines straight; and the projected figure was just as good an illustration of Desargues' theorem as the original one.

This story points up a remarkable feature of Desargues' theorem, which sets it apart from nearly all the theorems of classical Euclidean geometry. They all involve metric concepts such as distance, angle, or proportion; but Desargues' theorem involves none of these things. As a result it is — at first sight — invariant under projection from one plane onto another.

But we need to be a bit more careful. What exactly do we mean by projection? Idealizing the slide projector somewhat, we can represent the projection operation as follows. There are two fixed planes $\mathcal{P}$ and $\mathcal{Q}$ in three-dimensional space, and a fixed point O (the bulb) which is not on either plane. Projection gives a correspondence between points of $\mathcal{P}$ and points of $\mathcal{Q}$; the idea being that to a point $P \in \mathcal{P}$ corresponds the unique point of intersection of $\mathcal{Q}$ with $\overrightarrow{OP}$. This process is illustrated in Figure 2.10.

The problem is that this process does not quite match up each point of $\mathcal{P}$ with a point of $\mathcal{Q}$. There are some points of $\mathcal{P}$, such as the point P_1 in the figure, that have *no* image in $\mathcal{Q}$, because the line $\overrightarrow{OP_1}$ is parallel to $\mathcal{Q}$. A pair of lines in $\mathcal{P}$ that meet at the point P_1 will therefore be transformed by projection into a pair of lines in $\mathcal{Q}$ that do not meet at all; they must then be parallel. Conversely, there are some points of $\mathcal{Q}$, such as the point Q_1 in the figure, that are not the images of any points of $\mathcal{P}$, because the line $\overrightarrow{OQ_1}$ is parallel to $\mathcal{P}$. Two lines in $\mathcal{Q}$ meeting at Q_1 must be the images under projection of two parallel lines in $\mathcal{P}$. Both planes $\mathcal{P}$ and $\mathcal{Q}$ appear to have some points 'missing'; neither completely captures all the possibilities for the projection lines through O.

(2.4.1)   **Construction of a projective plane:** A *projective plane* is obtained from an ordinary or *affine* plane by adding in the 'missing' points. There are several ways to do this, and we will give a simple one. The disadvantage of our approach is that the extra 'points' that we add seem very different from the original points of the plane. This is an artefact of our construction; a more sophisticated approach would show that the extra points are on just the same footing as the ordinary ones.

(2.4.2)   **Definition:** *Let $\mathcal{P}$ be an affine plane. A* direction *in $\mathcal{P}$ is the class of all lines in $\mathcal{P}$ parallel to some given line.*

This definition makes it true to say that two lines in $\mathcal{P}$ are parallel if and only if they are in the same direction.

(2.4.3)   **Definition:** *The* projective plane $\overline{\mathcal{P}}$ *obtained from $\mathcal{P}$ is the set* $\mathcal{P} \cup \mathcal{D}$, *where $\mathcal{D}$ is the set of all directions in $\mathcal{P}$.*

In other words, $\overline{\mathcal{P}}$ has an extra point for each direction in the original plane $\mathcal{P}$. These extra points are referred to as *points at infinity*, and collectively they make up the *line at infinity*[4].

How is this concept related to our earlier ideas about projection? Let $\mathcal{P}$ be a plane in three-dimensional space, and let O be a point not on $\mathcal{P}$. Then there is an exact one-to-one correspondence between points of the projective plane $\overline{\mathcal{P}}$ and lines through O in the three-dimensional space[5]. Indeed, a line through O either meets $\mathcal{P}$ or is parallel to it; if it meets $\mathcal{P}$ then it corresponds to the point of intersection, and if it does not meet $\mathcal{P}$ then it corresponds to the point at infinity given by the set of all lines in $\mathcal{P}$ that are parallel to it. If $\mathcal{Q}$ is another plane not containing O, there is a similar one-to-one correspondence between the lines through O and the points of $\overline{\mathcal{Q}}$. Combining these two correspondences, we get a one-to-one 'projection' correspondence between $\overline{\mathcal{P}}$ and $\overline{\mathcal{Q}}$ — with no 'missing' points.

A *projective line* in the projective plane $\overline{\mathcal{P}}$ is defined to be either an ordinary line together with the point at infinity that represents its direction, or else the line at infinity. With this definition, the projection map gives a

---

[4]If you have studied complex analysis, you may be familiar with the idea of adding a *single* 'point at infinity' to the complex plane to obtain the *Riemann sphere*. The process whereby one obtains a projective plane is different; one adds a whole lot of points at infinity.

[5]This property is often used in projective geometry textbooks as a *definition* of projective space; a projective space of dimension n is defined as the set of all lines through the origin in a vector space of dimension $n + 1$.

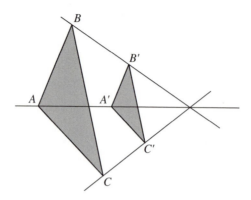

Fig. 2.11. Desargues' theorem: perspective line at infinity.

one-to-one correspondence between projective lines in $\overline{\mathcal{P}}$ and *planes* passing through the centre of projection O; the line at infinity corresponds to the unique plane through O parallel to $\mathcal{P}$. Therefore, projection from $\overline{\mathcal{P}}$ to $\overline{\mathcal{Q}}$ sends projective lines to projective lines. But just as projection may send a point at infinity to an ordinary point, so it may send the line at infinity to an ordinary line.

From the point of view of projective geometry, then, there is nothing special about the line at infinity. Now painting a picture can be thought of as a kind of projection (from physical space onto the plane of the canvas), and we find therefore that parallel horizontal lines in space should be represented in the painting as meeting on some 'vanishing line', the image of the line at infinity. It is this discovery, made by Renaissance artists in the early fifteenth century, which formed an important part of what art historian E. H. Gombrich [18] calls their 'conquest of reality'. He writes

It was said of Uccello [Paolo Uccello, 1379–1435] that the discovery of perspective had so impressed him that he spent days and nights drawing objects in foreshortening and setting himself ever new problems. His fellow artists used to tell that he was so engrossed in these studies that he would hardly look up when his wife called him to go to bed, and would exclaim 'What a sweet thing perspective is!'.

For a mathematical application of this idea, consider Desargues' theorem. Suppose that ABC and $A'B'C'$ are two triangles which are perspective from a point, and suppose that $\overleftrightarrow{AB}$ is parallel to $\overleftrightarrow{A'B'}$ and that $\overleftrightarrow{BC}$ is parallel to $\overleftrightarrow{B'C'}$. Then the intersections $\overleftrightarrow{AB} \cap \overleftrightarrow{A'B'}$ and $\overleftrightarrow{BC} \cap \overleftrightarrow{B'C'}$ are on the line at infinity, so by Desargues' theorem, the two triangles are perspective from this line. Therefore, $\overleftrightarrow{CA}$ is parallel to $\overleftrightarrow{C'A'}$ (Figure 2.11). The special case of Desargues' theorem that we used in Chapter 1 had the point O at infinity also.

(2.4.4)   **Duality:** There is a symmetry between points and lines in projective geometry which is not present in affine geometry. This is the *principle of duality*: *If T is a theorem of plane projective geometry, then the statement obtained from T by systematically interchanging the words 'point' and 'line' is also a theorem (called the* dual *of T).*

To illustrate this, notice that the line axiom is still true[6] in projective geometry: through any two points there is a unique line. But projective geometry also satisfies the less familiar 'dual' of the line axiom: On any two lines there is a unique point (*i.e.* any two projective lines meet at a point). For the line at infinity meets any ordinary line, and two ordinary lines are either parallel (in which case they meet at the point at infinity corresponding to their common direction) or else meet at an ordinary point.

One way to prove the principle of duality is the following. Think of $\mathcal{P}$ as a plane in some three-dimensional space, and let O be a point not in $\mathcal{P}$. Then we have seen that points of $\overline{\mathcal{P}}$ correspond to lines through O, whereas (projective) lines of $\overline{\mathcal{P}}$ correspond to planes through O. But there is also a one-to-one correspondence between lines through O and planes through O; each line (or plane) corresponds to the plane (or line) *perpendicular* to it. Combining these three one-to-one correspondences, we get a one-to-one correspondence between points and projective lines in $\overline{\mathcal{P}}$. This correspondence will transform any true statement into a true dual statement.

Here is another example of the principle of duality at work. Desargues' theorem can be stated as follows: if A, B, C, A', B', and C' are six points in a projective plane, no three of which are collinear, and if $\overline{AA'}$, $\overline{BB'}$, and $\overline{CC'}$ are concurrent, then the points of intersection $\overline{AB} \cap \overline{A'B'}$, $\overline{BC} \cap \overline{B'C'}$, and $\overline{CA} \cap \overline{C'A'}$ are collinear. The dual of Desargues' theorem is therefore the following: if $\mathcal{A}$, $\mathcal{B}$, $\mathcal{C}$, $\mathcal{A}'$, $\mathcal{B}'$, and $\mathcal{C}'$ are six lines in a projective plane, no three of which are concurrent[7], and if $\mathcal{A} \cap \mathcal{A}'$, $\mathcal{B} \cap \mathcal{B}'$, and $\mathcal{C} \cap \mathcal{C}'$ are collinear, then the lines $\overline{(\mathcal{A} \cap \mathcal{B})(\mathcal{A}' \cap \mathcal{B}')}$, $\overline{(\mathcal{B} \cap \mathcal{C})(\mathcal{B}' \cap \mathcal{C}')}$, and $\overline{(\mathcal{C} \cap \mathcal{A})(\mathcal{C}' \cap \mathcal{A}')}$ are concurrent.

But this dual result is just the *converse* of Desargues' theorem![8] The principle of duality has shown us that the converse of Desargues' theorem is also true, without our having to give a separate proof.

---

[6] One has to check three cases to verify this; for instance, if one point is an ordinary point and the other is at infinity, the unique line through them is the line through the given ordinary point in the direction given by the point at infinity.

[7] In forming the dual of a theorem we have also to interchange terms in which the concepts of point and line are *implicit*. For instance, 'collinear' means 'on the same line', and 'concurrent' means 'containing the same point': so in forming the dual it is necessary to interchange these two words.

[8] It is not always true that the dual of a theorem is the same as its converse.

2.5 Exercises

1. Let $\mathcal{P}$ be an affine plane, and let $\mathbf{i}, \mathbf{j}$ be a basis for the associated vector space $\vec{\mathcal{P}}$. Let A and B be the points whose position vectors, relative to some fixed origin O, are $3\mathbf{i} + 2\mathbf{j}$ and $6\mathbf{i} + 8\mathbf{j}$. Find the position vector of the point C on $\overleftrightarrow{AB}$ such that $AC : CB = 2 : 1$. Find also the position vector of the point D on that line such that C and D divide AB harmonically.

2. Three points in the plane $\mathcal{P}$ of the previous question have position vectors $x_1\mathbf{i} + y_1\mathbf{j}$, $x_2\mathbf{i} + y_2\mathbf{j}$, and $x_3\mathbf{i} + y_3\mathbf{j}$. Prove that they are collinear if and only if the determinant

$$\begin{vmatrix} 1 & x_1 & y_1 \\ 1 & x_2 & y_2 \\ 1 & x_3 & y_3 \end{vmatrix}$$

is equal to zero.

3. Let $\mathcal{S}$ be a three-dimensional affine space, and let $\mathbf{i}, \mathbf{j}, \mathbf{k}$ be a basis for the associated vector space $\vec{\mathcal{S}}$. Four points A, B, C, and D have position vectors relative to some fixed origin of $2\mathbf{i} + 3\mathbf{j} + 5\mathbf{k}$, $\mathbf{i} - \mathbf{j} + \mathbf{k}$, $\mathbf{i} + \mathbf{j} - 2\mathbf{k}$, and $2\mathbf{i} + 5\mathbf{j} + 2\mathbf{k}$ respectively. Show that they are coplanar. Which, if any, pairs of the six lines between the points A, B, C, and D are parallel to one another?

4. $P_1, \ldots, P_n$ are points in an affine space. Is there always a point Q such that

$$\overrightarrow{QP_1} + \cdots + \overrightarrow{QP_n} = \mathbf{0}?$$

5. Let A, B, C, and D be four points in an affine space. Show that the following three conditions are equivalent:

- The midpoints of AC and BD coincide;
- $\overrightarrow{AB} = \overrightarrow{DC}$;
- $\overrightarrow{AD} = \overrightarrow{BC}$.

(In this case one says that $ABCD$ is a parallelogram; see 1.4.1.)
   Show also that if the four points are not collinear, $\overleftrightarrow{AB}$ is parallel to $\overleftrightarrow{DC}$, and $\overleftrightarrow{AD}$ is parallel to $\overleftrightarrow{BC}$, then $ABCD$ is a parallelogram.

6. According to Ceva's theorem, the medians of a triangle ABC all meet in a point. Find the position vector of this point (called the *centroid*) in terms of the position vectors of A, B, and C.

7. Four points A, B, C, and D lie in an affine space. Let M_{AB} denote the midpoint of AB, and use a similar notation for other midpoints. Prove that the lines joining M_{AB} to M_{CD}, M_{AC} to M_{BD}, and M_{AD} to M_{BC}, all meet in a point.

8. ABC is a proper triangle. The points P on AB and Q on BC satisfy $AP : PB = CQ : QB$. Show that the point $X = AQ \cap CP$ must lie on a certain straight line passing through B.

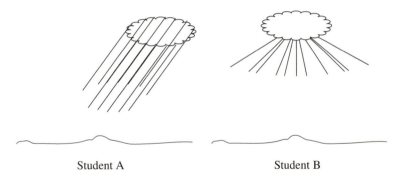

Student A   Student B

Fig. 2.12. The two students' pictures of sunlight through clouds.

9. Three points X, Y, and Z are given in some affine space. It is known that they are the midpoints of the sides of a triangle ABC. Express the position vectors of A, B, and C in terms of those of X, Y, and Z. Find three pairs of parallel lines among the lines joining these six points.

10. S is a three-dimensional affine space, and $\mathbf{i}, \mathbf{j}, \mathbf{k}$ is a basis for the associated space of vectors. Four points A, B, C, and D have position vectors $\mathbf{i}$, $\mathbf{j}$, $\mathbf{k}$, and $\mathbf{i} + \mathbf{j} + \mathbf{k}$ respectively. Show that the lines $\mathcal{L}_1 = \overrightarrow{AB}$ and $\mathcal{L}_2 = \overrightarrow{CD}$ are *skew* (not intersecting and not parallel). If P has position vector $\mathbf{i} + 2\mathbf{j} + 2\mathbf{k}$, show that there is exactly one line through P that meets $\mathcal{L}_1$ and $\mathcal{L}_2$, and find the position vectors of its points of intersection with these two lines.

11. Let ABC be a proper triangle; let X, Y, and Z be points on its sides, and use the same notation for the ratios as in Ceva's theorem. Prove *Menelaus' theorem*: X, Y, and Z are collinear if and only if $\alpha_1 \beta_1 \gamma_1 + \alpha_2 \beta_2 \gamma_2 = 0$. (See Exercise 1.6.11 for another proof.)

12. AB is a line segment in some affine space and C is its midpoint. P is a point not on AB. One can construct a line through P parallel to AB using a *straightedge alone*, as follows: Draw AP, let R be on AP produced, and let RC meet BP at S. Let AS produced meet RB at Q; then PQ is parallel to AB.

   Explain why the construction works. (The word 'produced' just means 'extended'; so, for instance, R lies on $\overrightarrow{AP}$ on the opposite side of P from A.)

13. Translate the affine version of Pappus' theorem (1.6.10) into projective terms by replacing the line at infinity with an ordinary line. Can you give a vector proof?

14. Let $\mathcal{L}$ and $\mathcal{M}$ be lines in some affine plane and let O be a point not on either line. Define *projection from O* to be the operation which takes a point X on $\mathcal{L}$ to the unique point Y on $\mathcal{M}$ such that O, X, and Y are collinear.

   Prove that if x and y are the coordinates of X and Y relative to fixed rulers on $\mathcal{L}$ and $\mathcal{M}$, then

$$y = \frac{\alpha x + \beta}{\gamma x + \delta}$$

where α, β, γ, and δ are constants.

15. Let A, B, C, and D be four points on an affine line $\mathcal{L}$. Their *cross ratio* $(A, B; C, D)$ is defined to be

$$\frac{(a - c)(b - d)}{(a - d)(b - c)}$$

where a, b, c, d are coordinates relative to some ruler. Prove that the cross ratio does not depend on the choice of ruler, and that it is invariant under projection from one line to another. Show also that C, D divide A, B harmonically if and only if $(A, B; C, D) = -1$.

16. Two students are arguing about how the rays of sunlight look when streaming through the clouds. Student A argues: 'The rays of the sun reaching the earth's surface are nearly parallel, so they must look parallel to us'. Student B argues: 'The rays of the sun come from the sun, so they must look as though they come from the sun'. (See Figure 2.12.) Which is right? Explain your answer in terms of projective geometry.

Now go outside on a fair but cloudy day and check your conclusions.

3

Congruence axioms

3.1 The length of a line segment

Affine geometry by itself does not capture the full richness of our intuitive picture of geometry. Its chief deficiency is that we are only able to compare lengths *along the same line* by means of the ruler axiom; and though we can extend this to compare lengths along two parallel lines (by drawing appropriate parallelograms), it makes no sense in affine geometry to compare lengths AB and AC if $\overrightarrow{AB}$ and $\overrightarrow{AC}$ are not parallel. Since the natural way of comparing lengths in two different directions is to draw a circle, another way of putting this is that there is no concept of a *circle* in affine geometry. Yet a third way of putting it is that in affine geometry one cannot say what it means for two triangles to be *congruent*.

In this chapter we will remedy the deficiency by introducing a concept of absolute distance. Affine geometry enriched by means of this additional structure is called *Euclidean* geometry. Formally speaking, we will obtain Euclidean geometry by adding one new undefined term and three new axioms to the ones already discussed in Chapter 1.

This chapter is related to Chapter 4 in the same way that Chapter 1 is related to Chapter 2. The axioms of Euclidean geometry will be used to define a vector concept, the *dot product* of two vectors. It turns out that the dot product encodes all the extra information needed to pass from affine to Euclidean geometry. Thus, the further development of Euclidean geometry can be based simply on the properties of the dot product, and we will take this approach from Chapter 4 onwards. If you are not interested in seeing how the properties of the dot product can be derived, then you can omit this chapter without loss of continuity.

The new undefined term that we will introduce is the *length* of a line segment AB; this length will be denoted $|AB|$. It will also be called the *distance* between the points A and B. We are assuming, therefore, that there

51

is some standard unit of length, with which *any* length, in any direction, can be compared.

Let $\mathcal{L}$ be a line. Among the various rulers on $\mathcal{L}$ there must be one whose scale is given by our standard unit of length, in other words one for which the distance between the point with coordinate 0 and the point with coordinate 1 is exactly one unit. Then we expect that the distance between the point with coordinate a and the point with coordinate b should be $|a - b|$. This amounts to a condition of compatibility between the old (affine) concept of distance ratio and the new (Euclidean) concept of absolute distance. We state it as our first axiom.

AXIOM 10 (RULER COMPATIBILITY AXIOM): For any line $\mathcal{L}$, there is at least one ruler on $\mathcal{L}$ which has the property that for all points A and B on $\mathcal{L}$,

$$|AB| = |a - b|$$

where a and b are the coordinates of A and B relative to the given ruler.

Such a ruler will be called a *standard ruler* on $\mathcal{L}$.

We can derive several consequences from this axiom. For instance, $|AB| = |BA|$, and $|AB| = 0$ if and only if $A = B$. Moreover, we see that affine distance ratios can be reconstructed from the Euclidean distance. In fact

(3.1.1)   **Proposition:** *Let A, B, and C be three collinear points. Then*

- *If $|AB| + |BC| = |AC|$, then $AB : BC = |AB| : |BC|$.*

- *Otherwise, $AB : BC = -|AB| : |BC|$.*

The point of having two cases is that there are *two* points B for which the ratio of the Euclidean distances $|AB| : |BC|$ is equal to some given ratio $x : y$, one between A and B and one not. This makes for some annoying case-splitting in the classical approach to Euclidean geometry.

Proof: Choose a standard ruler on the line through A, B, and C, and let a, b, and c be coordinates relative to this ruler. We can assume that $a \geq c$ (if not, just interchange A and C). Then the condition $|AB| + |BC| = |AC|$ translates to $|a - b| + |b - c| = |a - c|$, and this condition is fulfilled if and only if $a \geq b \geq c$.

If the condition is fulfilled, $a - b$ and $b - c$ are both positive, so

$$AB : BC = (a - b) : (b - c) = |a - b| : |b - c| = |AB| : |BC|.$$

If the condition is not fulfilled, one of $a - b$, $b - c$ is positive and the other is negative. So

$$AB : BC = (a - b) : (b - c) = -|a - b| : |b - c| = -|AB| : |BC|. \quad \square$$

Remark: One might feel that it is slightly pedantic to be so careful about the question of whether B is between A and C or not! However, it is easy to make mistakes by not being careful. Even Euclid himself did so. It was not until the nineteenth century that people noticed that certain statements about 'betweenness' had been implicitly assumed by Euclid, contrary to his philosophy which was to make all his assumptions as explicit as possible in the form of axioms. The most famous of these implicit assumptions is sometimes called *Pasch's axiom*: a line passing through a vertex and an interior point of a triangle meets the opposite side *between* the two opposite vertices.

(3.1.2)   **Definition:** *Let ABC and A′B′C′ be two triangles. They are said to be* congruent *if corresponding sides have the same length:* $|AB| = |A'B'|$, $|BC| = |B'C'|$, *and* $|CA| = |C'A'|$.

We allow the possibility that some (or even all) of the vertices of the triangle $A'B'C'$ are also vertices of the triangle ABC. For instance, if $|AB| = |AC|$ we may say that triangles ABC and ACB are congruent. (A triangle ABC of this sort is called *isosceles*.)

Our intuition tells us that the shape of a triangle is determined by the lengths of its sides, and that congruent triangles must therefore be 'the same shape'. For example, imagine making a model triangle from a framework of hinged metal rods. If you start with two rods and hinge them together, you get a rather wobbly structure. But as soon as you add a third side the wobbliness disappears; even though each joint is hinged, the triangle cannot change its shape. It is to take advantage of this rigidity that electricity pylons and similar structures are divided up into triangles by diagonal cross-bracing. We can express it axiomatically as follows.

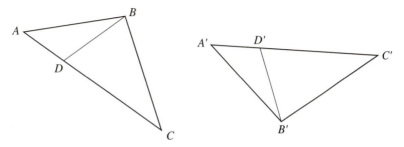

Fig. 3.1. The congruence axiom.

> AXIOM 11 (CONGRUENCE AXIOM): Let ABC and $A'B'C'$ be congruent triangles. Let D be a point on $\overrightarrow{AC}$ and D' a point on $\overrightarrow{A'C'}$. If $AD : DC = A'D' : D'C'$, then $|BD| = |B'D'|$.

The idea is that if the distances $|AB|$, $|BC|$, and $|CA|$ are fixed, then the distance from B to any point D on the side $\overrightarrow{AC}$ is fixed also; see Figure 3.1.

3.2 The length of a vector

Let $\mathbf{v}$ be a vector. We would like to define the *length* of $\mathbf{v}$ to be the distance through which $\mathbf{v}$ moves a general point: in other words, if $\mathbf{v}(X) = Y$, then the length of $\mathbf{v}$ should be $|XY|$. But this definition depends on the arbitrary choice of the point X, so there is a problem of well-definedness; would we have obtained the same result if we had started with a different X? The congruence axiom can be used to show that we would.

(3.2.1) **Proposition:** *Let $ABCD$ be a parallelogram. Then $|AB| = |CD|$ and $|AD| = |BC|$.*

Proof: (See Figure 3.2.) Let X be the midpoint of the parallelogram, and notice that since $BX : XD = 1 : 1$, $|BX| = |XD|$ by the ruler compatibility axiom. Using a standard ruler on the diagonal $\overrightarrow{AC}$, we can find points P and Q on $\overrightarrow{AC}$ such that $PX : XQ = 1 : 1$ and $|PX| = |QX| = |BX| = |DX|$. The triangle BXQ is then isosceles, so that BXQ is congruent to QXB; and $XP : PQ = XD : DB$. So by the congruence axiom, $|BP| = |DQ|$. Therefore, the triangles BXP and DXQ are congruent. Moreover, $PA : AX = QC : CX$. Applying the congruence axiom to triangles BXP and DXQ, we find that $|AB| = |CD|$, as required. Similarly we can prove that $|AD| = |BC|$. $\square$

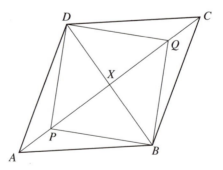

Fig. 3.2. Proof that opposite sides of a parallelogram have the same length.

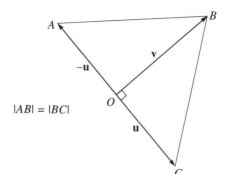

Fig. 3.3. Perpendicular vectors.

Let **v** be a vector. Then by definition 1.5.1, if X and X' are any two points, then $XX'\mathbf{v}(X')\mathbf{v}(X)$ is a parallelogram. So by the result we have just proved, the distance from X to $\mathbf{v}(X)$ is the same as the distance from X' to $\mathbf{v}(X')$. This allows us to make the following definition:

(3.2.2)   **Definition:** *The* length $|\mathbf{v}|$ *of a vector* **v** *is the distance* $|X\mathbf{v}(X)|$ *for any point* X. *A vector of length 1 is called a* unit *vector.*

In other words, $|\overrightarrow{XY}| = |XY|$. It is easy to check that $|\lambda\mathbf{v}| = |\lambda||\mathbf{v}|$, for any real number λ. In particular, $|-\mathbf{v}| = |\mathbf{v}|$.

(3.2.3)   **Definition:** *Let* **u** *and* **v** *be two vectors. We say that* **u** *and* **v** *are* perpendicular *if*

$$|\mathbf{u} + \mathbf{v}| = |\mathbf{u} - \mathbf{v}|$$

This definition is illustrated in Figure 3.3, where the vectors are shown as position vectors relative to some origin O. Notice that *any* vector is perpendicular to the zero vector.

(3.2.4)   **Proposition:** *Suppose that* **u** *and* **v** *are perpendicular vectors. Then* $\lambda\mathbf{u}$ *and* $\mu\mathbf{v}$ *are also perpendicular, for any real numbers* λ *and* μ.

Proof: The result is trivial if either vector is **0**. Otherwise, choose an origin O, and let A, B, and C be the points with position vectors $-\mathbf{u}$, **v**, and **u** respectively. Then $|AO| = |OC|$ and $|AB| = |CB|$, so the triangles AOB and COB are congruent. It follows from the congruence axiom, then, that *any* point P on $\overrightarrow{OB}$ has $|AP| = |CP|$. In particular, we could take P to be the point with position vector $\mu\mathbf{v}$, and this shows that **u** is perpendicular to $\mu\mathbf{v}$. Another application of the same argument then shows that $\lambda\mathbf{u}$ is perpendicular to $\mu\mathbf{v}$. □

Remember (2.1.4) that if $\mathcal{L}$ is a line, we said that a vector **v** is parallel to $\mathcal{L}$ if $\mathbf{v}(X) \in \mathcal{L}$ whenever $X \in \mathcal{L}$. The vectors parallel to a given line

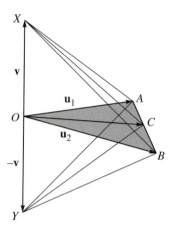

Fig. 3.4. The proof of lemma 3.2.5.

are all multiples of one another. We now define two lines $\mathcal{L}_1$ and $\mathcal{L}_2$ to be *perpendicular* if they intersect and every vector parallel to $\mathcal{L}_1$ is perpendicular to every vector parallel to $\mathcal{L}_2$. By the previous proposition, it comes to the same thing to say that the lines intersect and there is *just one* nonzero vector parallel to $\mathcal{L}_1$ which is perpendicular to *just one* nonzero vector parallel to $\mathcal{L}_2$.

(3.2.5) **Lemma:** *Let* $\mathbf{u}_1$, $\mathbf{u}_2$, *and* $\mathbf{v}$ *be vectors, and suppose that* $\mathbf{u}_1$ *and* $\mathbf{u}_2$ *are both perpendicular to* $\mathbf{v}$. *Then* $\frac{1}{2}(\mathbf{u}_1 + \mathbf{u}_2)$ *is also perpendicular to* $\mathbf{v}$.

Proof: (See Figure 3.4.) It is important in this proof to remember that the three vectors need not be in the same plane. Choose an origin O, and let X and Y have position vectors $\mathbf{v}$ and $-\mathbf{v}$. Let A have position vector $\mathbf{u}_1$, B have position vector $\mathbf{u}_2$, and C have position vector $\frac{1}{2}(\mathbf{u}_1 + \mathbf{u}_2)$. By the ratio theorem, A, B, and C are collinear, and $AC : CB = 1 : 1$.

Because $\mathbf{v}$ and $\mathbf{u}_1$ are perpendicular, $|XA| = |YA|$. Because $\mathbf{v}$ and $\mathbf{u}_2$ are perpendicular, $|XB| = |YB|$. Therefore the triangles XAB and YAB are congruent, so we may apply the congruence axiom to conclude that $|XC| = |YC|$, which implies that $\mathbf{v}$ and $\frac{1}{2}(\mathbf{u}_1 + \mathbf{u}_2)$ are perpendicular. $\square$

(3.2.6) **Proposition:** *Let* $\mathbf{v}$ *be a fixed vector. Then the set of all vectors that are perpendicular to* $\mathbf{v}$ *forms a* vector subspace *of the set of all vectors. This means that if* $\mathbf{u}_1$ *and* $\mathbf{u}_2$ *are perpendicular to* $\mathbf{v}$, *then so is* $\lambda_1\mathbf{u}_1 + \lambda_2\mathbf{u}_2$, *for any scalars* λ_1 *and* λ_2.

Proof: This follows from the two previous results; 3.2.5 deals with addition and 3.2.4 with scalar multiplication. Let $\mathbf{u}_1$ and $\mathbf{u}_2$ be perpendicular to $\mathbf{v}$. Then by 3.2.4, $2\lambda_1\mathbf{u}_1$ and $2\lambda_2\mathbf{u}_2$ are perpendicular to $\mathbf{v}$; and then by 3.2.5, so is $\lambda_1\mathbf{u}_1 + \lambda_2\mathbf{u}_2 = \frac{1}{2}(2\lambda_1\mathbf{u}_1 + 2\lambda_2\mathbf{u}_2)$. $\square$

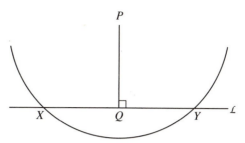

Fig. 3.5. Dropping a perpendicular.

3.3 Dropping a perpendicular

A famous construction in Euclidean geometry explains how to 'drop a perpendicular' from a point P to a line $\mathcal{L}$, it being assumed that P is not on $\mathcal{L}$. To carry out the construction, you set your compasses to a large radius and draw a circle whose centre is P and which intersects $\mathcal{L}$ at points A and B, say. Then you bisect the line segment AB. In this section, we will carry out a similar argument.

(3.3.1)   **Definition:** *Let P be a point in a plane $\mathcal{P}$ and let $r > 0$ be a real number. The* circle *in $\mathcal{P}$ with* centre *P and* radius *r is the set of all points $Q \in \mathcal{P}$ such that $|PQ| = r$.*

Our final axiom deals with the intersection of a circle and a line, and says that a 'sufficiently large' circle about any given point will meet any given line.

> AXIOM 12 (CIRCLE–LINE INTERSECTION AXIOM): Let O be a point and $\mathcal{L}$ a line. Then there is a circle with centre O which meets $\mathcal{L}$ in exactly two points.

(3.3.2)   **Lemma:** *Let $\mathcal{L}$ be a line and let P be a point not on $\mathcal{L}$. Suppose that there are two points X and Y on $\mathcal{L}$ with $|PX| = |PY|$. Let Q be the midpoint of XY. Then $\overrightarrow{PQ}$ is perpendicular to $\mathcal{L}$.*

Proof: (See Figure 3.5.) Let $\overrightarrow{PQ} = \mathbf{u}$ and $\overrightarrow{QX} = \mathbf{v}$. Then $\overrightarrow{QY} = -\mathbf{v}$. Therefore
$$\overrightarrow{PX} = \mathbf{u} + \mathbf{v}, \quad \overrightarrow{PY} = \mathbf{u} - \mathbf{v}.$$

Since $|PX| = |PY|$, $|\mathbf{u} + \mathbf{v}| = |\mathbf{u} - \mathbf{v}|$, so $\mathbf{u}$ and $\mathbf{v}$ are perpendicular. Moreover, $\mathbf{u}$ is parallel to $\overrightarrow{PQ}$ and $\mathbf{v}$ is parallel to $\mathcal{L}$. $\square$

(3.3.3)   **Proposition:** *Let $\mathcal{L}$ be a line and P a point. Then there is a unique point Q on $\mathcal{L}$ for which PQ is perpendicular to $\mathcal{L}$.*

Proof: To prove that such a point Q *exists*, we use the circle–line intersection axiom. This tells us that there are two points X and Y on $\mathcal{L}$ which are equidistant from P. Then lemma 3.3.2 says that the midpoint Q of XY is the foot of a perpendicular from Q to $\mathcal{L}$.

To prove that such a point Q is *unique*, suppose that there are two different points on $\mathcal{L}$, Q and Q', such that both PQ and PQ' are perpendicular to $\mathcal{L}$. Introduce a ruler on $\mathcal{L}$, and let small letters denote coordinates relative to this ruler as usual. Let $\varphi(z)$ denote the distance from P to the point $Z \in \mathcal{L}$ with coordinate z. Then since PQ is perpendicular to $\mathcal{L}$, $\varphi(z) = \varphi(2q - z)$, because q is the midpoint of z and $2q - z$. By similar reasoning, since PQ' is also perpendicular to $\mathcal{L}$, $\varphi(w) = \varphi(2q' - w)$ for any real number w. Put $w = 2q - z$ to obtain

$$\varphi(z) = \varphi(z + 2(q' - q)).$$

In this formula, z can be any real number, so the function φ is *periodic*, with period $2(q' - q)$. It follows that for any $r > 0$ there are either no points or infinitely many points on $\mathcal{L}$ at distance r from P. But this contradicts the circle-line intersection axiom, which tells us that for some value of r there are *exactly two* points on $\mathcal{L}$ at distance r from P. This contradiction shows that our initial hypothesis — that there were two distinct perpendiculars from P to $\mathcal{L}$ — must have been wrong. So there can only be one. $\square$

(3.3.4)   **Corollary:** *Coplanar lines that are perpendicular to the same line are parallel.*

Proof: If they met at a point P, then there would be two perpendiculars from P to the given line, contradicting the uniqueness of perpendiculars. $\square$

(3.3.5)   **Definition:** *An* altitude *of a triangle ABC is a perpendicular from a vertex to the opposite side, for example a perpendicular from A to $\overrightarrow{BC}$.*

(3.3.6)   **Proposition:** *Let ABC and A'B'C' be congruent triangles, and let $\overrightarrow{BQ}$ and $\overrightarrow{B'Q'}$ be the altitudes from B and B' to points Q and Q' on the sides $\overrightarrow{AC}$ and $\overrightarrow{A'C'}$. Then $AQ : QC = A'Q' : Q'C'$, and so $|BQ| = |B'Q'|$.*

Proof: By the circle–line intersection axiom, there are two points X and Y on $\overrightarrow{AC}$ that are equidistant from B; and by lemma 3.3.2, Q is the midpoint of XY. Now let X' and Y' be the corresponding points on $\overrightarrow{A'C'}$, *i.e.* those satisfying

$$A'X' : X'C' = AX : XC, \quad A'Y' : Y'C' = AY : YC.$$

Then by the congruence axiom, $|B'X'| = |BX| = |BY| = |B'Y'|$, so Q' is the midpoint of $X'Y'$. Now by the properties of ratios, $A'Q' : Q'C' = AQ : QC$. $\square$

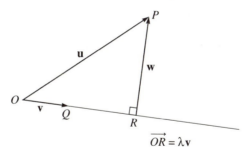

$$\overrightarrow{OR} = \lambda \mathbf{v}$$

Fig. 3.6. The proof of proposition 3.4.1.

3.4   The dot product of two vectors

The results of the previous section, about dropping perpendiculars, allow us to define what we mean by the *component* of one vector in the direction of another one.

(3.4.1)   **Proposition:** *Let* **u** *and* **v** *be vectors, with* **v** *nonzero. There is then a unique constant* λ *such that*

$$\mathbf{u} = \lambda \mathbf{v} + \mathbf{w}$$

with **w** *perpendicular to* **v**.

Proof: (See Figure 3.6.) Choose an origin O, and let P and Q be points such that $\overrightarrow{OP} = \mathbf{u}$ and $\overrightarrow{OQ} = \mathbf{v}$. Then $\lambda \mathbf{v} = \overrightarrow{OR}$, where R is a point on $\overrightarrow{OQ}$ such that $OR : OQ = \lambda : 1$. We have

$$\mathbf{w} = \mathbf{u} - \lambda \mathbf{v} = \overrightarrow{RP};$$

to say that this is perpendicular to **v** is just to say that R is the foot of a perpendicular from P to $\overrightarrow{OQ}$. Since there is exactly one such point R, there must be exactly one scalar λ with the required property. $\square$

The vector $\lambda \mathbf{v}$ appearing above is called the *component of* **u** *parallel to* **v**; the vector **w** is called the *component of* **u** *perpendicular to* **v**.

(3.4.2)   **Definition:** *Let* **u** *and* **v** *be vectors. Their* dot product **u** · **v** *is defined by*

$$\mathbf{u} \cdot \mathbf{v} = \lambda |\mathbf{v}|^2$$

where $\lambda \mathbf{v}$ *is the component of* **u** *in the direction of* **v**. *(If* **v** $= \mathbf{0}$, *define* **u** · **v** $= 0$.*)*

The dot product is a scalar (real number) obtained by 'multiplying' two vectors. For this reason it is often referred to as the *scalar product* of two vectors, but this can sometimes lead to confusion with the operation of

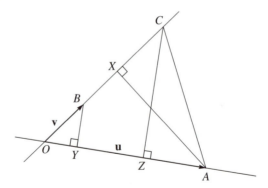

Fig. 3.7. Proof of the symmetry of the dot product.

multiplying a vector by a scalar, which is part of the structure of a vector space.

Here are the basic properties of the dot product which we will use in the subsequent discussion:

(3.4.3)   **Proposition:** *The dot product is* positive definite*: for any vector* **u**, *we have* **u** · **u** ≥ 0, *with equality if and only if* **u** = **0**.

Proof: One can write **u** = **u** + **0**, and **0** is perpendicular to **u**; so **u** is the component of **u** in the direction **u**. Therefore, by definition, **u** · **u** = $|\mathbf{u}|^2$; the result follows. □

(3.4.4)   **Proposition:** *The dot product is* symmetric*: for any two vectors* **u** *and* **v**,

$$\mathbf{u} \cdot \mathbf{v} = \mathbf{v} \cdot \mathbf{u}.$$

Proof: We assume that **u** and **v** are independent, for otherwise the result is easy (see Exercise 3.5.6). Choose an origin O, let $\overrightarrow{OA} = \mathbf{u}$ and $\overrightarrow{OB} = \mathbf{v}$. Let $\overrightarrow{OC} = \lambda \overrightarrow{OB}$, where $\lambda = |\mathbf{u}|/|\mathbf{v}|$. Thus $|OC| = |OA|$. Let X be the foot of the perpendicular from A to $\overrightarrow{OB}$, and let Y and Z be the feet of the perpendiculars from B and C to $\overrightarrow{OA}$ (see Figure 3.7).

The triangles OAC and OCA are congruent, and we have proved (3.3.6) that corresponding altitudes in congruent triangles divide the corresponding sides in the same ratio. Thus we find that $OZ : OA = OX : XC$. Also, the lines $\overleftrightarrow{ZC}$ and $\overleftrightarrow{YB}$ are parallel, since they are both perpendicular to $\overrightarrow{OA}$ and lie in the plane of O, A, and B. From the similarity axiom, therefore, $OZ : OY = OC : OB = |\mathbf{u}| : |\mathbf{v}|$.

By definition of the dot product, $OX : OB = \mathbf{u} \cdot \mathbf{v} : |\mathbf{v}|^2$; since $OC : OB = |\mathbf{u}| : |\mathbf{v}|$, it follows that

$$OX : OC = \mathbf{u} \cdot \mathbf{v} : |\mathbf{u}||\mathbf{v}|.$$

But also, $OY : OA = \mathbf{v} \cdot \mathbf{u} : |\mathbf{u}|^2$, and $OZ : OY = |\mathbf{u}| : |\mathbf{v}|$, so

$$OZ : OA = \mathbf{v} \cdot \mathbf{u} : |\mathbf{u}||\mathbf{v}|.$$

As $OZ : OA = OX : OC$, it follows that $\mathbf{u} \cdot \mathbf{v} = \mathbf{v} \cdot \mathbf{u}$. $\square$

(3.4.5)   **Proposition:** *The dot product is* linear*:*

$$(\mu_1 \mathbf{u}_1 + \mu_2 \mathbf{u}_2) \cdot \mathbf{v} = \mu_1 \mathbf{u}_1 \cdot \mathbf{v} + \mu_2 \mathbf{u}_2 \cdot \mathbf{v}.$$

Remark: Of course, because of the symmetry, the dot product is linear in the second variable $\mathbf{v}$ as well.

Proof: We will use the fact, proved in 3.2.6, that the set of all vectors perpendicular to a given vector $\mathbf{v}$ forms a *vector subspace* of the space of all vectors. Now suppose that

$$\mathbf{u}_1 = \lambda_1 \mathbf{v}_1 + \mathbf{w}_1, \qquad \mathbf{u}_2 = \lambda_2 \mathbf{v}_2 + \mathbf{w}_2$$

are the decompositions of $\mathbf{u}_1$ and $\mathbf{u}_2$ into components parallel and perpendicular to $\mathbf{v}$. Then

$$\mu_1 \mathbf{u}_1 + \mu_2 \mathbf{u}_2 = (\mu_1 \lambda_1 + \mu_2 \lambda_2)\mathbf{v} + (\mu_1 \mathbf{w}_1 + \mu_2 \mathbf{w}_2)$$

is such a decomposition of $\mu_1 \mathbf{u}_1 + \mu_2 \mathbf{u}_2$ — the second term is perpendicular to $\mathbf{v}$ because of 3.2.6. So by definition

$$\mathbf{u}_1 \cdot \mathbf{v} = \lambda_1 |\mathbf{v}|^2, \quad \mathbf{u}_2 \cdot \mathbf{v} = \lambda_2 |\mathbf{v}|^2, \quad (\mu_1 \mathbf{u}_1 + \mu_2 \mathbf{u}_2) \cdot \mathbf{v} = (\mu_1 \lambda_1 + \mu_2 \lambda_2)|\mathbf{v}|^2,$$

and the result follows. $\square$

3.5 Exercises

1. Show that a circle and a line cannot meet at more than two points.

2. Let P be a point on a circle with centre O. Show that the line $\mathcal{L}$ through P perpendicular to OP meets the circle only at P, but that any other line through P meets the circle again somewhere else.

   ($\mathcal{L}$ is called the *tangent line* to the circle at P.)

3. Show that if two circles meet in two points, then the line joining their points of intersection is perpendicular to the line joining their centres. Deduce that it is impossible for two circles to meet in three or more points.

4. Let $ABCD$ be a parallelogram. Show that the triangles ABC and CDA are congruent.

   Conversely, suppose that A, B, C, and D are four points in a plane, and that the triangles ABC and CDA are congruent. Must $ABCD$ be a parallelogram? If not, what else can happen?

5. $\mathcal{L}_1$ and $\mathcal{L}_2$ are two lines in a plane, meeting at O. $X \neq O$ is a point on one of the lines and Y is the foot of the perpendicular from X to the other line. Prove that the fraction $|OY|/|OX|$ does not depend on the choice of X.

6. Let $\mathbf{u}$ be a unit vector, $\mathbf{v} = \lambda \mathbf{u}$ and $\mathbf{w} = \mu \mathbf{u}$. Prove directly from the definition of the dot product that $\mathbf{v} \cdot \mathbf{w} = \lambda \mu$. Hence verify the symmetry of the dot product in the case of dependent vectors.

4

Euclidean geometry

4.1 The dot product

Let S be an affine space. We saw in the last chapter that if we also assume that S has the extra structure of an absolute *distance* $|AB|$ between two points A and B, then it is possible to define a *dot product* $\mathbf{u} \cdot \mathbf{v}$ between any two vectors $\mathbf{u}$ and $\mathbf{v}$ in $\vec{S}$. This dot product has the four properties of linearity, symmetry, positivity, and definiteness, which are summarized in Figure 4.1.

(4.1.1)   **Definition:** *An affine space whose associated vector space is equipped with a dot product operation that satisfies the four properties listed in Figure 4.1 is called a* Euclidean space.

The linearity property is stated in Figure 4.1 with reference only to the first vector in the dot product. But because the dot product is symmetric (the second property) it must be linear in the second variable as well; that is, we must have $\mathbf{u} \cdot (\mu_1 \mathbf{v}_1 + \mu_2 \mathbf{v}_2) = \mu_1 \mathbf{u} \cdot \mathbf{v}_1 + \mu_2 \mathbf{u} \cdot \mathbf{v}_2$.

Just as we did in Chapter 2 with reference to affine geometry, we can ask whether the process of going from geometrical axioms (the axioms of congruence) to algebraic structure (the dot product) is reversible. Given a Euclidean space S, is it always possible to define the distance $|AB|$ between points of S in such a way that the axioms of Chapter 3 are satisfied? It turns out that this *is* always possible; so the classical approach to Euclidean geometry by way of congruence axioms and the approach by way of the dot product are completely equivalent. Once again, any theorem that can be proved by one method can be proved by the other. Moreover, since any vector space can be equipped with a dot product, we will have a proof that our geometrical axioms, augmented now by the axioms of congruence, are still consistent relative to the theory of the real numbers.

To carry out this process we need first of all to define *distance* purely in terms of the dot product.

$$(\lambda_1 \mathbf{u}_1 + \lambda_2 \mathbf{u}_2) \cdot \mathbf{v} = \lambda_1 \mathbf{u}_1 \cdot \mathbf{v} + \lambda_2 \mathbf{u}_2 \cdot \mathbf{v}.$$
$$\mathbf{u} \cdot \mathbf{v} = \mathbf{v} \cdot \mathbf{u}.$$
$$\mathbf{u} \cdot \mathbf{u} \geq 0.$$
$$\mathbf{u} \cdot \mathbf{u} = 0 \Longleftrightarrow \mathbf{u} = \mathbf{0}.$$

Fig. 4.1. Axioms for the dot product.

(4.1.2) **Definition:** *Let S be a Euclidean space. If $\mathbf{v} \in \vec{S}$ is a vector, we define the* length *$|\mathbf{v}|$ of $\mathbf{v}$ to be $\sqrt{\mathbf{v} \cdot \mathbf{v}}$ (where we take the positive square root). If A and B are points of S, we define the* distance *$|AB|$ to be the length of the vector $\overrightarrow{AB}$.*

This definition is obviously compatible with the definition of the length of a vector in terms of distance given in the previous chapter (3.2.2). From the linearity of the dot product we obtain the 'semilinearity' property of the length of a vector: $|\lambda \mathbf{v}| = |\lambda||\mathbf{v}|$. From this property it follows that any vector $\mathbf{v}$ can be written as $|\mathbf{v}|\mathbf{u}$, where $\mathbf{u}$ is a *unit* vector, that is a vector of length 1.

(4.1.3) **Example:** Let $\mathbf{u}$ and $\mathbf{v}$ be vectors; can we calculate the length of $\mathbf{u} - \mathbf{v}$? We may write $|\mathbf{u} - \mathbf{v}|^2$ as $(\mathbf{u} - \mathbf{v}) \cdot (\mathbf{u} - \mathbf{v})$, and then expand this expression using the linearity of the dot product to get

$$|\mathbf{u} - \mathbf{v}|^2 = \mathbf{u} \cdot (\mathbf{u} - \mathbf{v}) - \mathbf{v} \cdot (\mathbf{u} - \mathbf{v}) = |\mathbf{u}|^2 - 2\mathbf{u} \cdot \mathbf{v} + |\mathbf{v}|^2.$$

It is sometimes useful to rearrange this equation in the form

(4.1.4)
$$\mathbf{u} \cdot \mathbf{v} = \frac{1}{2} \left(|\mathbf{u}|^2 + |\mathbf{v}|^2 - |\mathbf{u} - \mathbf{v}|^2 \right)$$

which is known as the *polarization identity*. This identity tells us that 'lengths determine dot products', and we will use it later in the chapter to show that any transformation that preserves distances (an *isometry*) must preserve angles as well.

(4.1.5) **Theorem:** *Let S be a Euclidean space, and define the term 'distance' as above. Then S is a model for the congruence axioms of Chapter 3.*

Proof: We must check that each of the axioms are satisfied.

Ruler compatibility axiom: On any line $\mathcal{L}$ there is a ruler such that the distance between points $A, B \in \mathcal{L}$ is given in terms of coordinates by $|a - b|$. Let $\mathcal{L}$ be a line; choose a point $X \in \mathcal{L}$ and a unit vector $\mathbf{u}$ parallel to $\mathcal{L}$.

Then for each point $P \in \mathcal{L}$ there is a scalar λ such that $\overrightarrow{XP} = \lambda\mathbf{u}$, and the assignment $P \mapsto \lambda$ gives a ruler on $\mathcal{L}$. This ruler satisfies the compatibility axiom, because if A and B are points on $\mathcal{L}$ and a and b their coordinates, then by definition

$$|AB| = |\overrightarrow{AB}| = |a\mathbf{u} - b\mathbf{u}| = |a - b||\mathbf{u}| = |a - b|.$$

Congruence axiom: Let ABC and $A'B'C'$ be congruent triangles, and let $D \in \overline{AC}$ and $D' \in \overline{A'C'}$ with $AD : DC = A'D' : D'C'$. Then $|BD| = |B'D'|$.

Let $\mathbf{a} = \overrightarrow{BA}$, $\mathbf{c} = \overrightarrow{BC}$, and define $\mathbf{a}'$ and $\mathbf{c}'$ similarly. By definition, the corresponding sides of congruent triangles are equal in length, so $|\mathbf{a}| = |\mathbf{a}'|$, $|\mathbf{c}| = |\mathbf{c}'|$, and $|\mathbf{a} - \mathbf{c}| = |\mathbf{a}' - \mathbf{c}'|$. Using the polarization identity (4.1.4), we find that $\mathbf{a} \cdot \mathbf{c} = \mathbf{a}' \cdot \mathbf{c}'$. Now suppose that $AD : DC = A'D' : D'C' = \lambda : \mu$. Then $\overrightarrow{BD} = (\lambda\mathbf{c} + \mu\mathbf{a})/(\lambda + \mu)$ by the ratio theorem (2.2.2), and so

$$|BD|^2 = \frac{1}{(\lambda + \mu)^2}\left(\lambda^2|\mathbf{c}|^2 + 2\lambda\mu\mathbf{a} \cdot \mathbf{c} + \mu^2|\mathbf{a}|^2\right).$$

This expression remains the same when $\mathbf{a}$ and $\mathbf{c}$ are replaced by $\mathbf{a}'$ and $\mathbf{c}'$, and so $|BD| = |B'D'|$.

Circle–line intersection axiom: Let O be a point and $\mathcal{L}$ a line; then there is a circle centre O intersecting $\mathcal{L}$ in exactly two points. Represent points by their position vectors relative to O as origin; then every point on $\mathcal{L}$ can be represented by a position vector

$$\mathbf{x} + \lambda\mathbf{u},$$

where $\mathbf{x}$ is the position vector of a fixed point $X \in \mathcal{L}$ and $\mathbf{u}$ is a vector parallel to $\mathcal{L}$. The condition for such a point to lie on the circle of radius r and centre O is $|\mathbf{x} + \lambda\mathbf{u}| = r$; squaring this, we get the quadratic equation

$$|\mathbf{u}|^2\lambda^2 + 2\mathbf{x} \cdot \mathbf{u}\lambda + |\mathbf{x}|^2 - r^2 = 0$$

for λ. If $r > |\mathbf{x}|$, then the left-hand side of this equation is negative for $\lambda = 0$ and positive for λ large; so the quadratic equation will have two real roots, which correspond to two points of intersection of the line and the circle. $\square$

4.2 Two classic theorems

The dot product allows us to give quick proofs of some classic theorems of Euclidean geometry. Here are two examples. Throughout, we are working in some fixed Euclidean space $\mathcal{S}$.

(4.2.1)   **Definition:** *Vectors* **u** *and* **v** *are* perpendicular *(or* orthogonal*) if* $\mathbf{u} \cdot \mathbf{v} = 0$.

We should check that this definition agrees with the one we used in the previous chapter. Suppose that **u** and **v** are vectors. Then

$$|\mathbf{v} + \mathbf{u}|^2 - |\mathbf{v} - \mathbf{u}|^2 = 4\mathbf{v} \cdot \mathbf{u},$$

and so $|\mathbf{v} + \mathbf{u}| = |\mathbf{v} - \mathbf{u}|$ if and only if $\mathbf{u} \cdot \mathbf{v} = 0$. We used the first of these properties as the definition of perpendicularity in 3.2.3.

We will also say that lines $\mathcal{L}$ and $\mathcal{M}$ are *perpendicular* if vectors parallel to them are perpendicular; and that a proper triangle ABC is *right-angled* if $\overrightarrow{AB}$ is perpendicular to $\overrightarrow{BC}$.

(4.2.2)   **Example:** Let A and B be two points in a plane $\mathcal{P}$, and let O be the midpoint of AB. There is a unique line $\mathcal{L}$ through O that is perpendicular to AB; this line $\mathcal{L}$ is called the *perpendicular bisector* of AB. The perpendicular bisector can also be characterized in terms of distances; $\mathcal{L}$ is the set of all points X such that $|AX| = |BX|$. To prove this, take position vectors relative to origin O. Let $\overrightarrow{OA} = \mathbf{u}$, so that $\overrightarrow{OB} = -\mathbf{u}$, and let $\overrightarrow{OX} = \mathbf{v}$. Then $\overrightarrow{AX} = \mathbf{v} - \mathbf{u}$ and $\overrightarrow{BX} = \mathbf{v} + \mathbf{u}$, so $|AX| = |BX|$ if and only if **v** and **u** are perpendicular, which is to say that X lies on the line $\mathcal{L}$.

(4.2.3)   **Theorem:** (PYTHAGORAS' THEOREM) *Let ABC be a right-angled triangle. Then*

$$|AB|^2 + |BC|^2 = |AC|^2.$$

Remark: AC is called the *hypotenuse* of the right-angled triangle ABC, so this theorem is usually stated in the form 'the square on the hypotenuse is equal to the sum of the squares on the other two sides'.

Proof: We start from the vector identity $\overrightarrow{AC} = \overrightarrow{AB} + \overrightarrow{BC}$. Taking the 'dot product square' of this and expanding out gives us

$$|\overrightarrow{AC}|^2 = |\overrightarrow{AB}|^2 + 2\overrightarrow{AB} \cdot \overrightarrow{BC} + |\overrightarrow{BC}|^2.$$

But $\overrightarrow{AB} \cdot \overrightarrow{BC} = 0$ because the triangle is right-angled, so

$$|AC|^2 = |AB|^2 + |BC|^2$$

as required. $\square$

(4.2.4)   **Corollary:** *Let $\mathcal{L}$ be a line and P a point not on $\mathcal{L}$. Then the point on $\mathcal{L}$ nearest to P is the unique point Q such that $\overrightarrow{PQ}$ is perpendicular to $\mathcal{L}$.*

The tradition that ascribes this theorem to Pythagoras derives from the commentary written by Proclus (410–485 AD) on the first book of Euclid. Since Pythagoras probably lived around 550 BC, the ascription is by no means

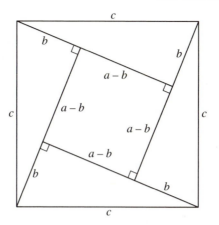

Fig. 4.2. The Indian proof of Pythagoras' theorem.

certain; Proclus himself writes, 'If we listen to those who wish to recount ancient history, we may find some of them referring this theorem to Pythagoras and saying that he sacrificed an ox in honour of his discovery'. Proclus was very impressed by Euclid's proof of the theorem, which depended on a clever addition and subtraction of areas; he calls it 'a most lucid demonstration'.

You may be more familiar with another proof that depends on the addition and subtraction of areas, the 'Indian' method. There are several versions of this, but they all depend on arranging four copies of the given triangle symmetrically around a square. A version given by Bhāskara (born 1114 AD) depends on the diagram of Figure 4.2. Each of the four triangles in the figure has sides a, b, and c, with hypotenuse c; we see from the figure and familiar properties of areas[1] that

$$(a - b)^2 + 4 \cdot \frac{ab}{2} = c^2$$

which gives $a^2 + b^2 = c^2$. Though Bhāskara is late, there is evidence that the theorem was known in India as early as the fifth century BC.

Yet another proof of Pythagoras' theorem makes use of similar triangles (Figure 4.3). Let ABC be a right-angled triangle, and let D be the foot of the perpendicular from B to the hypotenuse AC. Then triangles ADB and ABC are similar[2], and so the ratio $|AB| : |AC|$ is equal to $|AD| : |AB|$. Therefore

---

[1] We have not yet discussed area in this book; we will do so later (Chapter 9).

[2] We have not yet studied similar triangles in this degree of generality; our similarity axiom applies only to triangles cut off by two parallel lines. We only want to illustrate the proof of Pythagoras' theorem here, though.

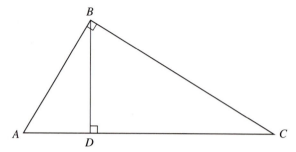

Fig. 4.3. The proof of Pythagoras' theorem by similar triangles.

$|AD| = |AB|^2/|AC|$, and by an analogous argument, $|DC| = |BC|^2/|AC|$. But $|AD| + |DC| = |AC|$, and so we obtain

$$|AB|^2 + |BC|^2 = |AC|^2$$

which is Pythagoras' theorem.

Thus we have three apparently independent methods of proving Pythagoras' theorem; using the dot product, using areas, or using similar triangles. It is not hard to see that the first and the third are more or less equivalent. In fact, our proof of the symmetry of the dot product (3.4.4) already makes use of similar triangles in a geometric situation very like Pythagoras' theorem. But why should there be a relation between similar triangles and area? The surprising answer is that the similarity axiom can in fact be proved as a theorem on the basis of the geometric properties of areas (see Exercise 9.7.11). This fact was already known to Euclid, where the result equivalent to our similarity axiom does not appear until Book 6, after a thorough discussion of areas and proportions. Pythagoras' theorem appears in Book 1.

(4.2.5)   **Theorem:** (THE ANGLE IN A SEMICIRCLE) *Let A, B, and C be three points and let O be the midpoint of AC. Suppose that $|OA| = |OB| = |OC|$. Then the triangle ABC is right-angled.*

Figure 4.4 should make it clear why this theorem is usually stated in the form 'the angle in a semicircle is a right angle'.

Proof: Let $\overrightarrow{OB} = \mathbf{u}$ and $\overrightarrow{OC} = \mathbf{v}$; then $\overrightarrow{OA} = -\mathbf{v}$. Therefore $\overrightarrow{AB} = \mathbf{v} + \mathbf{u}$ and $\overrightarrow{BC} = \mathbf{v} - \mathbf{u}$, so

$$\overrightarrow{AB} \cdot \overrightarrow{BC} = (\mathbf{v} + \mathbf{u}) \cdot (\mathbf{v} - \mathbf{u}) = |\mathbf{v}|^2 - |\mathbf{u}|^2 = |OC|^2 - |OB|^2 = 0$$

and the triangle is right-angled. □

Another ancient author, Diogenes Laertius, also tells the story of a great mathematician sacrificing an ox in honour of an important discovery. But

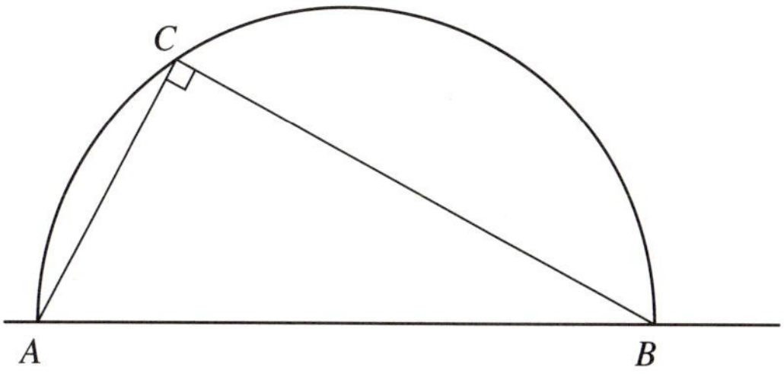

Fig. 4.4. Angle in a semicircle.

according to him, the mathematician in question was Thales, and the discovery was the result about the angle in a semicircle. At any rate, both this result and Pythagoras' theorem were thought significant enough for tales to be told about their origins.

4.3 Angles

There are several ways of introducing the concept of *angle* into geometry. One is to make it a new undefined term, governed by its own axioms; another is to use the theory of area, so that the angle between two lines is defined to be the area of a sector of a unit circle cut off between them. In this book, though, we will define angle in terms of the dot product; we will use the familiar identity $\mathbf{u} \cdot \mathbf{v} = |\mathbf{u}||\mathbf{v}| \cos \theta$ not as a definition of $\mathbf{u} \cdot \mathbf{v}$, but as a definition of θ.

This approach leads most quickly to the geometrical results that we want, but it does assume that we already know about the trigonometric functions $\sin \theta$ and $\cos \theta$. How is it possible to define these functions *without* mentioning geometrical concepts such as angle? In fact there are several ways of doing exactly this; one of the simplest is to use the power series expansions

$$\sin \theta = \sum_{n=0}^{\infty} \frac{(-1)^n \theta^{2n+1}}{(2n+1)!}, \quad \cos \theta = \sum_{n=0}^{\infty} \frac{(-1)^n \theta^{2n}}{(2n)!}$$

as a definition. The details (which really belong to analysis rather than to geometry) may be found in textbooks such as Binmore [5, Chapter 16] or Ebbinghaus *et al.* [13]. All the results about trigonometric functions that will be needed in this book can be proved fairly directly from the power series; they are set out in Appendix A.

(4.3.1)   **Proposition:** (CAUCHY–SCHWARZ INEQUALITY) *Let* $\mathbf{u}$ *and* $\mathbf{v}$ *be any two vectors; then*

$$-|\mathbf{u}||\mathbf{v}| \leq \mathbf{u} \cdot \mathbf{v} \leq |\mathbf{u}||\mathbf{v}|.$$

Moreover, these inequalities are strict ($\leq$ can be replaced by $<$) unless **u** *and* **v** *are linearly dependent (one being a multiple of the other).*

Proof: We can suppose that neither of **u** and **v** is zero, since otherwise the inequality is obvious. Let $\lambda = \mathbf{u} \cdot \mathbf{v}/|\mathbf{v}|^2$; then

$$0 \leq |\mathbf{u} - \lambda\mathbf{v}|^2 = |\mathbf{u}|^2 - 2\lambda\mathbf{u} \cdot \mathbf{v} + \lambda^2|\mathbf{v}|^2 = |\mathbf{u}|^2 - \frac{(\mathbf{u} \cdot \mathbf{v})^2}{|\mathbf{v}|^2}.$$

This implies that $(\mathbf{u} \cdot \mathbf{v})^2 \leq |\mathbf{u}|^2|\mathbf{v}|^2$, which is what we wanted. Moreover, in the case of equality we find that $\mathbf{u} - \lambda\mathbf{v} = 0$, so **u** and **v** are linearly dependent. $\square$

Remark: The value of λ used in this proof is not plucked from thin air; in fact, it is chosen so that $\lambda\mathbf{v}$ is the *component* (3.4.1) of **u** in the direction of **v**. The proof is really just an application of Pythagoras' theorem to say that this component of **u** is shorter than **u** itself.

(4.3.2) **Proposition:** (TRIANGLE INEQUALITY) *Let* **u** *and* **v** *be any two vectors; then* $|\mathbf{u} + \mathbf{v}| \leq |\mathbf{u}| + |\mathbf{v}|$. *Equality holds if and only if one vector is a non-negative multiple of the other.*

Proof: By the Cauchy–Schwarz inequality,

$$|\mathbf{u} + \mathbf{v}|^2 = |\mathbf{u}|^2 + 2\mathbf{u} \cdot \mathbf{v} + |\mathbf{v}|^2 \leq |\mathbf{u}|^2 + 2|\mathbf{u}||\mathbf{v}| + |\mathbf{v}|^2 = (|\mathbf{u}| + |\mathbf{v}|)^2.$$

We have equality if and only if there is equality in Cauchy–Schwarz, $\mathbf{u} \cdot \mathbf{v} = |\mathbf{u}||\mathbf{v}|$, and this happens if and only if one vector is a positive multiple of the other. $\square$

An equivalent form of the triangle inequality is that *in a proper triangle, the sum of any two sides is greater than the third.*

Let **u** and **v** be nonzero vectors. By the Cauchy–Schwarz inequality, the quantity $\mathbf{u} \cdot \mathbf{v}/(|\mathbf{u}||\mathbf{v}|)$ lies between -1 and 1. Now for any c between -1 and 1 inclusive, there is a unique θ between 0 and π (radians[3]) such that $\cos\theta = c$. We can therefore make the following definition:

(4.3.3) **Definition:** *Let* **u** *and* **v** *be nonzero vectors. The* angle[4] *between* **u** *and* **v** *is defined to be the unique θ between 0 and π such that*

$$\cos\theta = \frac{\mathbf{u} \cdot \mathbf{v}}{|\mathbf{u}||\mathbf{v}|}.$$

For points A, B, and C we define the angle $\widehat{ABC}$ to mean the angle between $\overrightarrow{BA}$ and $\overrightarrow{BC}$.

[3] In this book, as in most advanced mathematics books, angles will always be measured in radians; this leads to the simplest formulae for the sine and cosine functions. Remember that π radians are equal to 180 degrees.

[4] This is sometimes called the *unoriented angle*, to distinguish it from the *oriented angle* that will be introduced in the next chapter.

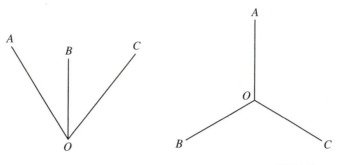

Fig. 4.5. (i) Addition of angles.   (ii) Failure.

Notice that two vectors are perpendicular if and only if the angle between them is $\frac{\pi}{2}$. Notice also that the angle between **u** and **v** is the same as the angle between $\lambda\mathbf{u}$ and $\mu\mathbf{v}$, for any scalars λ and μ that are either both positive or both negative.

Remark: What has become of 'reflex' angles (between π and 2π)? The answer is that in general we have no way of defining them. What looks like an angle of $\frac{3\pi}{2}$ viewed from above looks like an angle of $\frac{\pi}{2}$ viewed from below. We can only say what we mean by a 'reflex' angle if we are in two dimensions (so we can't go from 'above' to 'below'), and we agree on a choice of 'positive direction of rotation' or *orientation*. We will discuss orientation in the next chapter.

Angles add up: in Figure 4.5(i), the angle $\widehat{AOC}$ is the sum of angles $\widehat{AOB}$ and $\widehat{BOC}$. But we need to be careful with this statement; in Figure 4.5(ii), none of the angles is equal to the sum of the other two (they are all equal to $\frac{\pi}{3}$). The difficulty is that our angles are like absolute distances — they don't have signs — and so we have to be careful about 'betweenness', just as we had to for distances in proposition 3.1.1. The correct definition is the following:

(4.3.4)   **Definition:** *Let O, A, B, and C be four points in a plane. We say that $\overrightarrow{OB}$ is between $\overrightarrow{OA}$ and $\overrightarrow{OC}$ if it can be written*

$$\overrightarrow{OB} = \lambda\overrightarrow{OA} + \mu\overrightarrow{OC}$$

with $\lambda, \mu > 0$.

(4.3.5)   **Proposition:** (ANGLE ADDITION LAW) *Let O, A, B, and C be four points in a plane, with $\overrightarrow{OB}$ between $\overrightarrow{OA}$ and $\overrightarrow{OC}$. Then*

$$\widehat{AOC} = \widehat{AOB} + \widehat{BOC}.$$

Proof: Let **u** be a unit vector in the direction $\overrightarrow{OA}$ (that is, $\mathbf{u} = \overrightarrow{OA}/|\overrightarrow{OA}|$), and let **v** be a unit vector in the direction $\overrightarrow{OC}$. Let **w** be a unit vector in the direction $\overrightarrow{OB}$; because $\overrightarrow{OB}$ is between $\overrightarrow{OA}$ and $\overrightarrow{OC}$, we may write $\mathbf{w} = \lambda\mathbf{u} + \mu\mathbf{v}$ with $\lambda, \mu > 0$. Let $\theta = \widehat{AOC}$, $\varphi = \widehat{AOB}$, and $\psi = \widehat{BOC}$. Calculating the dot products, we get

$$\cos\theta = \mathbf{u}\cdot\mathbf{v}, \quad \cos\varphi = \mathbf{u}\cdot\mathbf{w} = \lambda + \mu\cos\theta, \quad \cos\psi = \mathbf{w}\cdot\mathbf{v}\lambda\cos\theta + \mu.$$

Since $\lambda\mathbf{u} + \mu\mathbf{v}$ is a unit vector, $\lambda^2 + \mu^2 + 2\lambda\mu\cos\theta = 1$. So

$$\sin^2\varphi = 1 - \cos^2\varphi = 1 - \lambda^2 - 2\lambda\mu\cos\theta - \mu^2\cos^2\theta$$
$$= \mu^2(1 - \cos^2\theta) = \mu^2\sin^2\theta.$$

Therefore (since $\mu > 0$), $\sin\varphi = \mu\sin\theta$. Similarly, $\sin\psi = \lambda\sin\theta$. Now we can calculate

$$
\begin{aligned}
\cos(\varphi + \psi) &= \cos\varphi\cos\psi - \sin\varphi\sin\psi \\
&= \lambda\mu(1 + \cos^2\theta - \sin^2\theta) + (\lambda^2 + \mu^2)\cos\theta \\
&= (2\lambda\mu\cos\theta + \lambda^2 + \mu^2)\cos\theta \\
&= \cos\theta, \\
\sin(\varphi + \psi) &= \sin\varphi\cos\psi + \cos\varphi\sin\psi \\
&= (\mu(\lambda\cos\theta + \mu) + \lambda(\lambda + \mu\cos\theta))\sin\theta \\
&= (2\lambda\mu\cos\theta + \lambda^2 + \mu^2)\sin\theta \\
&= \sin\theta.
\end{aligned}
$$

From these facts it follows that $\varphi + \psi$ equals θ plus some integer multiple of 2π (see A.1.7). Since, however, all three angles are between 0 and π, this integer multiple must be zero; so we have proved that $\varphi + \psi = \theta$, as required. □

Remark: Suppose that $\overrightarrow{OA} = -v\overrightarrow{OC}$ for some positive constant v, so that $\overrightarrow{OA}$ and $\overrightarrow{OC}$ are in exactly opposite directions. Then in general $\overrightarrow{OB}$ is not between $\overrightarrow{OA}$ and $\overrightarrow{OC}$ according to our definition. Nevertheless,

$$\cos\widehat{AOB} = -\cos\widehat{BOC}$$

and so it is still true that $\widehat{AOB} + \widehat{BOC} = \pi = \widehat{AOC}$. The angles $\widehat{AOB}$ and $\widehat{BOC}$ are called *supplementary* in this case.

Using the angle addition law we can prove the result that we discussed at the very beginning of this book.

(4.3.6)   **Theorem:** (ANGLE SUM OF A TRIANGLE) *The three angles of a proper triangle add up to π radians.*

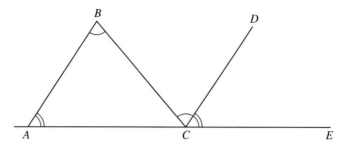

Fig. 4.6. The angle sum of a triangle.

Proof: Let ABC be our triangle. As in Figure 4.6, let D and E be points such that $\overrightarrow{CD} = \overrightarrow{AB}$ and $\overrightarrow{CE} = \overrightarrow{AC}$. Then $\widehat{ABC} = \widehat{BCD}$ (because $\overrightarrow{AB} = -\overrightarrow{DC}$ and $\overrightarrow{CB} = -\overrightarrow{BC}$) and $\widehat{CAB} = \widehat{ECD}$ (because $\overrightarrow{CA} = \overrightarrow{EC}$ and $\overrightarrow{BA} = \overrightarrow{DC}$). Furthermore, $\overrightarrow{CD}$ is between $\overrightarrow{CB}$ and $\overrightarrow{CE}$ (since in fact it is the sum of these two vectors), and so by the angle addition law $\widehat{BCE} = \widehat{BCD} + \widehat{ECD} = \widehat{ABC} + \widehat{CAB}$. Therefore

$$\widehat{CAB} + \widehat{ABC} + \widehat{BCA} = \widehat{ECB} + \widehat{BCA} = \pi,$$

because $\widehat{ECB}$ and $\widehat{BCA}$ are supplementary. □

Remark: This is essentially Euclid's proof as given in the *Elements*. Underlying it is the fact that an angle (such as $\widehat{CAB}$) is unchanged by translation (through the vector $\overrightarrow{AC}$, for example). This property that 'rotation is independent of translation' is equivalent to the parallel axiom, and it fails in non-Euclidean geometry. In Chapter 12 we will study the concepts of *holonomy* and *Gaussian curvature*, which measure the extent to which rotation and translation interact.

4.4 Trigonometry

'Trigonometry' (which literally means 'triangle-measurement') is the study of the relationships between the lengths of the sides and the sizes of the angles in a triangle. Let ABC be a proper triangle in some Euclidean space. It is conventional to denote the length of the side BC by a and to abbreviate the angle $\widehat{CAB}$ to A, and to use similar notation for the other sides and angles; see Figure 4.7. We will also use the notation s for the 'semiperimeter', $s = (a + b + c)/2$.

Our first trigonometrical result is the *cosine rule*,

(4.4.1) $$a^2 = b^2 + c^2 - 2bc \cos A.$$

(There are of course similar formulae for b^2 and c^2.) To prove this we let $\mathbf{b} = \overrightarrow{AB}$ and $\mathbf{c} = \overrightarrow{AC}$. Then we expand

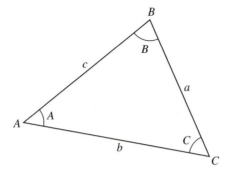

Fig. 4.7. Standard notation in a triangle.

$$a^2 = |\mathbf{b} - \mathbf{c}|^2 = |\mathbf{b}|^2 + |\mathbf{c}|^2 - 2\mathbf{b} \cdot \mathbf{c}$$
$$= b^2 + c^2 - 2bc \cos A.$$

From the cosine rule we can obtain the formulae for $\cos A$ and $\sin A$,

(4.4.2)
$$\cos A = \frac{b^2 + c^2 - a^2}{2bc}, \quad \sin A = \frac{2\sqrt{s(s-a)(s-b)(s-c)}}{bc}.$$

To obtain the formula for $\sin A$, calculate

$$\sin A = \sqrt{1 - \cos^2 A} = \frac{\sqrt{4b^2c^2 - (a^2 - b^2 - c^2)^2}}{2bc}$$

and factorize the expression under the square root sign. When using the formula for $\sin A$, one must be careful to note that $\sin A$ does not determine the angle A uniquely. Since $\sin A = \sin(\pi - A)$, there are two possibilities for A.

The *sine rule*,

(4.4.3)
$$\frac{a}{\sin A} = \frac{b}{\sin B} = \frac{c}{\sin C}$$

is an easy consequence of the formula for $\sin A$ (together with the corresponding formulae for $\sin B$ and $\sin C$).

What happens if we apply these rules in a right-angled triangle, say with $B = \frac{\pi}{2}$? Since $\sin \frac{\pi}{2} = 1$, the sine rule tells us that $\sin A = a/b$. Also, since the triangle is right-angled, $a^2 = b^2 - c^2$ by Pythagoras' theorem. Substituting for a^2 in the cosine rule and dividing through by $2bc$, we get $\cos A = c/b$. Thus we recover the familiar definitions of sine and cosine in terms of the ratios of the sides in a right-angled triangle.

In old-fashioned geometry texts a triangle was said to be *solved* if all its sides and angles were known. It is natural to ask how much information you need in order to solve a triangle, or to put it another way, in what respects you need to know that two triangles are equal in order to be sure that they are congruent to one another. For instance, the formula for the cosine makes it clear that you can solve the triangle if you know the three sides a, b, and c. Other cases in which you can solve the triangle are

- if you know two sides (say b and c) and the *included* angle A; then you can find the missing side a from the cosine rule;

- if you know two angles and any side; you can find the third angle from the theorem about the angle sum (4.3.6), and you can then find all the sides from the sine rule.

When thought of as conditions for two triangles to be congruent, these are often referred to as the 'cases of congruence', and are abbreviated by the letters SSS (three sides), SAS (two sides and the included angle), and AAS/ASA (two angles and a side). The discussion of these three cases is found in Proclus' commentary on Euclid's work.

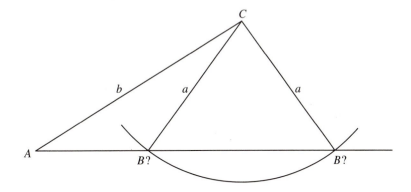

Fig. 4.8. The ambiguous case.

One thing that Proclus surprisingly failed to discuss is the 'ambiguous case', two sides and a non-included angle (memorably abbreviated to ASS). The point is that, if angle A and sides a and b are given, there may be *two* possibilities for the triangle (see Figure 4.8), provided that the given angle A is less than $\frac{\pi}{2}$. To see this algebraically, notice that from the given data one can find $\sin B$ using the sine rule, but this gives two possibilities for B, and either one can occur.

(4.4.4)   **Example:** Let LMN be a triangle, and let $\widehat{LMN} = \theta$. Suppose that it is known that there are two angles φ and ψ with $\varphi + \psi + \theta = \pi$ and

$$\frac{|LM|}{\sin \psi} = \frac{|MN|}{\sin \varphi}.$$

Then the angles $\widehat{NLM}$ and $\widehat{MNL}$ are equal to φ and ψ respectively. The simplest way to see this is to imagine that another triangle $L'M'N'$ has been constructed with angles θ, φ, and ψ and with $|L'M'| = |LM|$; then by the sine rule $|M'N'| = |MN|$, and so the two triangles are congruent (two sides and the included angle).

All the results that we have discussed so far in this chapter were known to the ancients. Here is a famous and beautiful example of a geometric theorem about triangles which was discovered only around 1899. Its discoverer, F. Morley, found it as a special case of a general theory that he was working on, and never published an independent proof[5].

To state Morley's theorem, we need to introduce a little more terminology, which is probably familiar to you anyway. A triangle is called *equilateral* if its sides are all equal, or equivalently if its angles are all $\frac{\pi}{3}$. And if $\widehat{AOB}$ is an angle, the *trisectors* of the angle are the two lines through O between $\overrightarrow{OA}$ and $\overrightarrow{OB}$ which divide the angle into three equal parts. Each part is then (by the angle addition theorem) equal to one-third of the original angle.

(4.4.5)   **Theorem:** (MORLEY'S THEOREM) *The three points of intersection of the adjacent trisectors of the three angles of any triangle form an equilateral triangle.*

See Figure 4.9: ABC is the original triangle, and P, Q and R are the points of intersection of the trisectors. The theorem says that PQR is equilateral.

Proof: Use the standard notation a, b, c, A, B, C for the sides and angles of the triangle ABC. By the sine rule,

$$\frac{a}{\sin A} = \frac{b}{\sin B} = \frac{c}{\sin C} = d, \text{ say.}$$

Let $\alpha = A/3$, $\beta = B/3$, and $\gamma = C/3$. Then

$$|AB| = c = d \sin C = d \sin(3\alpha + 3\beta).$$

(We have used the fact that the sum of the angles of triangle ABC is equal to π.) So by the sine rule in triangle ABR, together with the fact that the sum of the angles of this triangle is π,

---

[5]The first elementary proof of Morley's theorem was published in 1914. For references, see Coxeter [11, pages 23–25]. Our proof is taken from Berger [4].

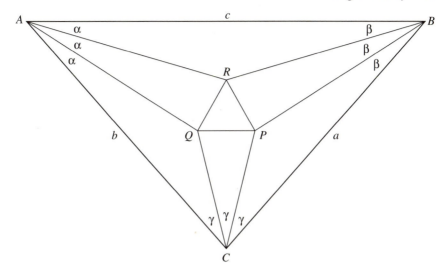

Fig. 4.9. Morley's theorem.

$$|AR| = |AB|\frac{\sin \beta}{\sin(\alpha + \beta)} = d\frac{\sin \beta \sin(3\alpha + 3\beta)}{\sin(\alpha + \beta)}.$$

Now it is easy to prove that for any angle θ,

$$\frac{\sin 3\theta}{\sin \theta} = 4 \sin \left(\frac{\pi}{3} - \theta\right) \sin \left(\frac{2\pi}{3} - \theta\right).$$

Making use of this identity with $\theta = \alpha + \beta$, and remembering that $\alpha + \beta + \gamma = \frac{\pi}{3}$, we find that

$$|AR| = 4d \sin \beta \sin \gamma \sin \left(\frac{\pi}{3} + \gamma\right).$$

Similarly

$$|AQ| = 4d \sin \beta \sin \gamma \sin \left(\frac{\pi}{3} + \beta\right).$$

Also, $(\frac{\pi}{3} + \gamma) + (\frac{\pi}{3} + \beta) + \alpha = \pi$. Therefore, by 4.4.4, the angles of the triangle AQR are α, $\frac{\pi}{3} + \gamma$, and $\frac{\pi}{3} + \beta$, and by the sine rule

$$|RQ| = 4d \sin \alpha \sin \beta \sin \gamma.$$

By symmetry, this is also the value of $|RP|$ and $|PQ|$; so the triangle PQR is equilateral. □

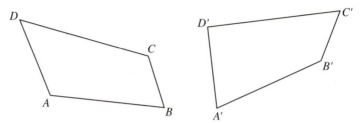

Fig. 4.10. Quadrilaterals with corresponding sides equal need not be congruent.

4.5 Isometries

Our definition of congruence for triangles (two triangles are congruent if corresponding sides are equal) is logically watertight; but in some ways it does not express very clearly the fundamental idea. For instance, our definition is specific for triangles; one would not want to call two *quadrilaterals* congruent if they had corresponding sides equal in length, because they might still not be the same shape (Figure 4.10). The fundamental idea seems to be that two geometrical figures are congruent if one can be 'moved without deformation' so as to coincide with the other. It is then a special property of triangles that this condition is equivalent to corresponding sides being equal.

This at any rate appears to be how Euclid felt. Congruence is introduced into Euclid's system by means of Common Notion 4, 'Things that coincide with one another are equal to one another'; and the Greek word translated 'coincide' here carries the meaning of one thing being moved without deformation onto the other.

But what exactly does the phrase 'moved without deformation' mean? In mathematical terms, a motion is a *mapping* T from a Euclidean space S to itself[6] (or even from one Euclidean space to another). To say that the motion is *without deformation* should mean that all geometric properties are preserved by the transformation, so that (for example) the distance between points A and B should always be equal to the distance between their images $T(A)$ and $T(B)$. A transformation of this kind is called an *isometry* (Greek for 'same distance').

(4.5.1)   **Definition:** *Let S and S' be Euclidean spaces, and let $T: S \rightarrow S'$ be*

---

[6]Each point A is transformed by the motion into a new point $T(A)$. We are thinking of this as some kind of instantaneous transformation. You might argue that when we talk of a *motion* we should be thinking of *continuous* rather than *instantaneous* transformation, so that each point A moves continuously to its new position $T(A)$. This leads to a slightly different concept of motion; see proposition 8.3.8.

a bijective map[7]. Then T is called an isometry *if it preserves distances:* $|AB| = |T(A)T(B)|$, *for all A and B in S.*

(4.5.2) **Example:** Suppose that **v** is a vector of some Euclidean space S. Then **v** defines a map $S \to S$, given by $X \mapsto \mathbf{v}(X)$; this map is called the *translation* by **v**, $T_{\mathbf{v}}$. It is easy to see that $T_{\mathbf{v}}$ is an isometry, since

$$\overline{AB} = \overline{T_{\mathbf{v}}(A)T_{\mathbf{v}}(B)}$$

by the parallelogram rule of vector addition.

Of course there are plenty of other examples including the familiar reflections, rotations, and so on. We will discuss these in the chapters on two- and three- dimensional geometry. But for the moment, what we want to do is to relate the concept of isometry to the intuitive notion of congruence, and to our particular definition of congruent triangles.

(4.5.3) **Lemma:** *Let A, B, C, and D be four points in an affine space. Then* $\overline{AB} = \overline{DC}$ *if and only if the midpoints of AC and BD are the same.*

Proof: This lemma will already be familiar to those who have read Chapter 1, when we used it 'in reverse' to define a parallelogram. To prove it, represent the four points by position vectors relative to some fixed origin, and notice that

$$\mathbf{b} - \mathbf{a} = \mathbf{c} - \mathbf{d} \iff (\mathbf{a} + \mathbf{c})/2 = (\mathbf{b} + \mathbf{d})/2. \qquad \square$$

(4.5.4) **Proposition:** *If T is an isometry, and* $\overline{AB} = \overline{DC}$, *then* $\overline{T(A)T(B)} = \overline{T(D)T(C)}$.

Proof: Let M be the common midpoint of AC and BD. The concept of 'midpoint' only depends on the distance, and an isometry preserves distances, so $T(M)$ is the common midpoint of $T(A)T(C)$ and $T(B)T(D)$. By the lemma, then, $\overline{T(A)T(B)} = \overline{T(D)T(C)}$. $\square$

So given an isometry $T: S \to S'$, we may define a map $\vec{T}: \vec{S} \to \vec{S}'$ by

$$\vec{T}(\overline{AB}) = \overline{T(A)T(B)}.$$

$\vec{T}$ is called the *vectorialization* of T or the transformation *induced* by T on the associated vector spaces. It is clear from the definition that $\vec{T}(\mathbf{u} \pm \mathbf{v}) = \vec{T}(\mathbf{u}) \pm \vec{T}(\mathbf{v})$, and (because T preserves distances) that $|\vec{T}(\mathbf{u})| = |\mathbf{u}|$, for all vectors $\mathbf{u}, \mathbf{v} \in \vec{S}$.

---

[7]That is, a one-to-one correspondence.

(4.5.5)   **Proposition:** *The transformation $\vec{T}$ induced by an isometry is linear, and satisfies the equation*

$$\vec{T}\mathbf{u} \cdot \vec{T}\mathbf{v} = \mathbf{u} \cdot \mathbf{v}$$

for any vectors $\mathbf{u}$ and $\mathbf{v}$.

Proof: We know that

$$|\vec{T}(\mathbf{u})| = |\mathbf{u}| \quad \text{and} \quad |\vec{T}(\mathbf{u}) - \vec{T}(\mathbf{v})| = |\mathbf{u} - \mathbf{v}|$$

for all vectors $\mathbf{u}, \mathbf{v} \in \vec{S}$. Now we use the polarization identity (4.1.4), which says that

$$\mathbf{u} \cdot \mathbf{v} = \frac{1}{2}\left(|\mathbf{u}|^2 + |\mathbf{v}|^2 - |\mathbf{u} - \mathbf{v}|^2\right).$$

Applying this identity to $\mathbf{u} \cdot \mathbf{v}$ and also to $\vec{T}(\mathbf{u}) \cdot \vec{T}(\mathbf{v})$, we find that

(4.5.6) $$\vec{T}(\mathbf{u}) \cdot \vec{T}(\mathbf{v}) = \mathbf{u} \cdot \mathbf{v}$$

for all $\mathbf{u}, \mathbf{v} \in \vec{S}$.

We still have to prove that $\vec{T}$ is a linear transformation, which means that we must check the identity

$$\vec{T}(\lambda_1 \mathbf{u}_1 + \lambda_2 \mathbf{u}_2) = \lambda_1 \vec{T}(\mathbf{u}_1) + \lambda_2 \vec{T}(\mathbf{u}_2)$$

for all scalars λ_1, λ_2 and vectors $\mathbf{u}_1, \mathbf{u}_2$. Let $\mathbf{w}$ denote the difference between the left- and right- hand sides, so we want to prove that $\mathbf{w} = \mathbf{0}$. Consider the expression $\mathbf{w} \cdot \vec{T}(\mathbf{v})$, where $\mathbf{v}$ is some vector to be determined later. Using the identity 4.5.6, we find that

$$\mathbf{w} \cdot \vec{T}(\mathbf{v}) = (\lambda_1 \mathbf{u}_1 + \lambda_2 \mathbf{u}_2) \cdot \mathbf{v} - \lambda_1 \mathbf{u}_1 \cdot \mathbf{v} - \lambda_2 \mathbf{u}_2 \cdot \mathbf{v} = 0$$

by the linearity of the dot product. This applies whatever the vector $\mathbf{v}$ is. In particular, since T is bijective one can choose $\mathbf{v}$ so that $\vec{T}(\mathbf{v}) = \mathbf{w}$; then we get $\mathbf{w} \cdot \mathbf{w} = 0$, and so $\mathbf{w} = \mathbf{0}$ by definiteness. $\square$

The whole of Euclidean geometry can be reconstructed from the linear algebra of the vector space $\vec{S}$, the action of $\vec{S}$ on S, and the dot product. We have now seen that an isometry preserves these three structures. Two figures[8] that are related by an isometry will be geometrically indistinguishable, and this is exactly what was required by our intuitive idea of congruence. We may therefore define two figures $\Omega \subset S$ and $\Omega' \subset S'$ to be *congruent* if there is an isometry $S \rightarrow S'$ that maps Ω onto Ω'.

We still have to prove, though, that the new concept of congruence given by isometry coincides with the old concept that we introduced for triangles. That is the point of the next proposition.

---

[8] A *figure* just means any subset of a space, though of course we usually think of the traditional kind of figure made up of points, lines, curves, and so on.

(4.5.7)   **Proposition:** *Let ABC and A′B′C′ be proper triangles in a Euclidean space
S. Suppose that they are congruent in the old sense that* $|AB| = |A′B′|$,
$|BC| = |B′C′|$, *and* $|CA| = |C′A′|$. *Then they are congruent in the new
sense that there is an isometry* $T: S \to S$ *with* $T(A) = A′$, $T(B) = B′$, *and*
$T(C) = C′$.

Proof: To keep things simple we will suppose that the space S is two-
dimensional[9].

Let $\mathbf{a} = \overrightarrow{BA}$ and $\mathbf{c} = \overrightarrow{BC}$; let $\mathbf{a}′ = \overrightarrow{B′A′}$ and $\mathbf{c} = \overrightarrow{B′C′}$. Just as we did
when verifying that a Euclidean space models the congruence axiom (4.1.5),
we deduce from the fact that ABC and $A′B′C′$ are old-sense congruent that
$|\mathbf{a}| = |\mathbf{a}′|$, $|\mathbf{c}| = |\mathbf{c}′|$, and $\mathbf{a} \cdot \mathbf{c} = \mathbf{a}′ \cdot \mathbf{c}′$. Now the vectors $\mathbf{a}$ and $\mathbf{c}$ are linearly
independent, and so form a basis for S; thus any point $P \in S$ has a position
vector $\overrightarrow{BP} = \lambda\mathbf{a} + \mu\mathbf{c}$. Define $T(P)$ to be the point such that

$$\overrightarrow{B′T(P)} = \lambda\mathbf{a}′ + \mu\mathbf{c}′;$$

then T is an isometry mapping A, B, and C to $A′$, $B′$, and $C′$, as required. □

(4.5.8)   **Example:** In classical geometry the concept of congruence was used to
show that two figures which agreed in certain respects in fact agreed in all
respects. As an illustration of this, here is the proof given by Pappus of the
fifth proposition of Euclid's Book 1, 'The base angles of an isosceles triangle
are equal'. (An *isosceles triangle* is one with two equal sides, and the *base
angles* are the angles opposite the equal sides.)

Let ABC be an isosceles triangle, with $|AB| = |AC|$. Then we can think of
ABC and ACB as two separate triangles, and they are then congruent because
$|AB| = |AC|$ and $|CB| = |BC|$. Therefore the corresponding angles in them
are equal, that is $\widehat{ABC} = \widehat{ACB}$.

Euclid himself did not give this proof; apparently he did not like the
idea of the triangle being congruent to itself. Instead he gave a complicated
construction which later became known as *Pons Asinorum* — the Asses'
Bridge. Some say the name was given because at this point many decided
that they could go no further with the study of geometry!

(4.5.9)   **The Erlanger Program:** Clearly the identity transformation is an isometry,
the composition of two isometries is an isometry, and the inverse of an
isometry is an isometry. In the language of abstract algebra, this tells us that
the isometries from a Euclidean space S to itself form a *group*, called the
isometry group and denoted $\text{Iso}(S)$. We have proceeded from the geometry
to the group, but it is equally possible to work the other way round; starting
from the group, one can investigate what are its *invariants* (the things that

---

[9]In the general case, one constructs an isometry *in the planes of the triangles* just as in two
dimensions and then extends it to the whole space.

it leaves unchanged) and so recover the concepts of distance, angle, and so on. More generally one can look for *representations* of the group; vectors, for example, are not invariant under isometries but they 'transform' in a certain way[10]. As an approach to Euclidean geometry this may seem back to front, but the general idea that *a geometry is determined by its group of 'allowable' transformations* has been profoundly influential. Because this idea was formulated by Felix Klein in his inaugural lecture at the University of Erlangen in 1872, it is often known as the 'Erlanger Program'.

(4.5.10)   **Example:** Many problems in mechanics are concerned with the motion of 'rigid bodies'. As a top spins and wobbles on a table top, the distance between any two of its points remains the same; this is what is meant by 'rigidity'. From a mathematical point of view, what this says is that the position of the top at time t is related to its position at time 0 by an *isometry* T_t (depending on time t, of course). The motion of the top is completely described by the time-varying family of isometries T_t. To study the dynamics of rigid bodies, it is therefore necessary to have a good understanding of the isometry group $\mathrm{Iso}(\mathcal{S})$, at least when $\mathcal{S}$ is ordinary three-dimensional space!

4.6 Exercises

1. a and **b** are two unit vectors in some Euclidean space, and the angle between them is $\frac{\pi}{3}$. Show that $2\mathbf{b} - \mathbf{a}$ is perpendicular to **a**.

2. The vectors **i**, **j**, and **k** form an orthonormal basis for a three-dimensional vector space. Let $\mathbf{a} = 3\mathbf{i} + \mathbf{j}$, $\mathbf{b} = \mathbf{i} + 2\mathbf{j} + \mathbf{k}$, and $\mathbf{c} = \mathbf{i} - \mathbf{j} + 2\mathbf{k}$. Find a scalar λ such that $\mathbf{a} + \lambda\mathbf{b}$ is perpendicular to **c**. Could such a scalar λ be found for *every* possible choice of **c**?

3. $\mathcal{P}$ is a Euclidean plane. The vectors **i** and **j** form an orthonormal basis for $\vec{\mathcal{P}}$. The points A, B, and C in $\mathcal{P}$ have position vectors (relative to some origin) $\mathbf{i} + 2\sqrt{3}\mathbf{j}$, $5\mathbf{i} + 2\sqrt{3}\mathbf{j}$, and $4\mathbf{i} + \sqrt{3}\mathbf{j}$ respectively. Find the angles of the triangle ABC.

4. Find a vector **v** that bisects the angle between $\mathbf{v}_1 = 4\mathbf{i} + 3\mathbf{j}$ and $\mathbf{v}_2 = -5\mathbf{j}$ (in other words, find **v** between $\mathbf{v}_1$ and $\mathbf{v}_2$ making equal angles with each of them).

5. Use the dot product to prove that for any two vectors **u** and **v**, $|\mathbf{u} - \mathbf{v}|^2 + |\mathbf{u} + \mathbf{v}|^2 = 2(|\mathbf{u}|^2 + |\mathbf{v}|^2)$. Deduce that the sum of the squares of the diagonals of a parallelogram is equal to the sum of the squares of all four sides.

6. Show that the perpendicular bisectors of the sides of a triangle meet at a point (the *circumcentre*) which is the centre of a circle passing through the vertices of the triangle.

---

[10]Some books (an example is Bourne and Kendall [8]) actually use this transformation property as the *definition* of a vector.

7. Let ABC be a proper triangle. Show that the altitudes of ABC meet at a point (called the *orthocentre*):

(i) by proving and using the vector identity

$$\overrightarrow{AB} \cdot \overrightarrow{CD} + \overrightarrow{AC} \cdot \overrightarrow{DB} + \overrightarrow{AD} \cdot \overrightarrow{BC} = \mathbf{0};$$

(ii) by using trigonometry and Ceva's theorem;

(iii) by showing that the altitudes of ABC are the perpendicular bisectors of the sides of the triangle $A'B'C'$ each of whose sides is parallel to the corresponding side of ABC and passes through the opposite vertex.

8. Let $\mathbf{n}$ be a unit vector. For any vector $\mathbf{v}$ define a new vector $T(\mathbf{v})$ by the equation

$$T(\mathbf{v}) = \mathbf{v} - 2(\mathbf{v} \cdot \mathbf{n})\mathbf{n}.$$

Prove that $T(T(\mathbf{v})) = \mathbf{v}$ and that $T(\mathbf{u}) \cdot \mathbf{v} = \mathbf{u} \cdot T(\mathbf{v})$. Deduce that $|T(\mathbf{v})|^2 = |\mathbf{v}|^2$.
    Can you give a geometrical description of the transformation T?

9. Let $\mathbf{a}$ be a given vector, and let $\mathbf{b} = \mathbf{a}/(1 - \lambda^2)$, where λ is a constant between 0 and 1. Prove that for any vector $\mathbf{r}$,

$$\frac{|\mathbf{r} - \mathbf{a}|^2 - \lambda^2|\mathbf{r}|^2}{1 - \lambda^2} = |\mathbf{r} - \mathbf{b}|^2 - \lambda^2|\mathbf{b}|^2.$$

Deduce that if O and A are fixed points in a plane, the locus of all points X such that $|AX|/|XO| = \lambda$ is a circle, and find its centre and radius. (This is called the *circle of Apollonius*.)

10. ABC is a triangle, and D is a point on AB between A and B. Use the dot product to prove *Stewart's theorem*:

$$|AC|^2|BD| + |BC|^2|AD| - |CD|^2|AB| = |AB||AD||BD|.$$

11. Let f and g be continuous functions on an interval $[a, b]$. Define a number $f \otimes g$ by

$$f \otimes g = \int_a^b f(x)g(x) \, dx.$$

Prove that the operation $\otimes$ has the four fundamental properties of the dot product (symmetry, linearity, positivity, and definiteness). Deduce that

$$\left| \int_a^b f(x)g(x) \, dx \right| \leq \sqrt{\left(\int_a^b f(x)^2 \, dx \right) \left(\int_a^b g(x)^2 \, dx \right)}.$$

12. Let ABC be an isosceles triangle with angles $A = \frac{\pi}{5}$ and $B = C = \frac{2\pi}{5}$. Let $a = |BC|$ and $b = |CA|$. By considering the point D on AB such that CD bisects angle ACB, and comparing the triangles ABC and BCD, show that

$$\frac{a}{b-a} = \frac{b}{a}.$$

Deduce that b/a is equal to the 'golden ratio' $(\sqrt{5}+1)/2$. Show also that

$$\cos\frac{\pi}{5} = \frac{\sqrt{5}+1}{4}.$$

13. Let S and S' be Euclidean spaces, and let $T: S \to S'$ be a bijective map. The map T is called a *similarity* with *ratio* ρ if it scales distances by the factor ρ, that is $|T(A)T(B)| = \rho|AB|$ for all $A, B \in S$.

Show that a similarity can be vectorialized in the same way as an isometry, and that

$$\vec{T}\mathbf{u} \cdot \vec{T}\mathbf{v} = \rho^2\mathbf{u} \cdot \mathbf{v}$$

for all vectors $\mathbf{u}$ and $\mathbf{v}$.

14. Let $\mathcal{P}$ and $\mathcal{P}'$ be parallel planes in some three-dimensional Euclidean space S. Let O be a point not on either plane, and define a transformation $T: \mathcal{P} \to \mathcal{P}'$ by projection from O (see Chapter 2). Show that T is a similarity, and that its ratio is equal to the ratio of the perpendicular distances from O to $\mathcal{P}$ and $\mathcal{P}'$.

5

Coordinates and equations

5.1 Cartesian coordinates

One of the greatest steps forward in geometry since the time of the Greeks is associated with the name of René Descartes (1596–1650). As a philosopher, Descartes was fascinated by geometry because it seemed to offer a model of how certain and reliable knowledge might be obtained. In his most famous work, the *Discourse on Method*, he wrote

> Those long chains of reasonings, each step simple and easy, which geometers are wont to employ in arriving even at the most difficult of their demonstrations, have led me to surmise that all the things we human beings are competent to know are interconnected in the same manner, and that none are so remote as to be beyond our reach or so hidden that we cannot discover them.

Descartes' approach is recognizably derived from Euclid's. Start with things you cannot doubt (the famous 'I think, therefore I am' was one of Descartes' axioms); proceed by logical steps; check your reasoning. Philosophers still debate the issues that Descartes raised. But his 'method' also led him to mathematical insights. In the same paragraph from which we have just quoted he goes on to say that he wants to combine the visual and imaginative impact of geometry with the power of algebra to express complicated ideas in a compact form, and he adds

> In this way, I should be borrowing all that is best in geometry and algebra, and should be correcting all the defects of the one by the help of the other.

Descartes spelled out his ideas about the relationship between geometry and algebra in an appendix to the *Discourse on Method*. Implicit in his work is the idea, now so familiar to us, of representing a point by its *coordinates* (x, y) (in the plane), and of representing a curve by an *equation* relating these coordinates. Formerly, an equation might have been thought of as the result or summary of a series of pure geometrical constructions; now the growing power of algebra was to be used in the other direction, and propositions in geometry were to be obtained as the results of algebraic manipulations.

Descartes was not the only one to hit on the idea of representing points by coordinates and lines by equations. His contemporary Fermat (1601–1665) arrived at the same method at about the same time. It even seems that both were motivated in their researches by the same classical problem of Greek geometry. But Fermat's attention was soon diverted to number theory, the subject for which he is most famous. We will see later that this was not as great a change of direction as it may appear. There is a profound link between geometry and number theory, which present-day mathematicians are still exploring.

In our system, coordinates can be introduced as follows.

(5.1.1)   **Definition:** *Let S be a Euclidean space. A list $\mathbf{v}_1, \ldots, \mathbf{v}_n$ of vectors in $\vec{S}$ is called an* orthonormal set *if each vector in the list has length* 1 *and distinct vectors in the list are perpendicular.*

One can abbreviate this by writing

$$\mathbf{v}_i \cdot \mathbf{v}_j = \delta_{ij}$$

where the *Kronecker delta symbol* δ_{ij} is defined by

$$\delta_{ij} = \begin{cases} 0 & \text{if } i \neq j \\ 1 & \text{if } i = j. \end{cases}$$

An orthonormal set $\{\mathbf{v}_1, \ldots, \mathbf{v}_n\}$ is called an *orthonormal basis* if every vector can be expressed as a linear combination of $\mathbf{v}_1, \ldots, \mathbf{v}_n$: that is, for every $\mathbf{v}$ there are scalars λ_i such that

$$\mathbf{v} = \lambda_1 \mathbf{v}_1 + \cdots + \lambda_n \mathbf{v}_n = \sum_{i=1}^{n} \lambda_i \mathbf{v}_i.$$

Notice that the coefficients λ_i in this expression are uniquely determined by $\mathbf{v}$; in fact, taking the dot product of both sides of the equation with $\mathbf{v}_j$ we find that

$$\mathbf{v} \cdot \mathbf{v}_j = \sum_{i=1}^{n} \lambda_i \delta_{ij} = \lambda_j.$$

(In other words, an orthonormal set is linearly independent.) In general there are many different orthonormal bases; but they all have the same number n of elements, which is equal to the *dimension* (see 2.1.2) of the space of vectors.

Now let us fix an origin $O \in S$ and an orthonormal basis or *frame* $\mathbf{v}_1, \ldots, \mathbf{v}_n$ for $\vec{S}$. As in 2.2.1, the choice of origin and of frame is arbitrary,

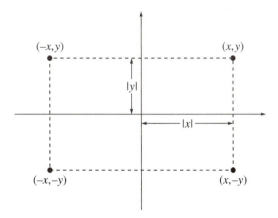

Fig. 5.1. Cartesian coordinates in two dimensions

but once it has been fixed we can represent each point P of S by a list of n real numbers or *coordinates* $x_1, \ldots, x_n$. These numbers are simply the components of the position vector $\overrightarrow{OP}$ relative to the orthonormal basis $\mathbf{v}_1, \ldots, \mathbf{v}_n$. That is,

(5.1.2)   **Definition:** *The* coordinates *of the point P (relative to the given origin O and frame* $\mathbf{v}_1, \ldots, \mathbf{v}_n$*) are the numbers* $x_1, \ldots, x_n$ *defined by*

$$\overrightarrow{OP} = x_1 \mathbf{v}_1 + \cdots + x_n \mathbf{v}_n.$$

Each point P now corresponds to a list $(x_1, \ldots, x_n)$ of coordinates, and each list of coordinates corresponds to a unique point. Thus we have set up a *one-to-one correspondence* between S and the set of all n-tuples of real numbers. This set of n-tuples is called *standard n-dimensional coordinate space*, and is denoted $\mathbf{R}^n$. In two or three dimensions it is customary to use the letters (x, y) or (x, y, z) instead of (x_1, x_2) or (x_1, x_2, x_3).

The *axes* of the coordinate system are the lines through O in the directions of $\mathbf{v}_1, \ldots, \mathbf{v}_n$. In two dimensions, the distance from the point with coordinates (x, y) to the x-axis is $|y|$, and the distance to the y-axis is $|x|$ (see Figure 5.1). The corresponding fact in three dimensions is that the distance from (x, y, z) to the plane containing the y and z axes is $|x|$. These facts are sometimes used to *define* coordinates in two and three dimensions. But one then has to fuss about the signs.

We can express the distance between two points P and Q in S in terms of their coordinates. To do this, we will need a formula for the dot product of two vectors in terms of components relative to an orthonormal basis. Suppose therefore that $\mathbf{v} = \sum_{i=1}^{n} \lambda_i \mathbf{v}_i$ and $\mathbf{u} = \sum_{i=1}^{n} \mu_i \mathbf{v}_i$ are two vectors; then their dot product can be written as the double sum $\sum_{i=1}^{n}\sum_{j=1}^{n}\lambda_i\mu_j\mathbf{v}_i \cdot \mathbf{v}_j$. But $\mathbf{v}_i \cdot \mathbf{v}_j = \delta_{ij}$; so the terms with $i \neq j$ vanish and the double sum collapses into a single one

(5.1.3)
$$\mathbf{u} \cdot \mathbf{v} = \sum_{i=1}^{n}\lambda_i\mu_i.$$

In particular, $|\mathbf{v}|^2 = \sum_{i=1}^{n}\lambda_i^2$. Now let P and Q have coordinates $(x_1,\ldots,x_n)$ and $(x_1',\ldots,x_n')$ respectively; then

$$\overrightarrow{PQ} = \overrightarrow{OQ} - \overrightarrow{OP} = (x_1' - x_1)\mathbf{v}_1 + \cdots + (x_n' - x_n)\mathbf{v}_n.$$

So we find the formula

$$|PQ|^2 = \sum_{i=1}^{n}(x_i - x_i')^2$$

for the distance in terms of coordinates.

It is often helpful to think of this formula in a different way. The standard coordinate space $\mathbf{R}^n$ has more structure than we have mentioned so far; it is in fact a Euclidean space in its own right, in which the distance d between points $(x_1,\ldots,x_n)$ and $(x_1',\ldots,x_n')$ is given by the formula

$$d^2 = \sum_{i=1}^{n}(x_i - x_i')^2.$$

For example, $\mathbf{R}^2$ is a Euclidean space whose associated vector space $\overrightarrow{\mathbf{R}^2}$ consists of all linear combinations of two orthonormal basis vectors $\mathbf{i}$ and $\mathbf{j}$, which operate on $\mathbf{R}^2$ by the equations

$$\mathbf{i}(x,y) = (x+1,y), \quad \mathbf{j}(x,y) = (x,y+1).$$

When $\mathbf{R}^n$ is thought of as a Euclidean space in this way, the one-to-one correspondence $S \rightarrow \mathbf{R}^n$ given by the choice of origin and frame becomes an *isometry*; the formulae for the distance in S and in $\mathbf{R}^n$ are the same. A coordinate system, therefore, is just an isometry between the given space S and the standard 'model space' $\mathbf{R}^n$.

(5.1.4)   **Example:** To see how useful this is, suppose that we have *two* coordinate systems for the same space S. Such a situation can easily arise; for example, the coordinate system in which it is most convenient to solve some problem may not be the same as the system in which the problem was originally stated.

The two coordinate systems then give rise to two isometries $T_1, T_2 \colon S \to \mathbf{R}^n$. Since the composite of two isometries is an isometry, $T_1 T_2^{-1}$ is an isometry from $\mathbf{R}^n$ to $\mathbf{R}^n$, which relates the coordinates of the *same* point of S in the two different coordinate systems. In other words, 'coordinate transformations' are simply isometries of the standard space $\mathbf{R}^n$.

This is often called the *alias interpretation* of an isometry: we are using the isometry to look at the same point 'under a different name'. It is contrasted with the *alibi interpretation* which we were using before, under which the isometry is thought of as relating one geometrical object with an identical object 'in a different place'.

5.2 Curves and equations

If Fermat and Descartes had confined themselves to labelling points by pairs of real numbers instead of by letters of the alphabet, their work would not have been especially interesting. Their greatest insight was expressed by Fermat as follows

Whenever in a final equation two unknown quantities are found, we have a locus... describing a line, straight or curved.

A *locus* is just the set of all points satisfying some geometrical condition; for instance, 4.2.2 describes the perpendicular bisector of a line segment AB as the locus of all points equidistant from A and B. Fermat was thinking in terms of plane geometry, so he is saying that an equation relating the coordinates x and y will describe a locus which is a curve in the plane. Two examples, with which you are probably familiar, are the equations of a straight line and of a circle.

(5.2.1)   **Example:** To find the equation of a straight line we may use the criterion 2.2.3 for collinearity. This tells us that a point (x, y) is collinear with two fixed points (x_1, y_1) and (x_2, y_2) if and only if there exist scalars α, β, γ, not all zero, such that

$$
\begin{aligned}
\alpha + \beta + \gamma &= 0 \\
\alpha x + \beta x_1 + \gamma x_2 &= 0 \\
\alpha y + \beta y_1 + \gamma y_2 &= 0.
\end{aligned}
$$

We can think of these as three simultaneous equations in the three unknowns $\alpha, \beta,$ and γ. In order that they should have a nonzero solution it is necessary and sufficient that the matrix of coefficients should be singular; in other words, that

$$\begin{vmatrix} 1 & 1 & 1 \\ x & x_1 & x_2 \\ y & y_1 & y_2 \end{vmatrix} = 0.$$

By expanding the determinant, we can write the equation of the line through (x_1, y_1) and (x_2, y_2) as

$$lx + my + n = 0$$

where $l = y_1 - y_2$, $m = x_2 - x_1$, and $n = x_1 y_2 - x_2 y_1$. Notice that at least one of l and m is nonzero; therefore, one may assume if necessary (by multiplying the whole equation by a constant factor) that $l^2 + m^2 = 1$.

(5.2.2)   **Example:** A circle is defined as the set of all points in the plane at a fixed distance from some given centre. If the centre is the point (a, b) and the distance is r, then the formula for the distance gives us $\sqrt{(x - a)^2 + (y - b)^2} = r$ or

$$(x - a)^2 + (y - b)^2 = r^2$$

as the equation of a circle.

(5.2.3)   **Example:** A slightly less familiar curve is the *folium of Descartes*, represented by the equation

$$x^3 + y^3 - 3xy = 0.$$

This curve was introduced by Descartes to show off the power of his new-fangled coordinate geometry. Unfortunately, he did not have a correct understanding of negative coordinates! A correct picture is given in Figure 5.2; Descartes noticed only the central loop.

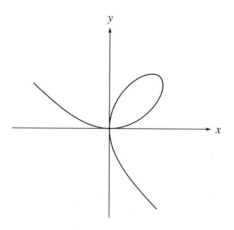

Fig. 5.2. The folium of Descartes

In general, the equation of a curve will have the form

$$f(x, y) = 0$$

where f is some function of the two variables x and y. But without some further conditions on the function f, this equation might define something quite un-curvelike. The difficulty is that the modern concept of 'function' is extremely general; any kind of rule that produces a number $f(x, y)$ from two numbers x and y might qualify as a function. We will need to impose some restrictions that take account of our intuition that a curve is a 'one-dimensional' object. We won't discuss this now, but merely note that it is a piece of unfinished business that will be taken up again in Chapter 7.

(5.2.4)   **Parametric equations:** So far we have understood the concept of a curve in a rather static way; it is the set of all points satisfying some condition expressed by the equation $f(x, y) = 0$. We can call this kind of equation a *locus equation*. But it is often more natural to think of a curve as being 'the path traced out by a moving point'; the coordinates x and y are given as functions of some other variable t, which it may be helpful to think of as the time, and the curve is described as the set of all points $(x(t), y(t))$ as t varies. For example, the circle $(x - a)^2 + (y - b)^2 = r^2$ can also be described as the set of all points

$$(a + r\cos t, b + r\sin t)$$

as t varies between 0 and 2π. Similarly, the line $lx + my + n$ can be described as the set of all points $(x_0 + mt, y_0 - lt)$ where (x_0, y_0) is some fixed point on the line and t varies between $-\infty$ and ∞. An equation of this kind is called the *parametric equation* of a curve.

(5.2.5)   **Example:** As one might expect, equations of this type tend to arise whenever one wants to talk about a mechanical system whose behaviour varies with time. The simplest possibility is that of a single particle moving in a plane and tracing out a curved path, such as a planet moving around the sun. But more complicated examples can be handled as well. Suppose that we want to study the motion of the whole solar system at once, nine planets all moving in three-dimensional space. We can make a list of the 27 coordinates of the planets, and this list will tell us where they all are at any one moment of time. Now we may think of this list as giving us the coordinates of a *single* point in some 27-dimensional 'configuration space'. As the planets move around the sun, the point will move along some curve in 27-dimensional space; and the parametric equation of this curve contains all the information about the motions of all the planets. This approach, which seems rather complicated at first, is in fact vital for the advanced study of mechanics. It gives one reason why it is important to study high-dimensional geometry, even though we live in a three-dimensional space.

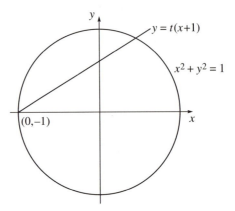

Fig. 5.3. Proof that the circle is a rational curve

(5.2.6)   **Definition:** *A curve is called* rational *if it can be represented by a parametric equation* $x = x(t)$, $y = y(t)$ *where* $x(t)$ *and* $y(t)$ *are* rational functions[1] *of* t.

Rational curves are important and rather rare; but they include many of the most familiar examples from elementary geometry. Let us prove, for example, that any circle is a rational curve. To keep things simple, we do the calculation only for the special case of the circle centre $(0,0)$ and radius 1; you will easily be able to generalize the argument. The equation of our circle is

$$x^2 + y^2 = 1.$$

We will find a rational parameterization by looking at the points of intersection of our circle with the line $y = t(x+1)$, which passes through the point $(0, -1)$ (see Figure 5.3). To find the other point of intersection, substitute into the equation of the circle, obtaining

$$x^2 + t^2(x + 1)^2 = 1$$

which implies that

$$(x + 1)((1 + t^2)x - (1 - t^2)) = (1 + t^2)x^2 + 2t^2x - (1 - t^2) = 0.$$

The roots of this equation are $x = -1$ and $x = (1 - t^2)/(1 + t^2)$. Therefore, the parametric equations

---

[1] A *rational function* is one polynomial divided by another, like

$$\frac{t^2 + 5t - 9}{t^5 - 2}.$$

$$x = \frac{1 - t^2}{1 + t^2}, \quad y = t(x + 1) = \frac{2t}{1 + t^2}$$

represent a point on the circle, and any point on the circle except $(-1, 0)$ can be so obtained. We recognize the familiar trigonometrical formulae giving $\sin\theta$ and $\cos\theta$ in terms of $\tan(\theta/2)$; the half angles enter in because of proposition 6.4.3, that the angle at the centre is twice that at the circumference.

(5.2.7)   **Example:** *Pythagorean triads* are triples of integers such as $3, 4, 5$ or $5, 12, 13$ which represent the lengths of the sides of a right-angled triangle; $3^2 + 4^2 = 5^2$, $5^2 + 12^2 = 13^2$, and so on. A couple of larger examples are $65^2 + 72^2 = 97^2$ and $119^2 + 120^2 = 169^2$; these examples are remarkable because they appear on a Babylonian clay tablet (Plimpton 322) dated to around 2000 BC. Perhaps you have experimented with Pythagorean triads at some stage and attempted to construct a formula for them. Such a formula can easily be found by considering the rational parameterization of the circle.

Suppose that a, b, c is a Pythagorean triad. Then $x = a/c$ and $y = b/c$ are rational numbers, and (x, y) is a point on the circle $x^2 + y^2 = 1$. Conversely, if (x, y) is a point on the circle with rational coordinates, we can recover Pythagorean triads from it by multiplying x and y by any common denominator. The problem of finding Pythagorean triads is therefore equivalent to the problem of finding points on the circle with rational coordinates.

But our parametric equations solve this problem! If x and y are rational, then $t = y/(x + 1)$ is rational; and if t is rational, then x and y are rational. So the rational points on the circle are exactly those with rational t-values. If we put $t = p/q$ in lowest terms, then

$$x = \frac{p^2 - q^2}{p^2 + q^2}, \quad y = \frac{2pq}{p^2 + q^2}.$$

The lowest common denominator is $p^2 + q^2$. Thus we discover that the general formula for all Pythagorean triads is

$$a = r(p^2 - q^2), \quad b = 2pqr, \quad c = r(p^2 + q^2)$$

for integers p, q, and r.

We investigated a number-theoretic problem about integers by reducing it to a problem about rational points on curves and then applying geometrical methods. This general idea has a long and honourable history in mathematics, and it is still the subject of active research under the rather formidable title of *arithmetic algebraic geometry*. On this whole area of interaction, one modern author[2] has written

---

[2]John Stillwell, in his book *Mathematics and its History* [38, Chapter 1].

Pythagoras' theorem was the first hint of a hidden, deeper relationship between arithmetic and geometry, and it has continued to hold a key position between these two realms throughout the history of mathematics. This has sometimes been a position of cooperation and sometimes one of conflict, as followed the discovery that $\sqrt{2}$ is irrational. It is often the case that new ideas emerge from such areas of tension.... The tension between arithmetic and geometry is, without doubt, the most profound in mathematics, and it has led to the most profound theorems.

Here is another example. Fermat is most famous, not for his contributions to coordinate geometry, but for his 'last theorem'. This 'theorem' asserts that it is impossible to find any whole numbers a, b, c, and $n \geq 3$ for which

$$a^n + b^n = c^n;$$

notice that Pythagorean triads provide plenty of examples when $n = 2$. Fermat wrote in the margin of a book that he was reading, 'I have a truly marvellous demonstration of this proposition, which however this margin is too narrow to contain'. Since Fermat's time, many mathematicians have tried to prove the 'theorem', but all in vain; it has become the most famous unsolved problem in all of mathematics. Sadly, most experts reject the romantic idea that Fermat really did have a proof which everyone else has missed.

Just as we studied Pythagorean triads by looking for rational points on the circle, one can study the Fermat problem by looking for rational points on the *Fermat curves* $x^n + y^n = 1$, $n \geq 3$. In Chapter 7 we will take a look at the Fermat problem from this point of view.

5.3   Coordinate form of an isometry

It is not only points and curves that can be represented by coordinates. In this section we will discuss isometries in coordinate terms. Because of 5.1.4, this will also allow us to understand the relationship between two different coordinate systems on the same space.

Let S be a Euclidean space, and suppose that we have fixed an orthonormal basis $\mathbf{v}_1, \ldots, \mathbf{v}_n$ for $\vec{S}$. If a vector $\mathbf{v} \in \vec{S}$ is equal to $\sum_{i=1}^{n} \lambda_i \mathbf{v}_i$, it is often convenient to represent $\mathbf{v}$ by the 'column vector' ($n \times 1$ matrix)

$$\mathsf{v} = \begin{pmatrix} \lambda_1 \\ \vdots \\ \lambda_n \end{pmatrix}$$

consisting of the coefficients λ_i. This representation allows one to express the dot product very simply; if $\mathbf{u} = \sum_{i=1}^{n} \mu_i \mathbf{v}_i$ is another vector, the dot product $\mathbf{u} \cdot \mathbf{v} = \sum_{i=1}^{n} \lambda_i \mu_i$ can be written in terms of the columns u and v as the matrix product $u^t v$. Here the superscript t denotes the transposed matrix of u (the $1 \times n$ matrix obtained by interchanging rows and columns); the matrix product is a 1×1 matrix, which we just think of as a real number.

It is a result of linear algebra that every linear transformation L acting on $\vec{S}$ can be represented (relative to the fixed basis $\mathbf{v}_1, \ldots, \mathbf{v}_n$) by a matrix L acting on column vectors. That is,

$$L(\mathbf{u}) = \mathbf{v} \quad \text{if and only if} \quad \mathsf{L} u = v,$$

where u and v are the column vectors representing $\mathbf{u}$ and $\mathbf{v}$. In fact, the (i,j)'th entry of the matrix L is just the dot product $(L\mathbf{v}_i) \cdot \mathbf{v}_j$. The columns of the matrix L, considered individually, are therefore the column vectors representing $L\mathbf{v}_1$, $L\mathbf{v}_2$, and so on.

Let us apply this theory to the study of isometries. The vectorialization $\vec{T}$ of an isometry is a linear transformation which satisfies the additional condition (4.5.6) that it preserves dot products. What does this condition become in matrix terms? The answer is provided by the next definition.

(5.3.1)   **Definition:** *An $n \times n$ matrix U of real numbers is called* orthogonal *if it is invertible and its transpose is equal to its inverse.*

We may write the condition for orthogonality in either of the two forms

$$\mathsf{U}^t \mathsf{U} = \mathsf{I} \quad \text{or} \quad \mathsf{U} \mathsf{U}^t = \mathsf{I},$$

where I denotes the $n \times n$ identity matrix. (The two forms are equivalent because of the linear algebra result which says that a left inverse for a square matrix is also a right inverse and vice versa.)

(5.3.2)   **Lemma:** *The following conditions on a matrix U are equivalent:*

(i) U *is orthogonal.*

(ii) *The columns of U form an orthonormal set.*

(iii) *The rows of U form an orthonormal set.*

Proof: Let c_i denote the i'th column of the matrix U. By definition of matrix multiplication, the entry in the (i,j) position of the product $\mathsf{U}^t \mathsf{U}$ is obtained by multiplying the i'th row of U^t by the j'th column of U, so it is equal to $c_i^t c_j$, which is the dot product of c_i and c_j. Thus

The columns $\{c_i\}$ are orthonormal $\iff c_i^t c_j = \delta_{ij} \iff \mathsf{U}^t \mathsf{U} = \mathsf{I}.$

This proves that the first two conditions are equivalent. The equivalence of the first and the last is proved similarly. $\square$

Simple as this lemma is, it is by no means easy to prove directly that the rows of a matrix form an orthonormal set if and only if its columns do. Try it in the 2×2 case!

(5.3.3) **Proposition:** *A linear transformation L preserves dot products if and only if its matrix* L *is orthogonal. In particular, the vectorization of an isometry is represented by an orthogonal matrix.*

Proof: The condition of preserving dot products,

$$L(\mathbf{u}) \cdot L(\mathbf{v}) = \mathbf{u} \cdot \mathbf{v},$$

translates into the matrix equation

$$(\mathsf{Lu})'\mathsf{Lv} = \mathsf{u}'\mathsf{L}'\mathsf{Lv} = \mathsf{u}'\mathsf{v}.$$

(Remember that transposition reverses the order of matrix multiplication!) Therefore, for any column vectors u and v,

$$\mathsf{u}'(\mathsf{L}'\mathsf{L} - \mathsf{I})\mathsf{v} = 0.$$

It follows that $\mathsf{L}'\mathsf{L} - \mathsf{I} = 0$, and so that L is orthogonal. $\square$

The determinant of the transpose of a matrix is equal to the determinant of the original matrix; and the determinant of a product of two matrices is equal to the product of their determinants. Using these facts, we find that if L is orthogonal, then

$$\det(\mathsf{L})^2 = \det(\mathsf{L}'\mathsf{L}) = \det(\mathsf{I}) = 1.$$

Thus $\det(\mathsf{L}) = \pm 1$. In particular, if T is an isometry, then $\vec{T}$ has determinant ± 1.

(5.3.4) **Definition:** *An isometry T is called* direct *if the determinant of $\vec{T}$ is $+1$, and* indirect *if it is -1.*

We now know that the vectorization of an isometry is represented by an orthogonal matrix. It is easy to deduce from this the coordinate description of a general isometry. Let S be a Euclidean space and $\mathbf{v}_1, \ldots, \mathbf{v}_n$ a frame, as above; suppose that we have also chosen an origin O. Each point X of S may then be represented by its coordinates $(x_1, \ldots, x_n)$, which we can bring together as the *coordinate vector*

$$\mathsf{x} = \begin{pmatrix} x_1 \\ \vdots \\ x_n \end{pmatrix}$$

of X. (In other words, x is the column vector that corresponds to the position vector $\overrightarrow{OX}$ of X.)

(5.3.5)   **Proposition:** *With the above notation, let $T: S \rightarrow S$ be an isometry. Then T maps the point X with coordinate vector* x *to the point X' with coordinate vector*

$$\mathsf{x'} = \mathsf{Ux} + \mathsf{c},$$

where U *is the orthogonal matrix representing the vectorialization of T, and* c *is the coordinate vector of $T(O)$.*

Proof: Let $C = T(O)$. By definition of $\vec{T}$,

$$\vec{T}(\overrightarrow{OX}) = \overrightarrow{CX'}.$$

Therefore

$$\mathsf{Ux} = \mathsf{x'} - \mathsf{c}$$

giving the result. □

An important special case arises if $T(O) = O$, when one says that *T fixes* the point O. Then $\mathsf{c} = \mathsf{0}$, and the action of T is represented simply by the orthogonal matrix U acting on coordinate vectors.

Remark: The *converse* of proposition 5.3.5 is also true; any transformation defined in terms of coordinate vectors by $\mathsf{x'} = \mathsf{Ux} + \mathsf{c}$ (where U is orthogonal) is an isometry. To prove this, suppose that X and Y are transformed to X' and Y'. Then in terms of coordinate vectors

$$\mathsf{x'} - \mathsf{y'} = \mathsf{U}(\mathsf{x} - \mathsf{y}).$$

Since an orthogonal matrix preserves dot products, it preserves lengths; so we find that $|X'Y'| = |XY|$ and the transformation is an isometry.

As an important example, let us find all the isometries of the coordinate plane $\mathbf{R}^2$. By proposition 5.3.5, they are all of the form $\mathsf{x} \mapsto \mathsf{Ux} + \mathsf{c}$, where U is a 2×2 orthogonal matrix. The next result gives all the possibilities for such a matrix U.

(5.3.6)   **Proposition:** *Let U be a 2×2 orthogonal matrix. If $\det \mathsf{U} = +1$, then there is an angle θ such that*

$$\mathsf{U} = \begin{pmatrix} \cos\theta & -\sin\theta \\ \sin\theta & \cos\theta \end{pmatrix} \qquad \text{(rotation matrix)},$$

and if $\det \mathsf{U} = -1$, then there is an angle θ such that

$$\mathsf{U} = \begin{pmatrix} \cos\theta & \sin\theta \\ \sin\theta & -\cos\theta \end{pmatrix} \qquad \text{(reflection matrix)}.$$

Proof: Let $U = \begin{pmatrix} a & b \\ c & d \end{pmatrix}$ be such a matrix. Orthogonality gives

$$a^2 + c^2 = b^2 + d^2 = 1, \quad ab + cd = 0.$$

Two real numbers whose squares add up to 1 can be represented as the cosine and sine of an appropriate angle (A.1.7). So we have

$$a = \cos\varphi, \; c = \sin\varphi, \; b = \sin\psi, \; d = \cos\psi$$

for some choice of φ and ψ, and the identity $ab + cd = 0$ translates to

$$\sin(\varphi + \psi) = \sin\varphi\cos\psi + \cos\varphi\sin\psi = 0.$$

Thus, $\varphi + \psi = k\pi$, where k is an integer. If k is even, then $\cos\psi = \cos\varphi$ and $\sin\psi = -\sin\varphi$, and U is a rotation matrix; if k is odd, then $\cos\psi = -\cos\varphi$ and $\sin\psi = \sin\varphi$, and U is a reflection matrix.

The determinant $ad - bc$ is equal to

$$\cos\varphi\cos\psi - \sin\varphi\sin\psi = \cos(\varphi + \psi) = \cos(k\pi),$$

which is $+1$ if k is even and -1 if k is odd. $\square$

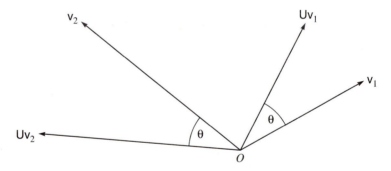

Fig. 5.4. The effect of a rotation matrix

The rotation and reflection matrices are so called because of the geometrical nature of the transformations that they produce on $\mathbf{R}^2$. Consider first a rotation matrix U with angle θ. Let $v = \begin{pmatrix} x \\ y \end{pmatrix}$ be a unit vector. The dot product between v and Uv is given by $v'Uv$. Working out this matrix product we get

$$x(x\cos\theta - y\sin\theta) + y(x\sin\theta + y\cos\theta) = \cos\theta.$$

Thus the angle between v and Uv is always θ; U rotates all vectors through θ (see Figure 5.4).

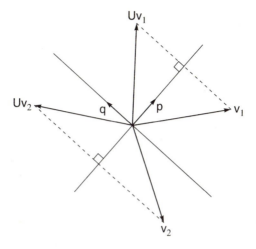

Fig. 5.5. The effect of a reflection matrix

The geometry of a reflection matrix is quite different. If U is the reflection matrix with angle θ, let p and q be the unit vectors

$$p = \begin{pmatrix} \cos\theta/2 \\ \sin\theta/2 \end{pmatrix}, \quad q = \begin{pmatrix} -\sin\theta/2 \\ \cos\theta/2 \end{pmatrix}.$$

Working out the matrix products, we find that

$$Up = \begin{pmatrix} \cos\theta\cos\theta/2 + \sin\theta\sin\theta/2 \\ \sin\theta\cos\theta/2 - \cos\theta\sin\theta/2 \end{pmatrix} = p,$$

and similarly $Uq = -q$. Thus U leaves the p-component of a vector unchanged, but reverses the q-component; it is a reflection in a line in the direction of p (see Figure 5.5).

Remark: Vectors such as p and q, which are transformed into multiples of themselves by a matrix U, are called *eigenvectors* for the matrix. They will appear several times in this book, and their study forms one of the central themes of any abstract linear algebra course. Their importance stems from the fact that they allow us to 'decompose' the multi-dimensional action of the matrix into a number of independent one-dimensional actions. Notice that p and q form an orthonormal basis; this is a very special case of a general theorem we will prove later (10.2.1), which says that given any symmetric matrix, one can find an orthonormal basis consisting of eigenvectors for it.

The addition of the constant vector c in the formula 5.3.5 just has the effect of a *translation*. Thus we can summarize our results as follows.

(5.3.7)   **Corollary:** *Any isometry of the coordinate plane* $\mathbf{R}^2$ *that leaves fixed the origin O can be expressed as a rotation (if it's direct) or a reflection (if it's indirect). A general isometry of* $\mathbf{R}^2$ *can be expressed as such a rotation or reflection followed by a translation.*

5.4   Change of coordinates

We have seen (5.1.4) that two different coordinate systems on the same Euclidean space are related by an isometry of the standard coordinate space $\mathbf{R}^n$. We have also seen (5.3.5) how to describe such an isometry of $\mathbf{R}^n$ in matrix terms. Putting these two facts together, we get a description of all possible coordinate changes, as follows:

(5.4.1)   **Proposition:** *Let S be an n-dimensional Euclidean space, equipped with two different coordinate systems. Then there exist an $n \times n$ orthogonal matrix* U *and an $n \times 1$ column vector* c, *such that the coordinate vectors* x *and* x' *of a point in the two coordinate systems are related by*

$$\mathsf{x} = \mathsf{U}\mathsf{x}' + \mathsf{c}.$$

Notice how similar this equation is to the one appearing in the ruler comparison axiom (1.3.1). We will refer to it as the transformation equation *from* the x system *to* the x' system of coordinates. This terminology seems illogical at first — shouldn't 'from' and 'to' be the other way around? — but it may seem more sensible if you note that c gives the coordinates in the x system of the origin of the x' system, and that the columns of U (thought of in the x system) are the basis vectors of the x' system.

To get familiar with this notion, we will work out some numerical examples in the plane. Suppose that

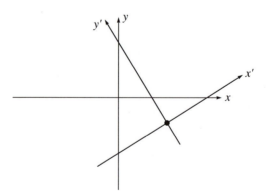

Fig. 5.6. Example of coordinate transformation

$$U = \frac{1}{2} \begin{pmatrix} \sqrt{3} & -1 \\ 1 & \sqrt{3} \end{pmatrix}, \quad c = \begin{pmatrix} 2 \\ -1 \end{pmatrix},$$

so that the transformation equations between $x = \begin{pmatrix} x \\ y \end{pmatrix}$ and $x' = \begin{pmatrix} x' \\ y' \end{pmatrix}$
are

$$x = \frac{\sqrt{3}x' - y'}{2} + 2, \quad y = \frac{x' + \sqrt{3}y'}{2} - 1.$$

The axes of the two coordinate systems are shown in Figure 5.6.

(5.4.2)   **Example:** To find the equation in the (x', y') coordinate system of the line $3x + 4y = 3$, we use the expressions for x and y in terms of x' and y' given by the transformation equations. Substituting them into the equation of the line, we get

$$3 \left(\frac{\sqrt{3}x' - y'}{2} + 2 \right) + 4 \left(\frac{x' + \sqrt{3}y'}{2} - 1 \right) = 3$$

which simplifies to

$$(3\sqrt{3} + 4)x' + (4\sqrt{3} - 3)y' = 2.$$

(5.4.3)   **Example:** Suppose we want to solve the inverse problem: we are given an equation in the (x', y') coordinate system and want to find the corresponding equation in the (x, y) system. Then we need to invert the transformation equations. This is most easily done in matrix terms. Multiplying the transformation equation by the transpose U' of U (which is the same as the inverse because U is orthogonal), we obtain

$$x' = U'x - U'c.$$

In our example the inverse transformation equations are

$$x' = \frac{\sqrt{3}x + y}{2} - \sqrt{3} + \frac{1}{2}, \quad y' = \frac{-x + \sqrt{3}y}{2} + 1 + \frac{\sqrt{3}}{2}.$$

You might like to check that you do recover the x, y equation of the line in our previous example by substituting these inverse transformation equations into its x', y' equation.

(5.4.4)   **Example:** The general form of the equation of a line is the same after coordinate transformation, although the values of the coefficients are changed. For more complicated curves, however, a suitable coordinate transformation may make them much simpler. For instance, consider the curve whose equation is $y = x^2$, which you will probably recognize as a parabola. In the other coordinate system its equation is

$$3x'^2 + y'^2 - 2\sqrt{3}x'y' + (8\sqrt{3} - 2)x' - (8 + 2\sqrt{3})y' + 20 = 0$$

which you may not find so transparent. If we were confronted with this second form, it would be worth our knowing that we could change coordinates to obtain $y = x^2$. But how would we know what coordinate change to make? We will investigate this problem in Chapter 7.

5.5 Exercises

1. Three points A, B, and C in a four-dimensional Euclidean space have coordinates $(1, 2, 3, 4)$, $(2, 3, 4, 6)$, and $(3, 5, 7, 9)$ respectively. Find the coordinates of a fourth point D so that $ABCD$ shall be a parallelogram. Also find two linearly independent vectors that are perpendicular to the plane of $ABCD$.

2. Let a, b, and c be positive. Find the coordinates of the points of intersection of the two circles

$$x^2 + y^2 = a^2, \quad (x - c)^2 + y^2 = b^2.$$

From your solution deduce that in order that the two circles should intersect it is necessary and sufficient that the sum of any two of a, b, and c should be at least as great as the third. Explain this condition geometrically.

3. The points P and Q in $\mathbf{R}^2$ have coordinates $(-1, 0)$ and $(1, 0)$ respectively. Find the equation of the locus of all points X for which $|PX| + |QX| = c$, where c is a constant greater than 2.

4. The points P and Q in $\mathbf{R}^2$ have coordinates $(-1, 0)$ and $(1, 0)$ respectively. Find the equation of the locus of all points X for which $|PX||QX| = c$, where c is a constant. Show that the locus consists of one piece when $c > 1$, but of two separate pieces when $c < 1$. What happens when $c = 1$?

(These figures are called the *ovals of Cassini*.)

5. A line $\mathcal{L}$ in $\mathbf{R}^2$ is described by the equation $lx + my + n = 0$, where $l^2 + m^2 = 1$. Show that the perpendicular distance from the point (x_0, y_0) to $\mathcal{L}$ is $|lx_0 + my_0 + n|$.

6. The *witch of Agnesi* is the plane curve defined as follows. A straight line $\mathcal{L}$ passes through the origin O and intersects the line $y = 2a$ in a point A and the circle $x^2 + (y - a)^2 = a^2$ in the point B. A point P is the intersection of a vertical line through A with a horizontal line through B. As $\mathcal{L}$ varies, P traces out a curve, and this curve is the 'witch'.

Prove that the Cartesian equation of the witch of Agnesi is

$$y = \frac{8a^3}{x^2 + 4a^2}.$$

7. The circle $\mathcal{C}$ has equation $(x - a)^2 + (y - b)^2 = r^2$. A point X with coordinates (x_0, y_0) does not lie on $\mathcal{C}$, and a line $\mathcal{L}$ passing through X meets $\mathcal{C}$ at A and B.

Show that $\overrightarrow{XA} \cdot \overrightarrow{XB} = (x_0 - a)^2 + (y_0 - b)^2 - r^2$. This quantity, which does not depend on the line $\mathcal{L}$, is called the *power* of X relative to the circle $\mathcal{C}$.

8. Let $\mathcal{C}_1$ and $\mathcal{C}_2$ be two circles. Show that the locus of all points having the same power with respect to each circle is a straight line perpendicular to the line joining the two centres. (This locus is called the *radical axis* of the two circles.)

9. Find the coordinates of the points of intersection of the folium of Descartes $x^3 + y^3 - 3xy = 0$ with the straight line $x + y = \frac{1}{2}$.

10. There are two possible isometries of the standard plane $\mathbf{R}^2$ that leave the origin fixed and send the point $(0, 5)$ to $(3, 4)$. Find their matrices and describe them geometrically.

11. Let T be an isometry of a Euclidean space $\mathcal{S}$. Prove that if the vectorialization $\vec{T}$ is the identity, then T is a translation.

(For the rest of this question you need some knowledge of group theory.) Show that the translations form a normal subgroup T of the isometry group $\mathrm{Iso}(\mathcal{S})$, and that the quotient group $\mathrm{Iso}(\mathcal{S})/T$ is isomorphic to the group of orthogonal matrices.

12. Let $\mathcal{L}$ be the line given in the standard coordinate system on $\mathbf{R}^2$ by $2x + 3y + 4 = 0$. A new coordinate system (x', y') has its origin at $(3, -2)$ and has basis vectors $\frac{1}{\sqrt{2}}(\mathbf{i} + \mathbf{j})$ and $\frac{1}{\sqrt{2}}(-\mathbf{i} + \mathbf{j})$. Find the equation of $\mathcal{L}$ in terms of the new coordinate system.

Find also a third coordinate system (x'', y'') relative to which $\mathcal{L}$ has the equation $y'' = $ constant.

13. The equation $x^2 + y^2 + z^2 = 9$ represents a sphere $\mathcal{S}$ of radius 3 centred at the origin in three-dimensional Euclidean space. The equation $x + y - z = 1$ represents a plane $\mathcal{P}$. Find a new coordinate system (x', y', z') relative to which the plane $\mathcal{P}$ has equation $z' = 0$. By transforming to this new coordinate system, find the equation (in terms of x' and y') of the curve in $\mathcal{P}$ given by its intersection with the sphere $\mathcal{S}$. What kind of curve is it?

14. Let U be a 2×2 reflection matrix and let $\mathbf{c}$ be a column vector. Show that the isometry of $\mathbf{R}^2$ given by

$$\mathbf{y} = \mathsf{U}\mathbf{x} + \mathbf{c}$$

has a fixed point if and only if $\mathsf{U}\mathbf{c} = -\mathbf{c}$.

15. Let $\mathcal{S}$ be a Euclidean space of dimension n, and let H by a *hyperplane*, that is an affine subspace of dimension $n - 1$. Show that there is just one non-identity isometry F_H of $\mathcal{S}$ that fixes every point of H; F_H is called the *hyperplane reflection* in H.

6

Plane geometry

6.1 Orientation in the plane

In this chapter we will explore in more detail the geometry of a Euclidean space $\mathcal{P}$ of dimension 2 or *Euclidean plane*. We will begin by looking at the concept of *orientation*, which allows us to distinguish between 'left-handed' and 'right-handed' coordinate systems.

Recall (4.5.1) that an *isometry* $T: \mathcal{P} \to \mathcal{P}$ is a transformation that preserves distances. In the previous chapter we have seen how isometries arise in geometry, both as congruences (the alibi interpretation) and as coordinate changes (the alias interpretation). Furthermore, we have obtained a complete classification of the isometries of a plane; in terms of coordinate vectors, any isometry has the form

$$\mathsf{x}' = \mathsf{U}\mathsf{x} + \mathsf{c},$$

where c is a fixed column vector and U is either a rotation matrix or a reflection matrix.

Let us try to express these isometries *without* using coordinates. To keep things simple, we will assume that $\mathsf{c} = \mathsf{0}$, so that the origin O is a fixed point of the isometry T in question. If T is indirect, so that U is a reflection matrix, the desired expression is easy to find:

(6.1.1)   **Proposition:** *Let $T: \mathcal{P} \to \mathcal{P}$ be an indirect isometry fixing the point O. Then there is a unit vector* $\mathbf{n}$ *such that T maps the point with position vector* $\mathbf{v}$ *to the point with position vector*

$$S_{\mathbf{n}}(\mathbf{v}) = \mathbf{v} - 2(\mathbf{v} \cdot \mathbf{n})\mathbf{n}.$$

Proof: We have already seen (5.3.6) that for a reflection matrix U we can find two orthonormal unit *eigenvectors* p and q satisfying $\mathsf{U}\mathsf{p} = \mathsf{p}$ and $\mathsf{U}\mathsf{q} = -\mathsf{q}$. The corresponding statement in terms of the transformation T is that there are two unit orthonormal vectors $\mathbf{p}, \mathbf{q} \in \vec{\mathcal{P}}$ such that $\vec{T}(\mathbf{p}) = \mathbf{p}$ and $\vec{T}(\mathbf{q}) = -\mathbf{q}$. Now $\mathbf{p}$ and $\mathbf{q}$ form an orthonormal basis for $\vec{\mathcal{P}}$, and therefore any vector $\mathbf{v} \in \vec{\mathcal{P}}$ can be written

$$\mathbf{v} = \lambda\mathbf{p} + \mu\mathbf{q}.$$

By linearity,

$$\vec{T}(\mathbf{v}) = \lambda\mathbf{p} - \mu\mathbf{q} = \mathbf{v} - 2\mu\mathbf{q} = \mathbf{v} - 2(\mathbf{v} \cdot \mathbf{q})\mathbf{q}.$$

Since the origin O is fixed, the effect of T on position vectors is given by its vectorialization $\vec{T}$. Thus we may take $\mathbf{n} = \mathbf{q}$ to obtain the required conclusion. □

Remark: The formula $\mathbf{v} \mapsto S_{\mathbf{n}}(\mathbf{v}) = \mathbf{v} - 2(\mathbf{v} \cdot \mathbf{n})\mathbf{n}$ also makes sense and defines an isometry in three dimensions or higher, in which case it is called reflection in a *plane* or *hyperplane*; see Exercises 4.6.8 and 5.5.15.

If we try to give a similar 'vector definition' for a *rotation* of the plane, though, we run into a difficulty. To specify a rotation we need to give not only its angle but also the *direction* — clockwise or anticlockwise — in which that angle is measured. Now this idea of choosing one out of the two possible directions of rotation is not at all intrinsic to the geometry of a Euclidean plane as we have defined it so far. All the definitions and theorems we have given so far would be unaffected if we replaced $\mathcal{P}$ by its mirror image, where 'clockwise' and 'anticlockwise' are interchanged. It follows that before we can discuss rotations we will need to enrich the geometry of $\mathcal{P}$ by adding an explicit choice of 'positive direction of rotation'. Such a choice is called an *orientation* of $\mathcal{P}$.

(6.1.2)   **Definition:** *An* orientation *for a Euclidean plane $\mathcal{P}$ is an operation which assigns to each vector $\mathbf{u} \in \vec{S}$ a new vector $\mathbf{u}^{\perp} \in \vec{S}$, having the following properties:*

(i) *The mapping $\mathbf{u} \mapsto \mathbf{u}^{\perp}$ is linear, in other words*

$$(\lambda_1\mathbf{u}_1 + \lambda_2\mathbf{u}_2)^{\perp} = \lambda_1\mathbf{u}_1^{\perp} + \lambda_2\mathbf{u}_2^{\perp}.$$

(ii) *The vector $\mathbf{u}^{\perp}$ is perpendicular to $\mathbf{u}$.*

(iii) *If $\mathbf{u}$ is a unit vector, then $\mathbf{u}^{\perp}$ is a unit vector also.*

The idea is that $\mathbf{u}^{\perp}$ specifies the result of rotating $\mathbf{u}$ through $\frac{\pi}{2}$ in the positive direction. Notice that the first and third properties imply that $|\mathbf{v}| = |\mathbf{v}^{\perp}|$ for *any* vector $\mathbf{v}$.

Examples of orientations are easy to find. If $\{\mathbf{i}, \mathbf{j}\}$ is an orthonormal basis, then the formula

$$(x\mathbf{i} + y\mathbf{j})^{\perp} = -y\mathbf{i} + x\mathbf{j}$$

defines an orientation, because $|x\mathbf{i} + y\mathbf{j}| = |-y\mathbf{i} + x\mathbf{j}|$ and

$$(x\mathbf{i} + y\mathbf{j}) \cdot (-y\mathbf{i} + x\mathbf{j}) = -xy + xy = 0.$$

Another orientation, different from the first one, is defined by the formula

$$(x\mathbf{i} + y\mathbf{j})^{\perp} = y\mathbf{i} - x\mathbf{j}.$$

In fact we can show that these are the only two possibilities. For $\mathbf{i}^{\perp}$ must be equal either to $\mathbf{j}$ or to $-\mathbf{j}$. If $\mathbf{i}^{\perp} = \mathbf{j}$, then $\mathbf{j}^{\perp} = -\mathbf{i}$; one can't have $\mathbf{j}^{\perp} = \mathbf{i}$ because in that case $(\mathbf{i} + \mathbf{j})^{\perp} = \mathbf{i} + \mathbf{j}$ by linearity, and this would contradict the perpendicularity requirement. Similarly, if $\mathbf{i}^{\perp} = -\mathbf{j}$, then $\mathbf{j}^{\perp} = \mathbf{i}$.

If an orientation has been chosen, we call a basis $\mathbf{i}, \mathbf{j}$ for which $\mathbf{i}^{\perp} = \mathbf{j}$ *right-handed*, and we call one for which $\mathbf{i}^{\perp} = -\mathbf{j}$ *left-handed*. A plane equipped with a choice of orientation is called an *oriented plane*.

Whichever orientation is chosen, $\mathbf{i}^{\perp\perp} = -\mathbf{i}$ and $\mathbf{j}^{\perp\perp} = -\mathbf{j}$. Because the orientation is linear, this implies that $\mathbf{v}^{\perp\perp} = -\mathbf{v}$ for any vector $\mathbf{v}$.

Fig. 6.1. The Möbius band: An example of a nonorientable surface

Remark: A plane can be regarded as a particularly simple kind of surface. Among more general surfaces there are examples that have *no* orientation. The famous *Möbius band* is such a surface (Figure 6.1): any right-handed basis comes back left-handed after it has travelled once round the band. Möbius discovered the concept of orientation in the following way. He had invented a rather complicated formula which gave the area of a *triangulated* surface (one made up of a lot of triangular pieces) as a sum of determinants. Also involved in his formula, however, were certain coefficients ± 1 which had to be chosen in a 'consistent' way for the formula to give the right answer. In fact these ± 1's told you whether each little triangle was left-handed or right-handed with respect to some assumed overall orientation. Möbius was therefore led to investigate whether these coefficients could always be chosen consistently, and he produced the example which now bears his name to show that sometimes they could not. We will discuss the orientation of surfaces in Chapter 12.

Suppose now that $\mathcal{P}$ is an oriented Euclidean plane, and let O be a point in $\mathcal{P}$ which we take as the origin.

(6.1.3)   **Proposition:** *Let $T: \mathcal{P} \to \mathcal{P}$ be a direct isometry fixing the point O. Then there is an angle θ such that T maps the point with position vector $\mathbf{v}$ to the point with position vector $R_\theta(\mathbf{v})$, where*

$$R_\theta(\mathbf{v}) = \cos\theta\,\mathbf{v} + \sin\theta\,\mathbf{v}^\perp.$$

Proof: Because O is fixed, we just need to find $\vec{T}$. Choose a right-handed orthonormal basis $\mathbf{i}, \mathbf{j}$ for $\vec{\mathcal{P}}$. Then by our classification of 2×2 orthogonal matrices (5.3.6), $\vec{T}$ is represented relative to this basis by a rotation matrix

$$\begin{pmatrix} \cos\theta & -\sin\theta \\ \sin\theta & \cos\theta \end{pmatrix}.$$

So $\vec{T}(\mathbf{i}) = \cos\theta\,\mathbf{i} + \sin\theta\,\mathbf{j} = R_\theta(\mathbf{i})$, and similarly $\vec{T}(\mathbf{j}) = R_\theta(\mathbf{j})$. Since the transformations $\vec{T}$ and R_θ are both linear and agree on $\mathbf{i}$ and $\mathbf{j}$, they must in fact be equal. $\square$

The argument can be reversed to show that the vector formula for R_θ defines an isometry whose matrix is a rotation matrix. An important property of rotations is the *composition law*:

(6.1.4) $$R_\theta R_\varphi = R_{\theta+\varphi}.$$

To prove this we write

$$R_\theta R_\varphi(\mathbf{v}) = \cos\theta(\cos\varphi\,\mathbf{v} + \sin\varphi\,\mathbf{v}^\perp) + \sin\theta(\cos\varphi\,\mathbf{v} + \sin\varphi\,\mathbf{v}^\perp)^\perp.$$

Using the fact that $\mathbf{v}^{\perp\perp} = -\mathbf{v}$, we may write this as

$$(\cos\theta\cos\varphi - \sin\theta\sin\varphi)\mathbf{v} + (\cos\theta\sin\varphi + \sin\theta\cos\varphi)\mathbf{v}^\perp,$$

which is equal to $R_{\theta+\varphi}(\mathbf{v})$ by familiar trigonometric identities.

Remark: You may have seen this calculation done the other way around; assuming that rotations add up as in 6.1.4, one can deduce the trigonometric identities. But there is no danger of circular reasoning here, because in this book we are assuming that all the properties of the trigonometric functions are obtained from some analytical definition (using power series, for instance). We are therefore free to make use of these properties in geometry.

(6.1.5)   **Example:** A particular example of a rotation is the *central inversion* ι with centre O, defined by $ι(\mathbf{v}) = -\mathbf{v}$. By definition, $ι = R_\pi$. Some people call this 'reflection in a point'.

6.2   Oriented angles

(6.2.1)   **Proposition:** *Let $\mathcal{P}$ be an oriented Euclidean plane, and let O, A, and B be three points in $\mathcal{P}$ with $|OA| = |OB| > 0$. Then there is exactly one rotation R_θ of $\mathcal{P}$ with centre O such that $R_\theta(A) = B$.*

Proof: Let $r = |OA| = |OB|$, and let $\mathbf{i} = \overrightarrow{OA}/r$ be a unit vector in the direction OA. Let $\mathbf{j} = \mathbf{i}^\perp$, so that $\mathbf{i}, \mathbf{j}$ forms a right-handed orthonormal basis. The vector $\overrightarrow{OB}/r$ is of unit length, so it can be written

$$\cos\theta\,\mathbf{i} + \sin\theta\,\mathbf{j}$$

where θ is uniquely determined modulo 2π. Then

$$R_\theta(\overrightarrow{OA}) = R_\theta(r\mathbf{i}) = r\cos\theta\,\mathbf{i} + r\sin\theta\,\mathbf{j} = \overrightarrow{OB},$$

and no other rotation can map $\overrightarrow{OA}$ to $\overrightarrow{OB}$. $\square$

Let $\mathbf{u}$ and $\mathbf{v}$ be nonzero vectors in an oriented Euclidean plane $\mathcal{P}$. It follows that there is a unique rotation R_θ such that

$$R_\theta\left(\frac{\mathbf{u}}{|\mathbf{u}|}\right) = \frac{\mathbf{v}}{|\mathbf{v}|}.$$

(6.2.2)   **Definition:** *This angle θ, which is defined up to a multiple of 2π, is called the oriented angle between $\mathbf{u}$ and $\mathbf{v}$.*

It is worth taking a little care over this business of 'defined up to a multiple of 2π'. To be precise, an oriented angle is not really 'measured' by a number, but by a whole class of numbers all differing from one another by multiples of 2π. (Sophisticated algebraists will recognize such a class as an element of the *quotient group* $\mathbf{R}/2\pi\mathbf{Z}$, and may enjoy working out the theory on this basis.) Each number in the class gives a different 'name' for the same angle. This means that any identity between the names of oriented angles must be thought of as holding only 'up to multiples of 2π', and to remind us of this we will introduce a special symbol $\equiv$; the equation

$$\theta \equiv \varphi$$

will mean that θ and φ are names for the same oriented angle, in other words that $\theta - \varphi = 2k\pi$, where $k \in \mathbf{Z}$. Notice that $\equiv$ is *not* the same as equality; if $2\theta \equiv \varphi$, for instance, it does *not* follow that $\theta \equiv \varphi/2$. (Why not?)

If A, B, and C are points in a Euclidean plane we will use the notation $\angle ABC$ for the oriented angle between $\overrightarrow{BA}$ and $\overrightarrow{BC}$.

The composition law for rotations immediately gives us the *addition law for oriented angles*:

(6.2.3)   **Proposition:** *Let* **u**, **v**, *and* **w** *be vectors in a plane. If the oriented angle between* **u** *and* **v** *is* θ, *the oriented angle between* **v** *and* **w** *is* φ, *and the oriented angle between* **u** *and* **w** *is* ψ, *then*

$$\theta + \varphi \equiv \psi.$$

Proof: We may assume without loss of generality that **u**, **v**, and **w** are unit vectors. Then we calculate that $R_{\theta+\varphi}(\mathbf{u}) = R_{\varphi}(R_{\theta}(\mathbf{u})) = R_{\varphi}(\mathbf{v}) = \mathbf{w}$, by the composition law for rotations (6.1.4). $\square$

A simple corollary is that the oriented angle between **v** and **u** is equal to *minus* the oriented angle between **u** and **v**.

How is this oriented angle related to the ordinary 'unoriented' angle between two vectors which we defined in 4.3.3? Notice that the unoriented angle only ranges from 0 to π, so the oriented angle is 'twice as informative' in some sense. The precise relationship is given in the next proposition.

(6.2.4)   **Proposition:** *Let* **u** *and* **v** *be vectors in an oriented Euclidean plane. Let* θ *denote the oriented angle between them,* φ *the unoriented angle between them. Then*

$$\theta \equiv \pm\varphi.$$

If $\mathbf{u}^{\perp} \cdot \mathbf{v} > 0$, *the positive sign should be selected and if* $\mathbf{u}^{\perp} \cdot \mathbf{v} < 0$ *the negative sign should be selected. If* $\mathbf{u}^{\perp} \cdot \mathbf{v} = 0$ *either sign may be selected, as then* $\varphi = 0$ *or* π.

Proof: We have, by definition of the oriented angle,

$$R_{\theta}\left(\frac{\mathbf{u}}{|\mathbf{u}|}\right) = \frac{\mathbf{v}}{|\mathbf{v}|}.$$

Therefore

$$\frac{\mathbf{v}}{|\mathbf{v}|} = \left(\frac{\mathbf{u}}{|\mathbf{u}|}\right)\cos\theta + \left(\frac{\mathbf{u}^{\perp}}{|\mathbf{u}|}\right)\sin\theta$$

and so

$$\cos\theta = \frac{\mathbf{u}\cdot\mathbf{v}}{|\mathbf{u}||\mathbf{v}|}, \qquad \sin\theta = \frac{\mathbf{u}^{\perp}\cdot\mathbf{v}}{|\mathbf{u}||\mathbf{v}|}.$$

Because of the definition of unoriented angle (4.3.3), the first equation says that $\cos\theta = \cos\varphi$, and this implies that $\theta = 2k\pi \pm \varphi$, $k \in \mathbf{Z}$. To see which sign we should take we look at the expression for $\sin\theta$. Since φ always lies between 0 and π, $\sin\varphi$ is always non-negative. Thus if $\sin\theta > 0$ we must take the positive sign, and if $\sin\theta < 0$ we must take the negative sign. $\square$

The expression $\mathbf{u}^{\perp} \cdot \mathbf{v}$, which came up in this proposition, is an interesting one; later we will see that it in fact gives the *area* of the parallelogram with sides specified by the vectors $\mathbf{u}$ and $\mathbf{v}$. We can calculate it by means of determinants. Choose a right-handed orthonormal basis $\{\mathbf{i}, \mathbf{j}\}$; then $\mathbf{u}$ and $\mathbf{v}$ are represented by column vectors u and v with respect to this basis. Let $(\mathsf{u}|\mathsf{v})$ denote the 2×2 matrix whose columns are u and v. Then

(6.2.5)
$$\mathbf{u}^{\perp} \cdot \mathbf{v} = \det(\mathsf{u}|\mathsf{v}).$$

To prove this, suppose that $\mathbf{u} = a\mathbf{i} + c\mathbf{j}$ and $\mathbf{v} = b\mathbf{i} + d\mathbf{j}$. Then

$$\mathbf{u}^{\perp} \cdot \mathbf{v} = (a\mathbf{j} - c\mathbf{i}) \cdot (b\mathbf{i} + d\mathbf{j}) = ad - bc = \det(\mathsf{u}|\mathsf{v}).$$

Let L be a linear transformation. Then $(L\mathsf{u})^{\perp} \cdot L\mathsf{v} = \det(L\mathsf{u}|L\mathsf{v})$, where L is the matrix of L. But by definition of matrix multiplication, $(L\mathsf{u}|L\mathsf{v}) = \mathsf{L}(\mathsf{u}|\mathsf{v})$. So, using the product rule for determinants, we find that

(6.2.6)
$$(L\mathbf{u})^{\perp} \cdot L\mathbf{v} = \det(L)\mathbf{u}^{\perp} \cdot \mathbf{v}.$$

A consequence of this is

(6.2.7)   **Lemma:** *Let T be an isometry of an oriented Euclidean plane. Then*

$$\vec{T}(\mathbf{v}^{\perp}) = \pm \left(\vec{T}(\mathbf{v}) \right)^{\perp}$$

where the + sign is to be taken if T is direct and the − sign if T is indirect.
Proof: Because $\vec{T}$ preserves the dot product,

$$\mathbf{u} \cdot \mathbf{v}^{\perp} = \vec{T}(\mathbf{u}) \cdot \vec{T}(\mathbf{v}^{\perp}).$$

By the previous calculation

$$\det\left(\vec{T} \right) \mathbf{u} \cdot \mathbf{v}^{\perp} = \vec{T}(\mathbf{u}) \cdot \left(\vec{T}(\mathbf{v}) \right)^{\perp}.$$

Comparing these equations, we find that

$$\left(\vec{T}(\mathbf{v}) \right)^{\perp} = \det\left(\vec{T} \right) \vec{T}(\mathbf{v}^{\perp})$$

as required. □

Remark: There is one subtle point in this argument. What it really shows is that the dot products of the left- and right- hand sides of the final equation with $\vec{T}(\mathbf{u})$ are equal. However, since $\mathbf{u}$ is arbitrary and $\vec{T}$ is invertible, $\vec{T}(\mathbf{u})$ can be any vector. Thus the two sides have the same dot product with any vector, and so they must be equal.

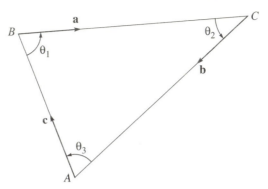

Fig. 6.2. The angle sum of a triangle

(6.2.8)   **Theorem:** (ANGLE SUM OF A TRIANGLE — ORIENTED VERSION) *Let A, B, and C be three distinct points in an oriented Euclidean plane. Then*

$$\angle ABC + \angle BCA + \angle CAB \equiv \pi.$$

Proof: Let $\theta_1 = \angle CAB$, $\theta_2 = \angle ABC$, and $\theta_3 = \angle BCA$. Also let **a**, **b**, and **c** be unit vectors in the directions of $\overrightarrow{BC}$, $\overrightarrow{CA}$, and $\overrightarrow{AB}$ respectively (see Figure 6.2). Then

$$R_{\theta_1}(\mathbf{b}) = \mathbf{c}, \quad R_{\theta_2}(\mathbf{c}) = -\mathbf{a}, \quad R_{\theta_3}(\mathbf{a}) = \mathbf{b}.$$

Thus $R_{\theta_1 + \theta_2 + \theta_3}(\mathbf{a}) = -\mathbf{a}$, and so $\theta_1 + \theta_2 + \theta_3 \equiv \pi.$ □

Remark: This proof is essentially equivalent to the 'total turning' argument which you may know from school geometry. Imagine a car driving round the triangle. At each vertex it turns through an angle equal to π minus the interior angle at that vertex. When it has finished its journey, it has turned through 2π. The result follows.

Isometries preserve unoriented angles; how do they affect oriented angles? The answer turns out to depend on whether the isometry is direct or indirect. Indirect isometries reverse the direction of rotation and so reverse the sign of oriented angles.

(6.2.9)   **Proposition:** *Let A, B, and C be points in an oriented Euclidean plane $\mathcal{P}$, and let $T: \mathcal{P} \to \mathcal{P}$ be an isometry. Then*

- *If T is direct, then $\angle T(A)T(B)T(C) \equiv \angle ABC$.*

- *If T is indirect, then $\angle T(A)T(B)T(C) \equiv -\angle ABC$.*

Proof: Because of the definition of oriented angles in terms of rotations, what we have to prove is simply that

$$R_\theta \vec{T} = \vec{T} R_{c\theta}$$

where $c = \det \vec{T}$. But this follows from the definition of a rotation (6.1.3) together with the fact that $(\vec{T}\mathbf{v})^\perp = c\vec{T}(\mathbf{v}^\perp)$ which was proved in 6.2.7. $\square$

As an illustration of the use of this proposition, here is an oriented version of the result about the base angles of an isosceles triangle:

(6.2.10)   **Proposition:** *Let ABC be an isosceles triangle in an oriented Euclidean plane, with $|AB| = |AC|$. Then $\angle ABC \equiv -\angle ACB$.*

Proof: We can use Pappus' proof (4.5.8), making the additional observation that the isometry which gives a congruence between ABC and ACB is in fact the reflection in the perpendicular bisector of BC, and so is indirect. $\square$

6.3 Polar coordinates

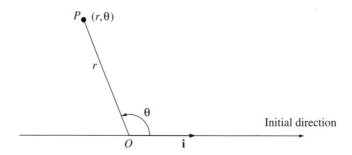

Fig. 6.3. Polar coordinates

Let $\mathcal{P}$ be an oriented Euclidean plane. If we fix an origin O and an *initial direction* given by a unit vector $\mathbf{i}$, then any point $P \neq O$ in the plane is uniquely determined by two data: the number $r = |OP| > 0$ and the oriented angle θ such that $R_\theta(r\mathbf{i}) = \overrightarrow{OP}$. The numbers (r, θ) are said to be the *polar coordinates* of the point P relative to the given origin, initial direction, and orientation (Figure 6.3).

Remark: Notice that, once again, the 'angle' coordinate θ is only defined up to a multiple of 2π. The simplest way of avoiding the multiple-valuedness of θ — always choose the 'principal value' so that $0 \leq \theta < 2\pi$ — has the serious disadvantage that the chosen coordinates can 'jump' as the point P moves continuously. The topological difficulties that this raises are discussed in books on complex analysis. A related problem is that the origin O is a 'singular' point of the polar coordinate system — the correspondence between points and their coordinates ceases to be one-to-one there, because if $r = 0$, θ can be anything.

From the explicit form of the rotation matrix we get the relationship between Cartesian and polar coordinates.

(6.3.1)   **Proposition:** *Let $\mathcal{P}$ be as above, and suppose that $P \in \mathcal{P}$ has polar coordinates (r, θ) (as defined above) and Cartesian coordinates (x, y) relative to the Cartesian coordinate system with origin O and basis vectors $\mathbf{i}$ and $\mathbf{i}^{\perp}$. Then*

$$x = r\cos\theta, \quad y = r\sin\theta.$$

When we defined polar coordinates we specified that $r > 0$. However, it is often convenient to consider polar coordinates with $r \leq 0$; they can still interpreted as Cartesian coordinates by means of the above formulae. The effect of this convention is that the polar coordinates (r, θ) and $(-r, \theta + \pi)$ represent the same point, so that negative values of r are measured 'in the opposite direction'.

It is possible to specify a curve by an equation $g(r, \theta) = 0$ relating the polar coordinates r and θ. We won't go into the general theory of the smoothness of such curves, but we will give two examples showing how it is possible to transform a Cartesian equation into a polar one using the relationship $x = r\cos\theta$, $y = r\sin\theta$ between the two coordinate systems.

(6.3.2)   **Example:** To find the *polar equation of a line*, we start with the Cartesian equation $lx + my + n = 0$, where we may assume that $l^2 + m^2 = 1$. Then there is a constant φ such that $l = \cos\varphi$, $m = \sin\varphi$. If we substitute $x = r\cos\theta$,

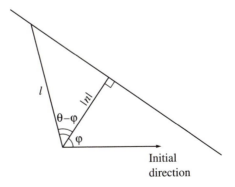

Fig. 6.4. Polar equation of a line

$y = r \sin \theta$ into the Cartesian equation, and use the expansion for $\cos(\theta - \varphi)$, we get

$$r \cos(\theta - \varphi) + n = 0$$

as the polar equation of the line. This form of the equation has a simple geometrical interpretation (Figure 6.4), which incidentally shows that $|n|$ is the perpendicular distance from the line to the origin.

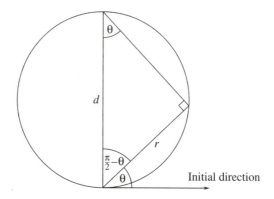

Fig. 6.5. Polar equation of a circle

(6.3.3)   **Example:** It is also useful to know the *polar equation of a circle*, when the origin is a point on the circle. Suppose for instance that the origin is at $(0, b)$ and the radius is b; then the Cartesian equation is $x^2 + (y - b)^2 = b^2$, or $x^2 + y^2 = 2by$. Substituting for x and y and cancelling a factor of r, we get

$$r = 2b \sin \theta = d \sin \theta$$

as the equation of the circle, where $d = 2b$ is the *diameter*. This also has a geometrical interpretation, using trigonometry and the fact that the angle in a semicircle is a right angle (Figure 6.5).

Notice that the entire circle is traced out as θ runs from 0 to π. As θ runs from π to 2π, r becomes negative and the circle is traced out a second time.

6.4   Circles in the plane

Circles have many interesting angular properties, mostly known to Euclid. In this section we will mention a few of the highlights. Throughout, we work in some oriented Euclidean plane $\mathcal{P}$.

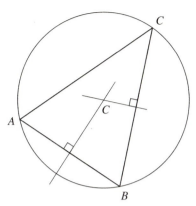

Fig. 6.6. The circumcircle of a triangle

(6.4.1)   **The circumcircle of a triangle:** Let ABC be a proper triangle. Then AB and BC are not parallel, and so their perpendicular bisectors are not parallel either. These perpendicular bisectors therefore meet in a point, O (Figure 6.6). Now $|OA| = |OB|$, because O is on the perpendicular bisector of AB, and $|OB| = |OC|$, because O is on the perpendicular bisector of BC. Therefore the three points A, B, C are equidistant from O, and so O is the centre of a circle passing through the three points. This circle is called the *circumcircle* of the triangle ABC.

The argument in fact shows that there is *only one* point O equidistant from A, B, and C, so the circumcircle is unique. In particular, the perpendicular bisector of AC must pass through O; thus, the three perpendicular bisectors of the sides of any triangle meet in a point, as you may already have proved in Exercise 4.6.6. (Compare this with the fact that the three medians must meet in a point, which we proved with the aid of Ceva's theorem.)

If A, B, and C are collinear then there is *no* circle through them. A circle can intersect a line in at most two points, because the points of intersection are given by the roots of a quadratic equation — see the verification of the circle-line intersection axiom in 4.1.5. But in this case there is of course a line through A, B, and C. This leads us to think of lines and circles as two species within the same genus, an idea which will recur several times. Sometimes it is expressed by saying that 'a line is a circle of infinite radius' — this is picturesque, and helps the imagination, but does not make precise sense.

(6.4.2)   **Tangent at a point:** Let $\mathcal{C}$ be a circle with centre O and radius r, and let P be a point on $\mathcal{C}$. It is a familiar fact that of all the lines through P there is only one that does not meet the circle again somewhere else; this unique line is called the *tangent line* at P, and it is perpendicular to the radius OP (Figure 6.7). To prove this, we may take position vectors relative to O. Let P have

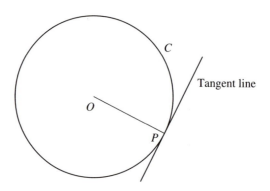

Fig. 6.7. Tangent to a circle

position vector **v**, and let **u** be any vector. The equation giving the points of intersection of the circle C with the line through P parallel to **u** is

$$|\mathbf{v} + \lambda \mathbf{u}|^2 = r^2$$

which reduces to

$$2\lambda \mathbf{u} \cdot \mathbf{v} + \lambda^2 |\mathbf{u}|^2 = 0;$$

and this has two distinct solutions unless $\mathbf{u} \cdot \mathbf{v} = 0$.

Using the tangent one can go on to make various other definitions. For instance, two circles meeting at a point P are said to be *tangent* at P if their tangent lines at P are the same. And one defines the *angle* between a circle and a line (or between two circles) at a point of intersection to be the angle between the tangent line(s).

Remark: By our definition, then, the angle between a circle and its tangent line is zero. This definition may seem natural to us, but it was the subject of intense controversy among geometers during the seventeenth century and earlier[1]. For them, an angle was a geometric figure bounded by two curves. Now, they argued, the angle between a circle and its tangent (for which there was a recognized name, the *cornicular* or 'horn-like' angle) cannot be *zero* — because the circle does not *coincide* with the line over any distance, however short — but on the other hand must be *less than any positive angle* — because any line that makes a positive angle with the tangent cuts the circle twice. It seems as though the cornicular angle is some kind of 'infinitesimal' quantity, but what sense does that make?

---

[1] Similar arguments probably took place among the Greek geometers, but there are only echoes of them in Euclid. Heath [22] remarks, 'The violence of the controversy [between two seventeenth-century geometers] will be understood from the fact that the arguments and counter-arguments cover 26 pages of small print'.

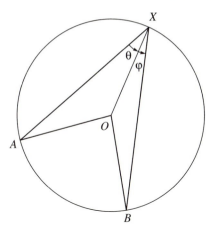

Fig. 6.8. The angle at the centre and at the circumference

You may know that Newton and Liebniz, who invented calculus, freely used the language of infinitesimals. But because of the paradoxes to which this language led, the rigorous analysis of nineteenth-century mathematicians like Cauchy and Weierstrass abolished the infinitesimals, and redefined everything in (rather complicated) finite terms. Recently, however, infinitesimals have been making a comeback. The subject of *nonstandard analysis* gives a rigorous (but still rather complicated) definition of infinitesimals, and can be used to give a retrospective justification of the free and easy arguments used by the founders of calculus. In one approach[2] to nonstandard analysis, infinitesimals and other 'nonstandard numbers' are defined by the 'rate' at which two functions (or curves) approach one another. From this point of view, the cornicular angle is indeed a good example of an infinitesimal quantity.

We have seen already (4.2.5) that the angle in a semicircle is a right angle. This is a special case of the following result, which establishes a close link between circles and oriented angles.

(6.4.3)   **Proposition:** *Let C be a circle with centre O, and let A and B be points on C. Then for any other point X on C, $\angle AOB \equiv 2\angle AXB$.*

One often states this as 'The angle at the centre is twice the angle at the circumference'. See Figure 6.8.

Proof: Let $\angle AXO \equiv \theta$ and $\angle OXB \equiv \varphi$. The triangle OAX is isosceles, and so its base angles are equal (6.2.10); so $\angle OAX \equiv \theta$. Hence, since the sum of

---

[2]For a brief introduction to nonstandard analysis from this point of view, see Ebbinghaus *et al.* [13, Chapter 12].

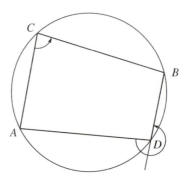

Fig. 6.9. Concyclic points

the oriented angles in the triangle OAX is π, $\angle XOA \equiv \pi - 2\theta$, and similarly $\angle BOX \equiv \pi - 2\varphi$. Therefore, by the addition law for oriented angles,

$$\angle BOA \equiv 2\pi - 2\varphi - 2\theta \equiv -2(\theta + \varphi),$$

and so $\angle AOB \equiv 2(\theta + \varphi) \equiv 2\angle AXB$. $\square$

(6.4.4)   **Proposition:** *Let A and B be points and let θ be an angle. Then the set of all points X such that $2\angle AXB \equiv \theta$ is a circle or a straight line, passing through the points A and B.*

    Watch out for the fact that $2\angle AXB \equiv \theta$ does *not* imply that $\angle AXB \equiv \theta/2$; it implies only that $\angle AXB \equiv \theta/2$ *or* $\pi + \theta/2$.

Proof: If $\theta \equiv 0$ then the condition that $2\angle AXB \equiv \theta$ just says that A, X, and B are collinear, so the set of all such points X is the line $\overleftrightarrow{AB}$. Otherwise, it is easy to see that there is a unique point O on the perpendicular bisector of AB for which $\angle AOB \equiv \theta$. Let $\mathcal{C}$ be the circle with centre O passing through A and B. Then any point $X \in \mathcal{C}$ satisfies $2\angle AXB \equiv \theta$, because the angle at the centre is twice the angle at the circumference (6.4.3). On the other hand, if X satisfies $2\angle AXB \equiv \theta$, then the centre of the circumcircle of the triangle AXB must be a point O' on the perpendicular bisector of AB with $\angle AO'B \equiv \theta$. The only such point is O, so the circumcircle must be $\mathcal{C}$, and then $X \in \mathcal{C}$. $\square$

(6.4.5)   **Definition:** *Four distinct points are said to be* concyclic *if there is a circle passing through all four of them.*

    It follows easily from the preceding proposition that four points A, B, C, and D are collinear or concyclic if and only if $2\angle ACB \equiv 2\angle ADB$. This is illustrated in Figure 6.9; notice particularly the directions in which the angles are measured. In classical geometry one worked with the *unoriented* angles $\widehat{ACB}$, $\widehat{ADB}$, which may be equal or may add up to π, depending on

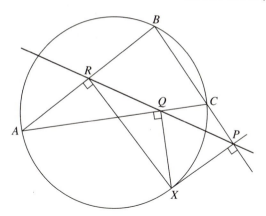

Fig. 6.10. The Simson line

whether C and D lie on the same or different arcs of the circle from A to B. In our treatment this is automatically taken care of by the conventions about orientation.

It is often possible to prove theorems about circles simply by using these and related results to write down long chains of equalities between angles, a procedure sometimes known as 'angle-chasing'. We will give one example of angle-chasing, the theorem on the so-called *Simson line*. Robert Simson (1687–1768) was an influential geometer whose edition of Euclid was widely studied, but it seems that he never proved this result which is named after him! It is in fact due to William Wallace (1768–1843), Professor of Mathematics at Edinburgh.

(6.4.6)   **Theorem:** *Let ABC be a proper triangle, and let X be another point. Let P, Q, and R be the feet of the perpendiculars from X to $\overline{BC}$, $\overline{CA}$, and $\overline{AB}$ respectively. Then the points P, Q, and R are collinear if and only if the point X lies on the circumcircle of the triangle ABC.*

The Simson line itself is the line PQR (see Figure 6.10).

Proof: Notice that $2\angle CQX \equiv 2\angle CPX \equiv \pi$, so the four points C, Q, P, X are concyclic. Similarly, the four points B, R, X, P are concyclic. Therefore $2\angle QPX \equiv 2\angle QCX \equiv 2\angle ACX$ and $2\angle RPX \equiv 2\angle RBX \equiv 2\angle ABX$. It follows that

$$A, B, C, X \text{ are concyclic} \iff 2\angle ABX \equiv 2\angle ACX$$
$$\iff 2\angle QPX \equiv 2\angle RPX$$
$$\iff 2\angle QPR \equiv 0.$$

But $2\angle QPR \equiv 0$ if and only if P, Q, and R are collinear, which completes the 'angle-chasing' and proves the result. $\square$

6.5 The use of complex numbers

The idea of representing the points of a plane by *complex numbers* was first proposed at the beginning of the nineteenth century by two amateur mathematicians, the Norwegian Caspar Wessel (who was a surveyor by trade) and the Swiss Jean Robert Argand (who was an accountant). This representation, nowadays familiar as the 'Argand diagram', was of great importance in clarifying the previously rather mysterious nature of complex numbers, and so contributed to their general acceptance among mathematicians. But the link, like so many between geometry and algebra, is two-way; we can also use our knowledge of complex numbers to prove theorems in geometry.

Let $\mathcal{P}$ be an oriented Euclidean plane. If we choose a unit vector $\mathbf{i}$, we can put the vectors of $\vec{\mathcal{P}}$ into one-to-one correspondence with the complex numbers $\mathbf{C}$. The vector $x\mathbf{i} + y\mathbf{i}^{\perp}$ corresponds to the complex number $x + iy$. If we also choose an origin O, each point of P can be represented by its position vector, and we get a correspondence between the points of $\mathcal{P}$ and the complex numbers. This correspondence can also be expressed in polar coordinates; the point with polar coordinates (r, θ) corresponds to the complex number $z = re^{i\theta}$ with modulus $|z| = r$ and argument $\operatorname{Arg} z \equiv \theta$. (We used the special $\equiv$ symbol because the argument of a complex number, like an oriented angle or a polar coordinate, is only defined up to multiples of 2π.)

In terms of this correspondence we can set up a 'dictionary' between geometry and complex numbers, as follows:

Vector, $\mathbf{v} = x\mathbf{i} + y\mathbf{j}$	Complex number, $z = x + iy$
Length of vector, $\|\mathbf{v}\|$	Modulus, $\|z\|$
Dot product, $\mathbf{v}_1 \cdot \mathbf{v}_2$	Real part of product, $\Re(\bar{z}_1 z_2)$
Oriented angle between $\mathbf{i}$ and $\mathbf{v}$	Argument, $\operatorname{Arg}(z)$
Orientation, $\mathbf{v} \mapsto \mathbf{v}^{\perp}$	Multiplication by i, $z \mapsto iz$
Translation, $\mathbf{v} \mapsto \mathbf{v} + \mathbf{u}$	Translation, $z \mapsto z + w$
Rotation, $\mathbf{v} \mapsto R_{\theta}(\mathbf{v})$	Multiplication by $e^{i\theta}$, $z \mapsto e^{i\theta}z$.
Reflection in x-axis	Complex conjugation, $z \mapsto \bar{z}$

Using this dictionary we could translate all that we have done so far into the language of complex numbers. Indeed, it seems that this is what Argand and Wessel were trying to do; they had no vector algebra, and they wanted to use complex numbers as vectors. More interesting to us, though, is the use of multiplication and division of complex numbers to suggest new types of geometrical transformations.

(6.5.1)   **Definition:** *A Möbius transformation is a mapping* $M: \mathbf{C} \to \mathbf{C}$ *given by a formula*

$$M(z) = \frac{az + b}{cz + d}$$

where a, b, c, and d are constants and ad − bc ≠ 0.

If one thinks of the Möbius transformation M as associated with the matrix $\begin{pmatrix} a & b \\ c & d \end{pmatrix}$, then it is easily checked that the effect of composing two Möbius transformations is to multiply the corresponding matrices. The composite of two Möbius transformations is therefore another Möbius transformation. Möbius transformations are invertible[3] (because $ad − bc ≠ 0$, which is the condition for the matrix to be invertible) and the inverse of a Möbius transformation is again a Möbius transformation. As usual, this implies that the Möbius transformations form a *group*.

All the direct isometries (rotations and translations) belong to this group, as one sees from the table above. (If we want to include indirect isometries as well, we have to throw in complex conjugation.) But the group also contains Möbius transformations which are neither rotations nor translations. According to the Erlanger Program (4.5.9), then, one should obtain a new kind of geometry by studying the invariants of the group of Möbius transformations. We will not develop this new geometry (called *inversive geometry*) very far, but we will give some examples to show how Möbius transformations can be used.

(6.5.2)   **Proposition:** *Let M be any Möbius transformation, and let* θ *be an angle. Then the equation*

$$2 \operatorname{Arg} M(z) \equiv \theta$$

defines a circle or straight line, and any circle or straight line can be defined by an equation of this form.

Proof: We must show that all those points Z represented by complex numbers z satisfying the equation lie on a circle or straight line. Suppose that $M(z) = \dfrac{az + b}{cz + d}$; then

$$\operatorname{Arg} M(z) \equiv \operatorname{Arg}\left(\frac{z - p}{z - q}\right) + \operatorname{Arg}\left(\frac{a}{c}\right)$$

where $p = -b/a$ and $q = -d/c$. Therefore, the equation $2 \operatorname{Arg} M(z) \equiv \theta$ is equivalent to

$$2 \operatorname{Arg}\left(\frac{z - p}{z - q}\right) \equiv \varphi$$

where $\varphi = \theta - 2 \operatorname{Arg}(a/c)$ is a constant. But this equation just says

---

[3] We have swept a problem under the carpet here. If $z = -d/c$, then $M(z)$ is not defined, and $M(z)$ can never take the value a/c. These difficulties may remind you of the problems we had with projection in Chapter 2, and again the remedy is to 'projectivize' by introducing suitable 'points at infinity'. In fact only one 'point at infinity' is needed, with $M(-d/c) = \infty$ and $M(\infty) = a/c$.

$$2\angle QZP \equiv \varphi$$

where Q, Z, P are the points represented by the complex numbers q, z, p; and this defines a circle or straight line by 6.4.4. Conversely, any circle or straight line can be represented in the form given in 6.4.4, and so can be given by an equation

$$2 \operatorname{Arg}\left(\frac{z - p}{z - q}\right) \equiv \varphi. \qquad \square$$

(6.5.3)   **Corollary:** *The image under a Möbius transformation of a circle or straight line is another circle or straight line.*

Proof: Suppose that the given circle or straight line is represented by the equation $2 \operatorname{Arg} N(z) \equiv \theta$ for some Möbius transformation N. Its image under the Möbius transformation M is the set of all complex numbers $w = M(z)$, where z satisfies the given equation; in other words, it is the set of all w such that

$$2 \operatorname{Arg} N(M^{-1}(w)) \equiv \theta.$$

But since NM^{-1} is again a Möbius transformation, this equation represents another straight line or circle. $\square$

It is quite possible for a Möbius transformation to turn a circle into a straight line or vice versa: we will see an example in a moment. But this theorem tells us that the combined concept 'circle-or-straight-line' is invariant under Möbius transformations (and so belongs to inversive geometry).

(6.5.4)   **Example:** About the simplest example of a Möbius transformation is given by $M(z) = 1/z$. We will show that this transformation maps some circles to lines (and, since it is its own inverse, this will show also that it maps some lines to circles). Let $a \in \mathbf{C}$ be nonzero, and consider the circle

$$|z - a| = |a|$$

of centre a and radius $|a|$ (which therefore passes through the origin 0). We may square and expand $|z - a|^2 = (z - a)(\bar{z} - \bar{a})$ to get the equation of the circle in the form $|z|^2 - 2\Re(\bar{z}a) = 0$. Let $w = 1/z = \bar{z}/|z|^2$; then the equation can be expressed as $\Re(aw) = \frac{1}{2}$. If now we put $w = x + iy$ (so that (x, y) are the Cartesian coordinates of the point w) and $a = k + il$ (where k, l are constants), then $\Re(aw) = kx - ly$ and the equation becomes

$$kx - ly - \tfrac{1}{2} = 0.$$

This is the equation of a straight line.

We might have expected this on the grounds that the origin 0 gets mapped to infinity, and so a circle through the origin should get mapped to 'a circle through infinity', that is a straight line.

The Möbius transformation $z \mapsto 1/z$ can be used to give a proof of a classical result on circles known as *Ptolemy's theorem*. Ptolemy (who worked in Alexandria around 150 AD) was a mathematician and astronomer. His most famous work, the *Almagest*, served as a textbook in astronomy until the time of Copernicus. Ptolemy used his theorem to compute $\sin \theta$ and $\cos \theta$ for the purposes of astronomy; a special case of the theorem (when AD is a diameter) is equivalent to the addition formula for $\sin(\theta + \varphi)$.

(6.5.5)   **Theorem:** (PTOLEMY'S THEOREM) *Let A, B, C, D be four concyclic points. Then, of the three numbers,*

$$|AB||CD|, \quad |AC||BD|, \quad |AD||BC|,$$

one is equal to the sum of the other two.

Proof: Represent points by complex numbers, with A as the origin, and let M denote the Möbius transformation $M(z) = 1/z$, which we have been considering. Notice that

$$\left| \frac{1}{z_1} - \frac{1}{z_2} \right| = \frac{|z_1 - z_2|}{|z_1||z_2|}$$

which gives

$$|M(B)M(C)| = \frac{|BC|}{|AB||AC|}$$

and similarly

$$|M(C)M(D)| = \frac{|CD|}{|AC||AD|}, \quad |M(B)M(D)| = \frac{|BD|}{|AB||AD|}.$$

But the points $M(B)$, $M(C)$, and $M(D)$ are collinear, because M sends the circle through them into a straight line. Therefore one of the distances between them is equal to the sum of the other two. Multiplying through by $|AB||AC||AD|$, we get Ptolemy's theorem. □

6.6 Straightedge and compasses

No account of plane geometry would be complete that did not discuss briefly the famous *construction problems* of the Greeks. Like some modern mathematicians, the Greeks seem to have felt that it is not enough just to give arguments showing that 'there exist' points or lines with certain properties;

one must also give explicit means for constructing these points or lines. Classically, such constructions had to be carried out with straightedge and compasses alone. Some constructions were very easy; for instance, Euclid's *Elements* begins with the construction of an equilateral triangle, one side of which is a given line segment *AB*. (Solution: Draw a circle centre *A* through *B*, and a circle centre *B* through *A*; let *C* be a point of intersection of the two circles. Then *ABC* is the desired equilateral triangle.) Other constructions were exceedingly difficult; for instance, the *Apollonian problem*, to construct a circle tangent to three given circles, or (from a later age) the *Malfatti problem*, to inscribe in a triangle three circles tangent to the sides of the triangle and to one another. But there were also some problems whose solution seemed not merely difficult but impossible; these were the *trisection of the angle* (to construct an angle equal to one-third of a given angle), the *duplication of the cube* (to construct a cube of twice the volume of a given cube), and the *squaring of the circle* (to construct a square whose area is equal to that of a given circle). In the nineteenth century, mathematicians using coordinate geometry were able to give an algebraic analysis of *all possible* constructions, and were able to prove that all three of the classical problems are in fact impossible to solve using straightedge and compasses alone.

To see why some constructions are impossible, it is necessary to look closely from the point of view of algebra at what can be achieved by construction. We therefore introduce a Cartesian coordinate system x, y in the plane. A straight line is represented by an equation

$$lx + my + n = 0$$

and a circle is represented by an equation

$$(x - a)^2 + (y - b)^2 = k.$$

The numbers l, m, n or a, b, k that determine a line or circle will be called the *coefficients* of that line or circle. To find out what points are constructible, we will examine the algebraic operations by which the coefficients of a line or circle are related to the coordinates of points through which it passes.

The coefficients of a line through two given points, or the coefficients of a circle with given centre passing through a given point, can be calculated from the coordinates of the two given points. For the line, this is just a matter of solving simultaneous equations, and for the circle we know that (a, b) is the centre and k is the square of the radius. In either case, the calculation only involves the four arithmetic operations of addition, subtraction, multiplication, and division; no other operations are required.

The coordinates of a point of intersection of two lines, a line and a circle, or two circles can be calculated from the coefficients using only the four arithmetic operations together with the extraction of square roots. There are three cases to consider here: line meets line, line meets circle, and circle meets circle. Line meets line is easy, and we have mentioned several times that line meets circle amounts to solving a quadratic equation, which of course requires extracting square roots only; we will do circle meets circle. Suppose then that we have two circles

$$(x - a_1)^2 + (y - b_1)^2 = k_1$$
$$(x - a_2)^2 + (y - b_2)^2 = k_2.$$

Subtracting we get

$$2x(a_1 - a_2) + 2y(b_1 - b_2) + (k_1 - a_1^2 - b_1^2) - (k_2 - a_2^2 - b_2^2) = 0.$$

(We have now reduced the problem to 'circle meets line'.) Use this equation to express y in terms of x, and then substitute into one of the original equations to get a quadratic for y, which can be solved using the four arithmetic operations and the extraction of square roots. Then x can be found in terms of y.

Now any straightedge-and-compass construction will consist of a succession of these five operations: drawing a line through two points, drawing a circle with centre one point and passing through another, finding the point of intersection of two lines, finding a point of intersection of a line and a circle, and finding a point of intersection of two circles. Each step can be expressed in terms of coordinates, and then will involve only the four arithmetic operations together with the extraction of square roots. Therefore

(6.6.1)   **Proposition:** *Suppose that n points $P_1, \ldots, P_n$ are given. Then the coordinates of any point P constructible from the given points by straightedge and compasses can be expressed in terms of the coordinates of $P_1, \ldots, P_n$ by repeated use of the four arithmetic operations together with the extraction of square roots.*

Repeated square roots are allowed, so expressions like $\sqrt{\sqrt{x} + \sqrt{y}}$ would be constructible. It is interesting that Euclid devoted the longest book (Book 10) of his *Elements* to an extensive discussion of 'quadratic irrationalities' of this type; did he anticipate the important algebraic rôle that they were later to play?

The duplication of the cube of course requires the solution of a *cubic* equation ($x^3 = 2$), and the trisection of the angle also involves the solution of a cubic equation coming from the trigonometrical identity $\cos 3\theta = 4\cos^3 \theta - 3\cos \theta$. We certainly feel that a cubic equation is a beast of a quite different character from a quadratic, and that it should therefore not be possible to solve a cubic just by extracting square roots, however often the process is repeated[4]. The problem is to make this precise. For example, the expression

$$\sqrt{\sqrt{\sqrt{\sqrt{26} + \sqrt{\sqrt{\sqrt{6}}}}}}$$

agrees with $\sqrt[3]{2}$ to an accuracy of 1 part in 10,000. How do we *know* that some more complicated expression cannot be exactly right?

To prove this it is helpful to arrange the constructible numbers in a hierarchy of increasing complexity; the level of a number in the hierarchy is related to the number of square roots that it contains. The formal definition of this hierarchy makes use of the algebraic concept of a *field*. By definition, a subset $\mathcal{K}$ of **R** is a *field* if it contains 0 and 1 and is closed under the four arithmetic operations, so that whenever $a \in \mathcal{K}$ and $b \in \mathcal{K}$, the numbers $a + b$, $a - b$, ab, and a/b are also in $\mathcal{K}$. A simple example of a field is the set **Q** of *rational numbers* $\frac{p}{q}$, p and q being integers.

Let $\mathcal{K}$ be a field and let $s \in \mathcal{K}$. Then $\mathcal{K}(\sqrt{s})$ is defined to be the set of all numbers of the form

$$a + b\sqrt{s}$$

where $a, b \in \mathcal{K}$. The most important algebraic fact that we will need is

(6.6.2)   **Lemma:** *If $\mathcal{K}$ is a field, then $\mathcal{K}(\sqrt{s})$ is a field also.*

Proof: It is obvious that $\mathcal{K}(\sqrt{s})$ is closed under the operations of addition, subtraction, and multiplication. The only tricky one is division, and here you can use the methods of school algebra ('rationalization of denominators') to write

$$\frac{1}{a + b\sqrt{s}} = \frac{a}{a^2 - sb^2} - \frac{b}{a^2 - sb^2}\sqrt{s}$$

and so to reduce division to multiplication by the reciprocal. $\square$

Now we can give an inductive definition of our hierarchy, as follows. The field of rational numbers is at level 0. A field $\mathcal{K}$ is at level 1 if it can be written as $\mathbf{Q}(\sqrt{s_1})$ for some $s_1 \in \mathbf{Q}$; a field $\mathcal{K}'$ is at level 2 if it can be written as $\mathcal{K}(\sqrt{s_2})$ for some $s_2 \in \mathcal{K}$, where $\mathcal{K}$ is at level 1; and in general a field $\mathcal{M}'$ is at level n if it can be written as $\mathcal{M}(\sqrt{s_n})$ for some $s_n \in \mathcal{M}$, where $\mathcal{M}$ is a field at level $n - 1$.

---

[4]Leonardo of Pisa, better known as Fibonacci, proved around 1200 AD that the roots of a certain cubic equation could not be irrational numbers of the type considered in Euclid's Book 10.

Any number which can be obtained from the rationals by repeating the four arithmetic operations together with extraction of square roots is at some finite level n. For instance, the number

$$\sqrt{\sqrt{5} + 8\sqrt{2 + \sqrt{3}}}$$

belongs to

$$\mathbf{Q}\left(\sqrt{3}\right)\left(\sqrt{2 + \sqrt{3}}\right)\left(\sqrt{5}\right)\left(\sqrt{\sqrt{5} + 8\sqrt{2 + \sqrt{3}}}\right)$$

and so is at level 4.

A cubic equation is called *irreducible* if it has rational coefficients but no rational roots; an example is the equation $x^3 - 2 = 0$. The result which will show that the trisection of the angle and the duplication of the cube are impossible is

(6.6.3)   **Proposition:** *Suppose that the number x satisfies an irreducible cubic equation*

$$x^3 + lx^2 + mx + n = 0.$$

Then x is not at any finite level.

Proof: Our strategy will be an 'infinite descent', a type of argument invented by Fermat to solve problems in number theory. Supposing that the equation has a root at level k, we will show that it must have a root at level $k - 1$, then at level $k - 2$, and so on. Finally, it will have to have a root at level 0, that is a rational number, and this will contradict our hypothesis that the equation is irreducible.

Suppose then that the equation *has* a root y at some level k. Then $y = a + b\sqrt{s}$, where a, b, and s are at level $k - 1$. We can substitute this expression for y into the equation and expand the powers of y to obtain

$$\left(a^3 + 3ab^2s + a^2l + b^2sl + ma + n\right) + \left(3a^2b + b^3s + 2abl + bm\right)\sqrt{s} = 0.$$

The two expressions in brackets are at level $k - 1$. If they are nonzero, then their quotient $\sqrt{s}$ is at level $k - 1$ and so y is at level $k - 1$ and we have achieved our 'descent'. Otherwise, both expressions are zero. But in this case, $z = a - b\sqrt{s}$ is another solution of the original equation, since $z^3 + lz^2 + mz + n$ equals

$$\left(a^3 + 3ab^2s + a^2l + b^2sl + ma + n\right) - \left(3a^2b + b^3s + 2abl + bm\right)\sqrt{s}$$

which is zero. The three roots of the original equation add up to $-l$, so another root is

$$-l - y - z = -l - 2a$$

which is at level $k - 1$. Thus, once again, we have achieved our descent. $\square$

We will use the next result to prove that some cubic equations are irreducible.

(6.6.4)   **Proposition:** *Let*
$$x^3 + lx^2 + mx + n = 0$$

be a cubic equation with integer coefficients l, m, n. If the equation has no integer roots, then it is irreducible (i.e. has no rational roots).

Proof: Let $x = r/s$ be a rational root, with r and s integers. We may assume that x is written in lowest terms, so that r and s have no common factor. Then

$$r^3 = -s(lr^2 + mrs + ns^2).$$

So any prime factor p of s is also a prime factor of r^3. But r^3 has the same prime factors as r, so p is a prime factor of r. This shows that $x = \frac{r}{s}$ was not in lowest terms after all. The only way to avoid this contradiction is to assume that s has no prime factors, so is equal to 1. □

(6.6.5)   **Example:** We can now show that it is impossible to duplicate the cube using straightedge and compasses. Duplication of the cube is equivalent to construction of the point $(0, \sqrt[3]{2})$ from the two points $(0,0)$ and $(0,1)$. If this construction were possible, then $\sqrt[3]{2}$ would be at some finite level. But it is a root of the cubic equation
$$x^3 - 2 = 0,$$

and this equation has no integer solutions and so must be irreducible. Therefore, by proposition 6.6.3, it has no solution at any finite level.

(6.6.6)   **Example:** Similarly, we can prove that it is impossible in general to trisect the angle using straightedge and compasses. In fact we will show that it is impossible to construct an angle of $\frac{\pi}{9}$ by this method. We know that it *is* possible to construct an angle of $\frac{\pi}{3}$ (which is just one of the angles of an equilateral triangle), so this will show that the angle of $\frac{\pi}{3}$ cannot be trisected.
    We use the trigonometric identity

$$\cos 3\theta = 4\cos^3 \theta - 3\cos \theta.$$

Since $\cos \frac{\pi}{3} = \frac{1}{2}$, this shows that $c = 2\cos \frac{\pi}{9}$ satisfies a cubic equation

$$c^3 - 3c - 1 = 0.$$

If we could construct an angle $\frac{\pi}{9}$, we could certainly construct the point $(0, c)$ from $(0,0)$ and $(0,1)$. But it is easy to check that the equation $c^3 - 3c - 1 = 0$ has no integer solutions[5], and so is irreducible. Arguing as before, we find that the point $(0, c)$ cannot be constructed.

---

[5] Any integer solution c would have $c(c^2 - 3) = 1$, so by factorization $c = \pm 1$; but neither $+1$ nor -1 is a solution.

The last of the classical problems is to square the circle. This is equivalent to constructing the point $(0, \sqrt{\pi})$ from the points $(0,0)$ and $(0,1)$. To show this to be impossible it is enough to show that the number π cannot be obtained from the rationals by repeatedly solving quadratic equations. Now it is a theorem proved by Lindemann in 1882 that π is a *transcendental number*; it cannot be obtained from the rationals by solving (repeatedly or not) any kind of algebraic equations of the form

$$x^n + a_{n-1}x^{n-1} + \cdots + a_0 = 0.$$

This result finally showed that it is impossible to square the circle using straightedge and compasses. The proof of Lindemann's theorem has been repeatedly simplified since it was first published, but it is still too difficult to include in this book. A reasonably accessible version may be found in Stewart [37].

Remark: One should not lose too much sleep over these facts. There are plenty of practical ways of trisecting an angle once we leave the artificially limited world of ruler and compasses: one can use special instruments like the *tomahawk* described by Eves [14], or special curves like the *trisectrix* (see Hollingdale [24]), or of course one can just measure the angle and divide by three. The remarkable thing is that it is possible to *prove* that certain procedures are not strong enough to solve the problem.

6.7 Exercises

1. *ABCD* is a parallelogram. Four squares are constructed external to the parallelogram, one on each of its sides. Show that the centres of the four squares themselves form a square (*Napoleon's theorem*).

2. Write down a formula (in terms of position vectors) for the action of the isometry obtained by rotation through oriented angle $\theta \neq 0$ followed by translation through vector **u**. Show that this isometry has a fixed point, and deduce that it is in fact a rotation through θ about that fixed point.

3. Let $A_1, \ldots, A_n$ be n points in a Euclidean plane. We think of them as forming a polygon, and make the convention that $A_0 = A_n$, $A_1 = A_{n+1}$, and so on. The polygon is said to be *convex* if, for every i and j, the vector $\overrightarrow{A_iA_j}$ is between $\overrightarrow{A_iA_{i-1}}$ and $\overrightarrow{A_iA_{i+1}}$. If this is so, prove that the unoriented angle sum

$$\sum_{i=1}^{n} \widehat{A_{i-1}A_iA_{i+1}}$$

is equal to $(n-2)\pi$.

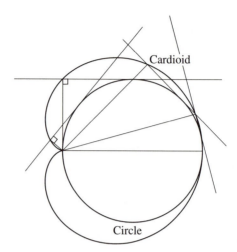

Fig. 6.11. The cardioid

4. Let $A_1, \ldots, A_n$ be n points in an oriented plane. We make the same convention as in the previous question, that $A_1 = A_{n+1}$ and so on. Prove that the *oriented* angle sum

$$\sum_{i=1}^{n} \angle A_{i-1} A_i A_{i+1}$$

is equal, as an oriented angle, to $(n-2)\pi$. How does this compare with the result of the previous question? In what ways is it stronger, and in what ways is it weaker?

5. Let $\mathcal{L}_1$ and $\mathcal{L}_2$ be two intersecting lines in a Euclidean plane; find the locus of all points P such that the perpendicular distances $|P\mathcal{L}_1|$ and $|P\mathcal{L}_2|$ are equal.

6. Let ABC be a triangle in a Euclidean plane. Show that the (internal) bisectors of its angles meet at a point (called the *incentre* of the triangle), and that this point is the centre of a circle (the *incircle*) which is tangent to the three sides $\overrightarrow{AB}$, $\overrightarrow{BC}$, and $\overrightarrow{CA}$ of the triangle. Show also that there are three other points (external to the triangle) which have this same property that a circle centred at any one of them is tangent to all three sides.

7. Let ABC be a triangle, and let X, Y, and Z be the points where the incircle touches BC, CA, and AB respectively. Show that the three lines AX, BY, and CZ meet in a point (the *Gergonne point* of the triangle).

8. Find the polar equation of the parabola whose Cartesian equation is $y = \frac{1}{4}x^2 - 1$.

9. The origin O of polar coordinates lies on a circle $\mathcal{C}$ of radius a. A new curve called the *cardioid* is defined as the locus of the feet of the perpendiculars from O to the tangent lines to $\mathcal{C}$ (see Figure 6.11). Find the polar equation of the cardioid.

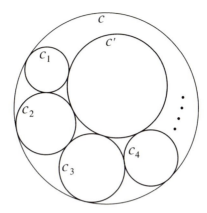

Fig. 6.12. Steiner's porism

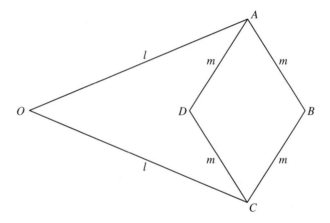

Fig. 6.13. Peaucellier's linkage

10. Show that, given any two nonintersecting circles C and C' in the complex plane, it is possible to find a Möbius transformation which converts them into two *concentric* circles.

    A *circle chain* between C and C' is a sequence of circles C_1, C_2 and so on, each of which is tangent to C, to C', and to the previous circle in the chain (see Figure 6.12). The circle chain is *closed* if $C_1 = C_n$ for some n. Show that any Möbius transformation takes circle chains to circle chains. Hence demonstrate *Steiner's porism*: if one circle chain is closed, then all circle chains are closed.

11. The arrangement of hinged rods shown in Figure 6.13 is called *Peaucellier's linkage*. It consists of six rods, four of length m (AB, BC, CD, and DA) and two of length $l > m$ (OA and OC). The point O is fixed and all the others are free to move.

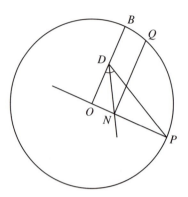

Fig. 6.14. Construction of a regular pentagon

The rods are hinged at O, A, B, C, and D.

Prove that $|OB||OD| = l^2 - m^2$. Hence show that if B moves on an arc of a circle passing through O, then D will move on a straight line. Thus Peaucellier's linkage is a *parallel motion* — an arrangement for converting circular into straight line movement. Such a parallel motion was eagerly sought after by James Watt and other early designers of steam engines.

12. A regular pentagon can be constructed by ruler and compasses as follows (see Figure 6.14). Draw a circle centre O and let P be a point on the circumference. Construct OB perpendicular to OP with B on the circle and bisect the segment OB at D. Bisect the angle $\widehat{ODP}$, and let the bisector meet OP at N. Construct NQ perpendicular to OP with Q on the circle. Then P and Q are two consecutive vertices of a regular pentagon.

Explain why the construction works.

13. Let z denote the complex number $\cos(2\pi/7) + i\sin(2\pi/7)$. Explain why the points $1, z, z^2, z^3, z^4, z^5, z^6$ represent the vertices of a regular heptagon in the complex plane, and why $z^7 = 1$.

Let $x = \cos(2\pi/7)$. Prove that $\Re z = \Re z^6 = x$, $\Re z^2 = \Re z^5 = 2x^2 - 1$, and $\Re z^3 = \Re z^4 = 4x^3 - 3x$.

By multiplying by $z - 1$, prove that $1 + z + z^2 + z^3 + z^4 + z^5 + z^6 = 0$. Hence prove that $y = 2x$ satisfies the irreducible cubic equation

$$y^3 + y^2 - 2y - 1 = 0.$$

Deduce that the regular heptagon cannot be constructed by ruler and compasses.

14. In lemma 6.6.2, we cleared denominators by multiplying top and bottom by $a - b\sqrt{s}$. Why was it legitimate to assume that this expression is not zero?

7

Conics and other curves

7.1 Curves and their tangents

In this chapter we will study in more detail some curves in the plane described by 'locus equations' of the form $f(x, y) = 0$, where f is a function of the two variables x and y. Our main attention will be devoted to *conics*, which are the curves that arise when $f(x, y)$ is a *quadratic* function of the variables x and y; at the end of the chapter we will also take a look at some examples where $f(x, y)$ is a *cubic* function of x and y. Before we start, though, we want to take up our 'unfinished business' from Chapter 5, and to investigate what conditions have to be imposed on the function f in order that the equation should define something that we would intuitively recognize as a curve. This investigation will turn out to be closely linked to the problem of finding the *tangent line* through a point on the curve. Indeed, our intuitive picture of a curve requires that through any point on it there should be a *unique* tangent line. We therefore begin by defining the concept of tangent line more precisely.

To give the general definition, we take our cue from the special case of the circle, which we discussed in 6.4.2. Let C be a circle and P a point on it. We found that a line $\mathcal{L}$ through P will usually meet the circle again at another point Q, but for one particular choice of $\mathcal{L}$, the points P and Q coalesce into a single point. This line $\mathcal{L}$ is the tangent line, and it can be described by saying that the equation which defines its intersection with the circle has a *repeated root* at the point P.

In general, then, let $P = (x_0, y_0)$ be a point on a curve given by the equation $f(x, y) = 0$, and consider the line $\mathcal{L}$ through P in the direction of the vector $p\mathbf{i} + q\mathbf{j}$. This line has parametric equation $x = x_0 + pt, y = y_0 + qt$, and so it meets the curve at those points given by parameter values t for which

$$f(x_0 + pt, y_0 + qt) = 0.$$

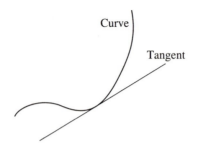

Fig. 7.1. Curve and tangent line.

We want to know whether this equation has a repeated root at $t = 0$. Now an equation of the form $h(t) = 0$ has a repeated root at a value $t = t_0$ if and only if both $h(t)$ and the derivative $h'(t)$ vanish at $t = t_0$. So we are led to make the following definition:

(7.1.1)   **Definition:** *The line $\mathcal{L}$ given parametrically by $x = x_0 + pt$, $y = y_0 + qt$ is tangent to the curve $f(x, y) = 0$ at the point (x_0, y_0) on the curve if*

$$\left. \frac{df(x_0 + pt, y_0 + qt)}{dt} \right|_{t=0} = 0.$$

(7.1.2)   **Example:** Let us calculate directly from this definition the condition for a line to be tangent to a curve which is given as the graph of a function g, $y = g(x)$. The relevant function $f(x, y)$ in this case is $g(x) - y$, and so

$$\left. \frac{df(x_0 + pt, y_0 + qt)}{dt} \right|_{t=0} = pg'(x_0) - q$$

by the chain rule. Setting this equal to zero we get $q/p = g'(x_0)$. This tells us, as we expected, that the gradient of the tangent line is given by the derivative $g'(x_0)$.

Although definition 7.1.1 can be applied as it stands, it is convenient to simplify it by using some ideas from the calculus of partial derivatives. The *chain rule for partial differentiation* allows one to calculate the derivative of f with respect to t, if f is a function of two variables x and y which themselves are functions of t. The chain rule is

$$\frac{df}{dt} = \frac{\partial f}{\partial x} \frac{dx}{dt} + \frac{\partial f}{\partial y} \frac{dy}{dt},$$

where the 'partial derivative' $\partial f / \partial x$ means the derivative of f with respect to x thinking of y as a constant, and $\partial f / \partial y$ means the derivative of f with respect to y thinking of x as a constant[1].

---

[1] See any book on advanced calculus, such as Apostol [3] or Kaplan [25].

We may apply the chain rule to the expression appearing in 7.1.1. We obtain

(7.1.3)
$$\frac{df(x_0 + pt, y_0 + qt)}{dt} = \frac{\partial f}{\partial x}p + \frac{\partial f}{\partial y}q.$$

This expression will equal zero at $t = 0$ if and only if the direction vector $p\mathbf{i} + q\mathbf{j}$ for the line $\mathcal{L}$ is perpendicular to the vector $(\partial f/\partial x)\mathbf{i} + (\partial f/\partial y)\mathbf{j}$ at the point (x_0, y_0). This vector is important enough to have a special name:

(7.1.4)   **Definition:** *Let $f(x, y)$ be a smooth function of the two variables x and y. Let $\mathbf{i}$ and $\mathbf{j}$ denote the standard basis vectors of the coordinate system. Then the* gradient *of f at the point (x, y) is defined to be the vector*

$$\nabla f(x, y) = \frac{\partial f}{\partial x}\mathbf{i} + \frac{\partial f}{\partial x}\mathbf{j}.$$

Remark: The gradient of f is an example of a *vector field*; a vector field is a mapping that associates a *vector* to every point of space. (Contrast this with an ordinary function like $f(x, y)$, which associates a *scalar* to every point of space.) Vector fields come up frequently in physics. An example is the *velocity vector field* of a moving fluid, which associates to each point of space the velocity with which a small element of the fluid at that point is moving. Under certain conditions (irrotational flow) the velocity field of a moving fluid can actually be written as ∇f for a suitable function f, and this fact is of importance in fluid dynamics[2].

The vector field ∇f points in the direction in which f is changing most rapidly. If you think of $z = f(x, y)$ as describing a surface in three-dimensional space (like a mountain range), then the vector $-\nabla f(x, y)$ points in the direction of steepest descent at the point (x, y). It is intuitively clear that this direction of steepest descent is always perpendicular to the contour lines, and that is the content of the next proposition.

(7.1.5)   **Proposition:** *A line through the point (x_0, y_0) on the curve given by $f(x, y) = 0$ is tangent to the curve at that point if and only if it is perpendicular to the gradient vector $\nabla f(x_0, y_0)$.*

Proof: Follows from equation 7.1.3. □

Having defined the tangent we are now in a position to say more precisely what kind of functions will be allowed in the definition of a curve. First of all, the functions will have to be *differentiable*, so that we can sensibly talk about partial derivatives. To avoid having to fuss over just how many orders of differentiability are needed, we will in fact require that f is *smooth*,

---

[2]For an introduction, see Acheson [2].

that is differentiable infinitely often. All the common functions of analysis of course satisfy this requirement. Second, we will require that our curves should have well-defined tangent lines; by proposition 7.1.5, the direction of a tangent line is well-defined provided that the gradient vector is nonzero. These requirements are summarized in the next definition.

(7.1.6)   **Definition:** *Let $f(x,y)$ be a smooth function of x and y, and let (x_0, y_0) be a point. We say that (x_0, y_0) is a* regular point *for f if $\nabla f(x_0, y_0) \neq \mathbf{0}$. Otherwise, (x_0, y_0) is a* singular point. *The locus*

$$C = \{(x, y) : f(x, y) = 0\}$$

is a curve *if all its points, with perhaps a finite number of exceptions, are regular. It is a* regular curve *if all its points (with no exceptions) are regular.*

(7.1.7)   **Example:** To get familiar with this, let us check that the circle is a regular curve according to our definitions, and that the tangent line is what we expect. The circle with centre (a, b) and radius r is given by the equation $f(x, y) = 0$ where

$$f(x, y) = (x - a)^2 + (y - b)^2 - r^2.$$

So by the usual rules for partial differentiation,

$$\nabla f(x, y) = 2(x - a)\mathbf{i} + 2(y - b)\mathbf{j}.$$

This is nonzero for all points except (a, b), and $f(a, b) \neq 0$. So the circle is a regular curve. By definition, the tangent line at (x, y) is perpendicular to $\nabla f(x, y)$; but this is just twice the *radius vector* from the centre to the point (x, y). So we recover the fact that the tangent is perpendicular to the radius vector.

Remark: We need to allow a finite number of singular points in order to handle examples like the folium of Descartes (5.2.3), which has a self-intersection or *double point* at the origin. The folium of Descartes is given by the equation $f(x, y) = x^3 + y^3 - 3xy$, so the gradient is

$$\nabla f(x, y) = 3(x^2 - y)\mathbf{i} + 3(y^2 - x)\mathbf{j}.$$

This vanishes at the point $(0, 0)$, which is therefore a singular point of the curve. We can see from Figure 5.2 that something must go wrong at $(0, 0)$; the curve crosses itself, so there is not much hope of defining a unique tangent line. One might say that *any* line through a singular point is a tangent line there, but we will usually just say that the tangent line at a singular point is *undefined*. There are 'worse' singularities than double points, such as the *cusp* exhibited by the equation $y^2 = x^3$ at the origin, but we will not consider them in this book.

We have now seen how the tangent line to a curve given by a *locus* equation $f(x, y) = 0$ can be found, and this study has allowed us to say more precisely when such an equation defines a good curve. Our next task is to carry out the same programme for *parametric* equations $x = x(t)$, $y = y(t)$. To find the tangent line through the point $(x(t_0), y(t_0))$, we once again look for a double root of the equation giving the intersection of the curve with a fixed line. Now, however, since the *curve* is given by a parametric equation, we use the *locus* equation $lx + my + n = 0$ of a line, where the constant n is chosen so that $(x(t_0), y(t_0))$ lies on the line. The equation that needs to be solved to find the points of intersection is

$$lx(t) + my(t) + n = 0$$

and this will have a double root at $t = t_0$ if and only if

$$lx'(t_0) + my'(t_0) = 0.$$

This implies that the tangent line $lx + my + n = 0$ must be parallel to the *tangent vector* $x'(t_0)\mathbf{i} + y'(t_0)\mathbf{j}$. If we think of $(x(t), y(t))$ as the coordinates of a moving particle (with t being the time), then $x'(t_0)\mathbf{i} + y'(t_0)\mathbf{j}$ is the *velocity vector* of the particle at time t_0.

In giving the formal definitions it is convenient to amalgamate $x(t)$ and $y(t)$ into a single map $\gamma(t) = (x(t), y(t))$; γ is therefore a smooth map from an interval $[a, b]$ of real numbers (the *parameter interval*) to the plane $\mathbf{R}^2$, and $x(t)$ and $y(t)$ are its components. The tangent vector $x'(t)\mathbf{i} + y'(t)\mathbf{j}$ will simply be denoted $\gamma'(t)$.

(7.1.8)   **Definition:** *Let $\gamma: [a, b] \to \mathbf{R}^2$ be a smooth map. We say that $t = t_0$ gives a* regular point *of γ if $\gamma'(t_0) \neq \mathbf{0}$, and a* singular point *otherwise. The map γ defines a* regular parameterized curve *if all its points are regular, and a* parameterized curve *if there are at most finitely many singular points. The tangent line* at a regular point $\gamma(t_0)$ *is the line through that point in the direction of the tangent vector $\gamma'(t_0)$.*

Remark: It is possible to prove that our two definitions of a curve are actually equivalent; any non-self-intersecting[3] curve that can be given by a (regular) locus equation can also be given by a (regular) parametric equation and vice versa. The proof of this fact is not especially important for us and we won't bother with it. But we ought at any rate to show that our two definitions of the tangent line are the same. Suppose that a curve is given both by a locus equation $f(x, y) = 0$ and by a parametric equation $x = x(t)$, $y = y(t)$; is

---

[3]This condition is necessary because our definition of a regular parameterized curve allows self-intersections; there might be two different parameter values t_0 and t_1 corresponding to the same point $\gamma(t_0) = \gamma(t_1)$.

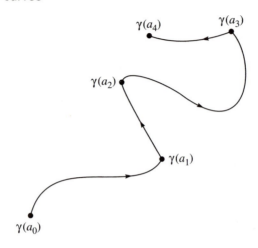

Fig. 7.2. A piecewise regular curve.

the locus-equation definition of the tangent line at a point the same as the parametric-equation definition? To prove that it is, we just have to show that $\nabla f(x(t), y(t))$ is perpendicular to the tangent vector $x'(t)\mathbf{i} + y'(t)\mathbf{j}$. But this follows from the chain rule; since $f(x(t), y(t)) = 0$ for all values of t,

$$\nabla f \cdot (x'(t)\mathbf{i} + y'(t)\mathbf{j}) = \frac{\partial f}{\partial x}\frac{dx}{dt} + \frac{\partial f}{\partial y}\frac{dy}{dt} = \frac{df}{dt} = 0.$$

(7.1.9)   **Piecewise regular curves:** It is often convenient to have a definition of 'parameterized curve' that is general enough to include the boundary of a triangle or polygon. Our current definition runs into problems at the corner points, because the natural parameterization is not differentiable there. We therefore define a *piecewise regular parameterized curve* to be a map $\gamma \colon [a, b] \rightarrow \mathbf{R}$ with the property that there are finitely many numbers $a = a_0 < a_1 < \cdots < a_n = b$ such that γ becomes a regular curve when restricted to each of the intervals $[a_i, a_{i+1}]$. The points $\gamma(a_i)$ are called the *vertices* of the curve. See Figure 7.2.

(7.1.10)   **Parameter changes:** It is sometimes necessary to be careful about the distinction between a parameterized curve $\gamma(t) = (x(t), y(t))$, which is a mapping from an interval $[a, b]$ of real numbers to $\mathcal{P}$, and its *trace* or *image*, which is the set of points γ^* in $\mathcal{P}$ defined by

$$\gamma^* = \{\gamma(t) : t \in [a, b]\}.$$

In physical terms, it is possible for two particles to traverse the same path but at different speeds; this corresponds to the fact that two different parameterized curves can have the same trace.

One way in which two curves γ_1 and γ_2 might have the same trace is for them to be related by an equation of the form

$$\gamma_2(t) = \gamma_1(\varphi(t))$$

where $\varphi \colon \mathbf{R} \to \mathbf{R}$ is a smooth function with nonvanishing derivative. (It can be proved that two regular curves with the same trace are always related in this way, but we won't use this fact.) We call the map φ a *parameter change map*. One can think of such a parameter change map as a generalized 'change of coordinate system' on the curve γ, analogous to the affine changes of coordinate system on straight lines introduced by the ruler comparison axiom in Chapter 1.

Since the derivative of a parameter change map is nonvanishing, it must be either always positive or always negative throughout its range. It is often convenient to restrict attention to parameter change maps whose derivative is always positive, so that the curve is traced out in the same direction after the parameter change as before. When we want to draw attention to this restriction, we will speak of *oriented* parameterized curves and parameter changes.

7.2 The ellipse, hyperbola, and parabola

Circles and straight lines were not the only curves known to the Greek mathematicians. The next most simple in a certain sense are the *conic sections* or *conics*, obtained by cutting a cone with a plane as shown in Figure 7.3. No old-fashioned mathematics classroom would be complete without a polished wooden model of a cone which could be taken apart in various ways to show

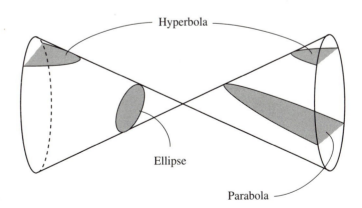

Fig. 7.3. Conic sections.

the different kinds of conic section. As the angle of the plane varies we get three kinds of curve, which the Greeks called *ellipse, parabola,* and *hyperbola.* These words come from the Greek terms for 'falling short', 'laying alongside', and 'exceeding'; it is interesting that all three have passed into the English language as names for figures of speech (ellipsis, parable, hyperbole) as well as for mathematical curves.

Apollonius (born about 260 BC) worked out the theory of conics systematically in a treatise of eight books. When asked about the practical value of his results, he is said to have replied

> They are worthy of acceptance for the sake of the demonstrations themselves, in the same way as we accept many other things in mathematics for this and for no other reason.

It has often turned out in the history of mathematics that subjects which were first explored for their own intrinsic beauty are later found to be exactly the ones needed to understand the physical world; one twentieth-century writer even went so far as to speak of the 'unreasonable effectiveness' of mathematics in science. So it was for the Greek study of conics. In 1609, Kepler published his laws of planetary motion, based on the astronomical observations of Tycho Brahe. The first law states that planets move around the sun in ellipses with the sun at one focus. Conic sections therefore proved to be the key to understanding the motions of the heavenly bodies. When Isaac Newton explained Kepler's laws in terms of his theory of universal gravitation, he made heavy use of Apollonius' results. We will discuss Newton's work in Section 11.2.

The Greek definition of the conics has the disadvantage of depending on a topic in *three*-dimensional geometry. In this book we will define conics directly by their equations in two dimensions, and we will make the connection with the Greek definition in 10.4.11, after we have discussed quadric surfaces. This kind of approach seems to have been first used in the book on conics by John Wallis (1616–1703), contemporary and friend of Isaac Newton. It was not universally popular; Thomas Hobbes, who as we have seen disapproved of algebraic methods, described Wallis' book as a 'scab of symbols'.

We will define the conics by an equation in polar coordinates.

(7.2.1)   **Definition:** *Let $\mathcal{P}$ be a Euclidean plane. A (nondegenerate) conic in $\mathcal{P}$ is a curve that can be represented, relative to a suitable polar coordinate system (r, θ), by the equation*

$$r = \frac{k}{1 - e \cos \theta}$$

where $k > 0$ and $e \geq 0$ are constants.

The constant e, which controls the shape of the conic, is called the *eccentricity.* The constant k is simply an overall scale factor. If $e > 1$, then r

will become negative for some values of θ; the negative coordinates should be interpreted according to the usual convention that (r, θ) and $(-r, \pi + \theta)$ represent the same point.

(7.2.2)   **Proposition:** *Let (u, v) be Cartesian coordinates[4] in $\mathcal{P}$, related to the polar coordinates (r, θ) by the usual formula $u = r\cos\theta$, $v = r\sin\theta$. Then (u, v) lies on the conic defined by the polar equation 7.2.1 if and only if $u^2 + v^2 = (k + eu)^2$.*

Proof: The equation $u^2 + v^2 = (k + eu)^2$ is equivalent to $r^2 = (k + er\cos\theta)^2$, which has solutions $r = \pm(k + er\cos\theta)$. If $r = +(k + er\cos\theta)$ we find immediately that

$$r = \frac{k}{1 - e\cos\theta}$$

and so (r, θ) lies on the given conic. If $r = -(k + er\cos\theta)$ then

$$-r = \frac{k}{1 + e\cos\theta} = \frac{k}{1 - e\cos(\theta + \pi)}.$$

Since $(-r, \theta + \pi)$ and (r, θ) represent the same point, we again find that (r, θ) lies on the given conic. $\square$

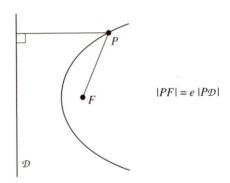

$|PF| = e\,|P\mathcal{D}|$

Fig. 7.4. Focus–directrix definition of a conic.

The result of this proposition can be described geometrically as follows. If P is the point (u, v), then $u^2 + v^2$ is just the square of the distance $|PF|$ from P to the origin (which we denote by F and call the *focus*). On the other hand, provided that $e \neq 0$, $(k + eu)^2 = e^2(k/e + u)^2 = e^2|P\mathcal{D}|^2$, where $|P\mathcal{D}|$ denotes the perpendicular distance from P to the line $\mathcal{D}$ (called the *directrix*)

---

[4]We use the letters u and v rather than x and y because we are later going to want to consider another Cartesian coordinate system with a different origin.

with equation $u = -k/e$. Thus the points P on the conic are described by the equation

$$|PF| = e|PD|.$$

Traditionally, one expresses this by saying that a conic is *the locus of those points P such that the distance of P from a fixed point F bears a fixed ratio e to the distance of P from a fixed line D.* See Figure 7.4.

Conics are classified into three types depending on the value of the eccentricity e.

(7.2.3)   **Definition:** *An* ellipse *is a conic with eccentricity $e < 1$.*

If we expand out the Cartesian equation 7.2.2 for an ellipse we obtain $u^2 + v^2 = k^2 + 2eku + e^2u^2$. Since $e \neq 1$ we can complete the square to write this as

$$(1 - e^2)\left(u - \frac{ek}{1 - e^2}\right)^2 + v^2 = \frac{k^2}{1 - e^2}.$$

The form of this equation suggests that we change to a new coordinate system (x, y) related to (u, v) by $x = u - ek/(1 - e^2)$, $y = v$. (Geometrically, this represents a shift of the origin to the right by $ek/(1 - e^2)$.) In terms of x and y the equation becomes

(7.2.4)
$$\frac{(1 - e^2)^2}{k^2}x^2 + \frac{1 - e^2}{k^2}y^2 = 1.$$

Since $e < 1$ the coefficients of x^2 and y^2 are both positive, so we may write them as $1/a^2$ and $1/b^2$ for positive numbers a and b. We then obtain the *standard form for the Cartesian equation of an ellipse:*

$$\frac{x^2}{a^2} + \frac{y^2}{b^2} = 1.$$

Conversely, given a and b, it is easy to recover the original parameters e and k. We will need to notice that $e = \sqrt{1 - b^2/a^2}$.

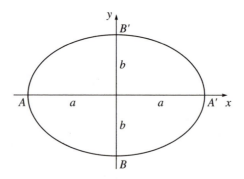

Fig. 7.5. An ellipse.

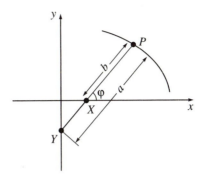

Fig. 7.6. The elliptical chuck.

If $e = 0$, the ellipse is just a circle of radius $a = b$. In general, an ellipse (see Figure 7.5) can be thought of as a circle that has been squashed by rescaling the y-direction by a factor b/a. Notice that since squares of real numbers are positive, an ellipse is a bounded figure: $|x|$ cannot exceed a, and $|y|$ cannot exceed b. The points $A = (-a, 0)$ and $A' = (a, 0)$ are therefore the extremes of the ellipse in the x-direction, and $B = (0, -b)$ and $B' = (0, b)$ are the extremes of the ellipse in the y-direction. The line segment AA' is called the *major axis* and the line segment BB' is called the *minor axis* of the ellipse.

A parametric equation for the ellipse, with parameter φ, is given by

$$x = a \cos \varphi, \quad y = b \sin \varphi.$$

Putting $t = \tan \varphi/2$ we obtain the *rational* parameterization

$$x = a\frac{1 - t^2}{1 + t^2}, \quad y = b\frac{2t}{1 + t^2}.$$

(7.2.5)   **Example:** Suppose that PXY is a straight line with $|PX| = b$ and $|PY| = a$, and that the point X is constrained to lie on the major axis AA' and the point Y is constrained to lie on the minor axis BB'. (See Figure 7.6.) If φ denotes the oriented angle between $\overrightarrow{XA'}$ and $\overrightarrow{XP}$, then by trigonometry,

$$\overrightarrow{MP} = a \cos \varphi \mathbf{i}, \quad \overrightarrow{LP} = b \sin \varphi \mathbf{j};$$

so P has coordinates $(a \cos \varphi, b \sin \varphi)$ and so lies on the ellipse.

This construction is put to use in the *elliptical chuck*, which is a device for cutting an elliptical figure on a lathe. PXY becomes a rod, fixed in direction, with a cutting tool at P and with X and Y sliding along perpendicular guides AA' and BB' which are arranged to turn along with the *work* (the piece of metal or whatever that is being machined). Then, as the work turns on the lathe, the rod PXY moves back and forward in such a way that P cuts out an ellipse on the work.

Notice that the ellipse is symmetrical about the origin of the (x, y) coordinate system. Therefore there are in fact two focus–directrix pairs with respect to which it obeys the definition of a conic: the original focus F and directrix $\mathcal{D}$, and a new focus F' and corresponding directrix $\mathcal{D}'$ obtained by inverting F and $\mathcal{D}$ in the origin. A simple calculation shows:

(7.2.6)   **Proposition:** *The coordinates of the foci of the ellipse defined above are* $(\pm ea, 0)$ *and the equations of the directrices are* $x = \pm a/e$.

(7.2.7)   **Proposition:** (DISTANCE PROPERTY OF THE ELLIPSE) *If the point P moves on an ellipse with foci F and F', then the sum* $|PF| + |PF'|$ *is a constant.*

Proof: From the polar equation of the ellipse, $|PF| + |PF'| = e|P\mathcal{D}| + e|P\mathcal{D}'| = e|\mathcal{D}\mathcal{D}'|$, which is constant. In the last step we used the fact that P always lies between $\mathcal{D}$ and $\mathcal{D}'$. $\square$

This proposition is the basis of the so-called *gardener's method* of constructing an ellipse. If you want to cut an elliptical flower bed in your lawn, insert two pegs at the foci and make a loop of string around the two pegs and your spade. As you move the spade around (keeping the string taut) it will cut out an ellipse, because the sum of its distances from the two foci is forced to be constant.

(7.2.8)   **Definition:** *A hyperbola is a conic with eccentricity e > 1.*

The argument leading to the Cartesian equation 7.2.4 remains valid for the hyperbola. But now, since $e > 1$, the coefficient of x^2 is positive while the coefficient of y^2 is negative. We may therefore write these coefficients as $1/a^2$ and $-1/b^2$ for positive numbers a and b, and so obtain the *standard form for the Cartesian equation of a hyperbola:*

$$\frac{x^2}{a^2} - \frac{y^2}{b^2} = 1.$$

Again, one can recover the original parameters e and k from a and b. This time we have $e = \sqrt{1 + b^2/a^2}$.

In contrast to the ellipse, the hyperbola is *not bounded*, for however large y may be, one can still solve the equation of the hyperbola for x. Moreover, the hyperbola is *not connected* — it is impossible for any point on the hyperbola to have $|x| < a$ (which would imply $y^2 < 0$), so the curve comes in two separate pieces. The equations

$$x = \pm a \cosh \varphi, \quad y = b \sinh \varphi$$

parameterize the hyperbola; the positive sign describes one 'branch' and the negative sign describes the other. The hyperbola also has a rational parameterization given by

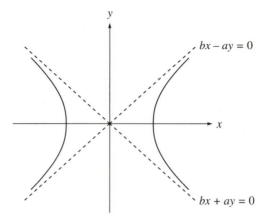

Fig. 7.7. A hyperbola and its asymptotes.

$$x = a\frac{1+t^2}{1-t^2}, \quad y = b\frac{2t}{1-t^2}.$$

A feature of the hyperbola that has no analogue in the case of the ellipse is the existence of *asymptotes*, that is straight lines that are approached more and more closely as the parameter φ becomes large. To investigate these lines, suppose that $(x, y) = (a\cosh\varphi, b\sinh\varphi)$ is a point on one branch of the hyperbola. Then

$$bx - ay = ab(\cosh\varphi - \sinh\varphi) = ab\,e^{-\varphi}.$$

As φ tends to $+\infty$, this quantity tends to zero, and so (x, y) draws very close to the line $bx - ay = 0$. By a similar argument, as φ tends to $-\infty$ the point (x, y) draws very close to the line $bx + ay = 0$. These two lines, which are more and more closely approached by the hyperbola as $|x|$ and $|y|$ become large, are called the *asymptotes* (see Figure 7.7).

In the language of projective geometry the directions of the asymptotes define two points at infinity, and these points at infinity are thought of as lying on the hyperbola. Indeed from the projective point of view there is no essential difference between a hyperbola and an ellipse; one merely happens to contain points at infinity whereas the other does not.

A hyperbola is called *rectangular* if its asymptotes are perpendicular to one another. From the discussion above we can see that a hyperbola is rectangular if $a = b$, so its equation is $x^2 - y^2 = a^2$. One can simplify the equation by using a new coordinate system whose axes are the asymptotes; this coordinate system (X, Y) is defined by $X = (x+y)/\sqrt{2}$ and $Y = (x-y)/\sqrt{2}$, and the equation becomes $XY = a^2/2$.

Like the ellipse, the hyperbola is symmetrical about the origin of the (x, y) coordinate system, and therefore has two foci and two directrices. Moreover, their coordinates can be calculated by the same formulae as for the ellipse: the foci are the points $(\pm ea, 0)$, and the directrices are the lines $x = \pm a/e$. Notice that none of the points of the hyperbola are between the two directrices.

(7.2.9)   **Proposition:** (DISTANCE PROPERTY OF THE HYPERBOLA) *If the point P moves on a hyperbola with foci F and F′, then the difference $|PF| - |PF'|$ is a constant on each branch of the hyperbola, positive for one branch and negative for the other.*

Proof: We may write $|PF| - |PF'| = e|P\mathcal{D}| - e|P\mathcal{D}'| = \pm e|\mathcal{D}\mathcal{D}'|$, using the fact that P does not lie between $\mathcal{D}$ and $\mathcal{D}'$. □

This property, which characterizes the hyperbola, has been put to use in a variety of navigational aids for ships and aircraft, such as LORAN and Decca Navigator. These work on the following principle. Two fixed base stations emit accurately synchronized radio signals, and a special receiver measures the difference in the arrival times of the signals from the base stations. Assuming that the signals travel in straight lines, the value of this difference defines a certain hyperbola with foci the two base stations, on which the receiver must lie. By comparing signals from three or more base stations the position of the receiver can be determined completely as the intersection of a number of hyperbolas. Special charts may be prepared on which the relevant hyperbolas are already drawn, though nowadays it is more likely that the calculations will be performed automatically by computer.

(7.2.10)   **Definition:** *A parabola is a conic with eccentricity $e = 1$.*

This time when we expand the Cartesian equation 7.2.2 we find that the terms in u^2 cancel, and we get $v^2 = k^2 + 2ku = 2k(\frac{1}{2}k + u)$. If we change to new coordinates (x, y) given by $x = u + \frac{1}{2}k$, $y = v$, and put $k = 2a$, we obtain the *standard form of the Cartesian equation for a parabola:*

$$y^2 = 4ax.$$

This parabola has the rational parameterization $x = at^2$, $y = 2at$, as t varies over the real numbers. Notice that the parabola is *not* symmetrical about the origin and therefore has only one focus $(a, 0)$ and one directrix $x = -a$.

7.3   More general conics

We have defined the ellipse, the hyperbola, and the parabola, and we have seen that each of them can be given by simple Cartesian equations. But in order to obtain these simple equations, we have had to choose our coordinate systems rather carefully. What do their equations look like in more general coordinate systems?

(7.3.1) **Example:** Consider the ellipse with equation $2x^2 + 4y^2 = 7$. Suppose that we make the coordinate transformation (a combination of a rotation and a translation) with equation

$$x = \frac{x' + y' + 1}{\sqrt{2}}, \quad y = \frac{-x' + y'}{\sqrt{2}}.$$

We substitute these expressions for x and y into the original equation and expand, obtaining

$$3x'^2 + 3y'^2 - 2x'y' + 2x' + 2y' = 6.$$

This equation therefore describes the same ellipse, but in a less suitable coordinate system. We would like a systematic way to find the best coordinate system.

It is clear that the equations obtained in the way illustrated above are all going to be of *degree* 2; in other words, the highest total power of x or y appearing is 2. (We count the term xy as being of degree 2 because it contains one x and one y.) This then suggests the following definition.

(7.3.2) **Definition:** *A (general)* conic *is a curve in a plane whose equation in some Cartesian coordinate system is*

$$Ax^2 + 2Bxy + Cy^2 + Dx + Ey + F = 0$$

where $A, \ldots, F$ are constants and not all of A, B, and C are zero. The conic is called a central conic *(relative to the given coordinate system) if $D = E = 0$.*

If (x, y) is on a *central* conic, then so is $(-x, -y)$, so that central inversion[5] is a symmetry of the curve. This is the reason for the phrase 'central conic'.

(7.3.3) **Theorem:** *Suppose that*

$$Ax^2 + 2Bxy + Cy^2 + F = 0$$

is a central conic. Then one can find a rotation matrix R_θ such that, in new coordinates x' and y' related to the old by

$$\begin{pmatrix} x \\ y \end{pmatrix} = \mathsf{R}_\theta \begin{pmatrix} x' \\ y' \end{pmatrix}$$

the equation of the conic becomes

$$A'x'^2 + C'y'^2 + F = 0.$$

---

[5] The map ι which reverses position vectors relative to the origin, defined in 6.1.5.

Proof: We start by noticing that the expression $Ax^2 + 2Bxy + Cy^2$ can be written in matrix form as

$$(x \quad y) \begin{pmatrix} A & B \\ B & C \end{pmatrix} \begin{pmatrix} x \\ y \end{pmatrix};$$

that is, as $x'Mx$ where x denotes the coordinate vector $\begin{pmatrix} x \\ y \end{pmatrix}$ and M the symmetric matrix $\begin{pmatrix} A & B \\ B & C \end{pmatrix}$. If new coordinates $x' = \begin{pmatrix} x' \\ y' \end{pmatrix}$ are related to the old by the rotation matrix R_θ, then

$$x'Mx = (R_\theta x')'MR_\theta x' = x''R_{-\theta}MR_\theta x'.$$

What we need to do, therefore, is to find a value of θ so that the matrix $R_{-\theta}MR_\theta$ is *diagonal*, having its top right- and bottom left- hand entries equal to zero.

We may calculate explicitly

$$R_{-\theta}MR_\theta = \begin{pmatrix} \cos\theta & \sin\theta \\ -\sin\theta & \cos\theta \end{pmatrix} \begin{pmatrix} A & B \\ B & C \end{pmatrix} \begin{pmatrix} \cos\theta & -\sin\theta \\ \sin\theta & \cos\theta \end{pmatrix}$$

$$= \begin{pmatrix} A\cos^2\theta + C\sin^2\theta + B\sin 2\theta & \frac{1}{2}(C - A)\sin 2\theta + B\cos 2\theta \\ \frac{1}{2}(C - A)\sin 2\theta + B\cos 2\theta & A\sin^2\theta + C\cos^2\theta - B\sin 2\theta \end{pmatrix}.$$

To make this matrix diagonal, we must ensure that $\frac{1}{2}(C - A)\sin 2\theta + B\cos 2\theta = 0$, and we can do this by choosing θ so that

$$\tan 2\theta = \frac{2B}{A - C};$$

where if $A = C$ we choose $\theta = \pi/4$.

By this rotation we have reduced the equation of the conic to the form $A'x'^2 + C'y'^2 + F = 0$, where A' and C' are given by

$$\begin{aligned} A' &= A\cos^2\theta + C\sin^2\theta + B\sin 2\theta \\ C' &= A\sin^2\theta + C\cos^2\theta - B\sin 2\theta. \end{aligned}$$

This is what was required. $\square$

The numbers A' and C' above determine what kind of curve our equation gives rise to. For instance, if $A' > 0$, $C' > 0$, and $F < 0$ the curve is an ellipse; if $A' > 0$ and $C' < 0$ it is a hyperbola; and if $A' < 0$, $C' < 0$, and $F < 0$ it is the empty set $\emptyset$. A' and C' can be calculated, by the method given above, from the coefficients A, B, and C of the original equation. We might wonder, however, whether there is some direct way of working out A' and C' which does not involve working out θ in between. Such a way can be found using the concept of *eigenvalues* of a matrix.

(7.3.4)   **Definition:** *Let* M *be a square matrix, and let* I *denote the identity matrix of the same size. The equation*

$$\det(\mathsf{M} - \lambda \mathsf{I}) = 0$$

is called the characteristic equation *of the matrix* M, *and its roots are the* eigenvalues *of* M.

For example, the characteristic equation of the 2×2 matrix M that we have been considering is

$$\begin{vmatrix} A - \lambda & B \\ B & C - \lambda \end{vmatrix} = \lambda^2 - (A + C)\lambda + (AC - B^2) = 0.$$

(7.3.5)   **Proposition:** *The numbers* A' *and* C', *defined above, are the eigenvalues of the matrix* M.

Proof: Let M' denote the diagonal matrix with entries A' and C', calculated as $\mathsf{M}' = \mathsf{R}_{-\theta} \mathsf{M} \mathsf{R}_{\theta}$. *Its* characteristic equation is

$$\begin{vmatrix} A' - \lambda & 0 \\ 0 & C' - \lambda \end{vmatrix} = (A' - \lambda)(C' - \lambda) = 0$$

whose roots are A' and C'. It is enough, therefore, to show that M and M' have the *same* characteristic equation, and this follows from the properties of determinants:

$$\det(\mathsf{M}' - \lambda \mathsf{I}) = \det(\mathsf{R}_{-\theta}(\mathsf{M} - \lambda \mathsf{I})\mathsf{R}_{\theta}) = \det((\mathsf{M} - \lambda \mathsf{I})\mathsf{R}_{\theta}\mathsf{R}_{-\theta}) = \det(\mathsf{M} - \lambda \mathsf{I}).$$

We used the fact that $\det(\mathsf{AB}) = \det(\mathsf{BA})$ for square matrices A and B. $\square$

Thus, A' and C' are roots of the characteristic equation

$$\lambda^2 - (A + C)\lambda + (AC - B^2) = 0.$$

Their product $A'C'$ is therefore equal to $AC - B^2$, the determinant of M, and their sum $A' + C'$ is equal to $A + C$, the so-called *trace* of M. From the signs of their product and their sum, their own signs can be deduced; for instance, we can argue that if $AC - B^2 < 0$, the two eigenvalues A' and C' have opposite sign and so the conic must be a hyperbola.

We can use the same kind of methods to look at the general conic

$$Ax^2 + 2Bxy + Cy^2 + Dx + Ey + F = 0$$

(where not all of A, B, and C are zero). We can write this equation in the matrix form

$$x'Mx + Nx + F = 0$$

where x and M are as above, and N is the one-rowed matrix $(\; D \;\; E \;)$. If new coordinates x' are related to the old by $x = R_\theta x'$, then the equation becomes

$$x''M'x' + N'x' + F = 0$$

where

$$M' = R_{-\theta}MR_\theta, \quad N' = NR_\theta.$$

As we have seen (7.3.3), the rotation angle θ can be chosen so as to make M' a diagonal matrix. In other words, *any* conic can be reduced by a rotation to the form

$$A'x'^2 + C'y'^2 + D'x' + E'y' + F = 0.$$

What will happen next? We know that at least one of A' and C' is nonzero; for if both of them were zero, the matrix M' would be zero, and then the matrix M would be zero as well, contrary to our original hypothesis that not all of A, B, and C were zero. If $A' \neq 0$, we can 'complete the square' to absorb the D' term, by writing

$$A'x'^2 + D'x' = A'\left(x' + \frac{D'}{2A'}\right)^2 - \frac{D'^2}{4A'}.$$

Similarly if $C' \neq 0$ we can complete the square to absorb the E' term. The upshot of this is that if *both A' and C'* are nonzero, we can rewrite the equation as

$$A'x''^2 + C'y''^2 + F'' = 0$$

(the standard form of the equation of a central conic) by changing the coordinates by a translation:

$$x' = x'' - \frac{D'}{2A'}$$
$$y' = y'' - \frac{E'}{2C'}$$

and then

$$F'' = F - \frac{D'^2}{4A'} - \frac{E'^2}{4C'}.$$

But this approach will not work if one of A' or C' is zero: these cases need special treatment. Here then is the complete description of all possible conics, taking into account special cases.

(7.3.6)   **Theorem:** *A general conic*

$$Ax^2 + 2Bxy + Cy^2 + Dx + Ey + F = 0$$

(where not all of A, B, and C are zero) represents one of the following eight types of loci: an ellipse, a hyperbola, a parabola, a pair of intersecting straight lines, a pair of parallel straight lines, a single straight line 'counted twice' (like a double root of an equation), a single point, or the empty set.

Moreover the cases that can occur are governed by the sign of $AC - B^2$, as follows:

- *If $AC - B^2 > 0$, the possibilities are an ellipse, a single point, or the empty set.*

- *If $AC - B^2 = 0$, the possibilities are a parabola, two parallel straight lines, a single straight line, or the empty set.*

- *If $AC - B^2 < 0$, the possibilities are a hyperbola or two intersecting straight lines.*

Remark: The first three cases (ellipse, parabola, or hyperbola) are referred to as *nondegenerate* conics; the others are *degenerate*.

Proof: We have seen above that the equation can be reduced by a rotation to the form

$$A'x'^2 + C'y'^2 + D'x' + E'y' + F = 0$$

where A' and C' are not both zero. Moreover, $A'C' = AC - B^2$.

Suppose first that *neither* of A' and C' are zero. Then, as we have seen, the equation can be further reduced by a translation (change of origin) to the form

$$A'x''^2 + C'y''^2 + F'' = 0.$$

If $F'' \neq 0$, this equation represents either an ellipse, a hyperbola, or the empty set. If $F'' = 0$, this equation represents the single point $x'' = y'' = 0$ if A' and C' have the same sign, and it represents the two intersecting straight lines

$$\frac{y''}{x''} = \pm\sqrt{-\frac{A'}{C'}}$$

if A' and C' have opposite signs.

Suppose now that *just one* of A' and C' is zero; we suppose without loss of generality that $C' = 0$. We may then complete the square in x' only to reduce the equation to the form

$$A'x''^2 + E'y' + F' = 0$$

where

$$x' = x'' - \frac{D'}{2A'}$$

and then

$$F' = F - \frac{D'^2}{4A'}.$$

There are now two subcases to consider:

- If $E' = 0$, the equation reduces to $x''^2 = $ constant. This represents two parallel lines, two coincident lines, or the empty set, according to whether the constant is positive, zero, or negative.

- If $E' \neq 0$, we can make a further translation

$$y' = y'' - \frac{F'}{E'}$$

to reduce the equation to the form

$$A'x''^2 + E'y'' = 0.$$

This is the equation of a parabola.

The conclusions about the sign of the determinant $AC - B^2$ follow if you trace through the above argument, using the fact if $AC - B^2 > 0$ then A' and C' have the same sign, if $AC - B^2 = 0$ then one of them is zero, and if $AC - B^2 < 0$ then they have different signs. $\square$

(7.3.7)   **Example:** Suppose we are faced with the following problem: *Reduce the conic*

$$31x^2 - 24xy + 21y^2 + 4x + 6y = 25$$

to the simplest possible form by means of a coordinate transformation.

To solve the problem, the first step is to write the equation as $\mathbf{x}'\mathbf{M}\mathbf{x} + \mathbf{N}\mathbf{x} + F = 0$, where

$$\mathbf{M} = \begin{pmatrix} 31 & -12 \\ -12 & 21 \end{pmatrix}, \quad \mathbf{N} = (\, 4 \quad 6 \,), \quad F = -25.$$

We need first to find a rotation matrix $\mathbf{R}_\theta$ such that $\mathbf{R}_{-\theta}\mathbf{M}\mathbf{R}_\theta$ is diagonal. By the calculations in theorem 7.3.3, we must take

$$\frac{2\tan\theta}{1 - \tan^2\theta} = \tan 2\theta = \frac{-24}{31 - 21} = -\frac{12}{5}.$$

This gives a quadratic equation for $\tan\theta$,

$$6\tan^2\theta - 5\tan\theta - 6 = 0$$

that is

$$(2\tan\theta - 3)(3\tan\theta + 2) = 0$$

so $\tan\theta = 3/2$ or $\tan\theta = -2/3$. We may choose either value (the two corresponding values of θ will differ by a right angle, so the only difference it makes is to which axis gets called the x' axis and which gets called the y' axis). We take $\tan\theta = -2/3$; then

$$\sin\theta = \frac{-2}{\sqrt{13}}, \quad \cos\theta = \frac{3}{\sqrt{13}}, \quad \mathbf{R}_\theta = \frac{1}{\sqrt{13}}\begin{pmatrix} 3 & 2 \\ -2 & 3 \end{pmatrix}.$$

We may calculate

$$\begin{aligned} \mathbf{R}_{-\theta}\mathbf{M}\mathbf{R}_\theta &= \frac{1}{13}\begin{pmatrix} 3 & -2 \\ 2 & 3 \end{pmatrix}\begin{pmatrix} 31 & -12 \\ -12 & 21 \end{pmatrix}\begin{pmatrix} 3 & 2 \\ -2 & 3 \end{pmatrix} \\ &= \begin{pmatrix} 39 & 0 \\ 0 & 13 \end{pmatrix} \end{aligned}$$

and

$$\mathbf{N}\mathbf{R}_\theta = \frac{1}{\sqrt{13}}(\, 4 \quad 6 \,)\begin{pmatrix} 3 & 2 \\ -2 & 3 \end{pmatrix} = \frac{1}{\sqrt{13}}(\, 0 \quad 26 \,).$$

So in new coordinates x' and y' related to the old by $\mathbf{R}_\theta$, the equation becomes

$$39x'^2 + 13y'^2 + \frac{26}{\sqrt{13}}y' - 25 = 0.$$

We may complete the square to get

$$39x'^2 + 13 \left(y' + \frac{1}{\sqrt{13}} \right)^2 - 26 = 0$$

or

$$\frac{x''^2}{2/3} + \frac{y''^2}{2} = 1$$

where $x' = x''$ and $y' = y'' - 1/\sqrt{13}$. We find therefore that the equation is that of an ellipse of eccentricity

$$\sqrt{1 - \frac{2/3}{2}} = \sqrt{\frac{2}{3}}$$

and that the coordinates in which it takes its simplest form are x'', y'' related to x, y by

$$
\begin{aligned}
x &= \frac{1}{\sqrt{13}}(3x' + 2y') &= \frac{1}{\sqrt{13}}(3x'' + 2y'') - \frac{2}{13} \\
y &= \frac{1}{\sqrt{13}}(-2x' + 3y') &= \frac{1}{\sqrt{13}}(-2x'' + 3y'') - \frac{3}{13}.
\end{aligned}
$$

7.4 Intersections of lines and conics

(7.4.1) **Proposition:** *Let C be a conic in a Euclidean plane. A straight line that is not contained in C can meet C in at most two points. In particular, a straight line can meet a nondegenerate conic in at most two points.*

Proof: If the straight line $\mathcal{L}$ passes through a point whose Cartesian coordinates are (x_0, y_0), then a general point on the line has coordinates

$$x = x_0 + kt, \quad y = y_0 + lt$$

where t is a real parameter and $k\mathbf{i} + l\mathbf{j}$ is a vector parallel to the line. If we substitute these expressions for x and y into the equation of a conic, we will in general obtain a quadratic equation, whose roots will be the parameter values t at the points of intersection of the line and conic. Since a quadratic has at most two roots, there are at most two points of intersection.

The only way we could have more than two points of intersection would be if we obtained the trivial equation $0 = 0$ when we substituted for x and y. (A quadratic polynomial with three distinct roots is identically zero.) But this would mean that *every* point of the line was on the conic, contrary to our hypothesis. $\square$

Of course the number of intersection points might be *less* than two. Typically there will be some lines that intersect the conic in two points, some that don't intersect it at all, and a few that intersect it only in one point. These last are in fact the *tangent lines* (see 7.1.5) to the conic, because the quadratic equation whose solutions define the points of intersection must have a double root in this case.

(7.4.2)   **Example:** Let us use this approach to calculate the equation of the tangent line to the parabola $y^2 = 4ax$ at the point $(at^2, 2at)$. Let $\mathcal{L}$ be the line through $(at^2, 2at)$ parallel to the vector $k\mathbf{i} + l\mathbf{j}$. The general point on this line has coordinates

$$x = at^2 + ku, \quad y = 2at + lu$$

where u is a real parameter. Substituting this into the equation of the parabola to find the points of intersection, we get $(2at + lu)^2 = 4a(at^2 + ku)$, which reduces to

$$l^2u^2 + 4a(tl - k)u = 0.$$

One root of this equation is $u = 0$, which corresponds to the fact that we arranged the line $\mathcal{L}$ to pass through the point $(as^2, 2as)$ on the parabola. We are interested in knowing when $u = 0$ is a *double* root, and the condition for this is clearly $k = tl$. Thus our tangent line $\mathcal{L}$ is the line through $(at^2, 2at)$ parallel to $t\mathbf{i} + \mathbf{j}$, in other words the line with equation

$$x - ty + at^2 = 0.$$

We have shown that a line and a nondegenerate conic can intersect in at most two points; in how many points can two nondegenerate conics intersect? The answer is given by

(7.4.3)   **Proposition:** *Two nondegenerate conics can intersect in at most four points.*

See Figure 7.8 for an example showing four points of intersection.

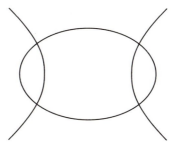

Fig. 7.8. Intersection of two conics.

Proof: We will need to use the fact that every nondegenerate conic is a rational curve which can be parameterized by rational functions of degree 2. We have checked this separately for the ellipse, parabola, and hyperbola; alternatively, Exercise 7.6.13 gives a general proof which imitates the argument used for the circle in 5.2.6. So let C_1 and C_2 be two distinct conics. Take a rational parameterization $(x(t), y(t))$ for C_1 and substitute it into the locus equation for C_2. After clearing the denominators, you will obtain a *quartic* (degree 4) equation for t, which can therefore have at most four distinct roots, corresponding to four possible points of intersection. □

This result is a special case of a general theorem called *Bézout's theorem*: a curve of degree m and a curve of degree n intersect in at most mn points. (Furthermore, there are exactly mn points of intersection if you take proper account of imaginary points, multiple points, and points at infinity!) We have checked Bézout's theorem for $m, n \le 2$. The general case is rather harder than these special cases might suggest. The point is that our proofs have made use of a rational parameterization of one of the curves, and such a rational parameterization is not available in general.

(7.4.4)   **Proposition:** *Let $P_1, \ldots, P_5$ be five points in a plane, no three of which are collinear. Then there is a unique nondegenerate conic passing through the five points.*

Proof: Suppose that the five points have coordinates (x_i, y_i) for $i = 1, \ldots, 5$. A general conic defined by the function

$$f(x, y) = Ax^2 + 2Bxy + Cy^2 + Dx + Ey + F$$

will pass through the five points if the equations

$$Ax_i^2 + 2Bx_iy_i + Cy_i^2 + Dx_i + Ey_i + F = 0$$

are satisfied for $i = 1, \ldots, 5$. Considered as equations for $A, \ldots, F$, these are five homogeneous linear equations in six unknowns, so by linear algebra they must have a nonzero solution. Thus there is at least one conic passing through the five points.

Any conic passing through the five points must be nondegenerate, since a degenerate conic is made up of one or two straight lines, and no three of the points are collinear. By the special case of Bézout's theorem proved above (7.4.3), two distinct nondegenerate conics can have at most four points in common. Thus there can be only one nondegenerate conic passing through the given five points, and the result is proved. □

7.5   A short tour around some higher-degree curves

So far we have studied curves of degree 1 (lines) and of degree 2 (conics). It is natural to suppose that the step to curves of higher degree will not involve anything harder than a more complicated notation, but this is far from being the case. The theory of higher-degree curves, which was begun by Newton in 1695 with a classification of cubics (degree 3) into 72 different types, and which developed in the nineteenth century through a profound link with complex analysis and number theory, is still an active and exciting area of research mathematics.

The first indication that something may go wrong when we pass to higher degrees comes from the fact that not all curves of degree 3 or more are rational. (Remember that a curve is *rational* if it can be given a parametric equation $x = x(t)$, $y = y(t)$, where x and y are rational functions of t.) In order to show that certain higher-degree curves are not rational we will make use of a result about polynomials known as *Mason's theorem*. Mason's theorem was proved as recently as 1983. My account of it is based on an article by Serge Lang [28].

To state Mason's theorem we need to introduce some notation. Let $p(t)$ be a polynomial (a function made up of finitely many powers of t, like $2t^6 + 5t^2 + 9$). We will use the result, known as the 'fundamental theorem of algebra', that every polynomial has complex roots, and so can be completely factorized as

$$p(t) = c \prod_{i=1}^{n} (t - \lambda_i)^{m_i}$$

where the complex numbers λ_i are the roots, and the positive integers m_i are their multiplicities. Let $d(p)$ be the *degree* (the highest power of t that appears, which is equal to the overall number of roots counted according to multiplicity), and let $d_0(p)$ be the number of *distinct* roots (counted without regard to multiplicity). For example, if $p(t)$ is the polynomial $(t - 5)^2(t - 2)^3(t - 3)$, then $d(p) = 6$, while $d_0(p) = 3$.

(7.5.1)   **Lemma:** (MASON'S THEOREM) *Let p, q, and r be nonconstant polynomials without any common factor, and suppose that $p + q = r$. Then*

$$\max\{d(p), d(q), d(r)\} \leq d_0(pqr) - 1.$$

Proof: Let $f = p/r$ and $g = q/r$; then f and g are rational functions, and

$$f(t) + g(t) = 1.$$

Differentiating with respect to t, we get

$$f'(t) + g'(t) = 0$$

so that

$$\frac{q(t)}{p(t)} = \frac{g(t)}{f(t)} = -\frac{f'(t)/f(t)}{g'(t)/g(t)}.$$

Now suppose that the polynomials p, q, and r can be factorized as

$$p(t) = \prod_{i=1}^{d_0(p)} (t - \alpha_i)^{k_i}, \quad q(t) = \prod_{i=1}^{d_0(q)} (t - \beta_i)^{l_i}, \quad r(t) = \prod_{i=1}^{d_0(r)} (t - \gamma_i)^{m_i}.$$

Then

$$\frac{f'(t)}{f(t)} = \frac{d\log f}{dt} = \frac{d}{dt}\left(\sum_{i=1}^{d_0(p)} k_i \log(t - \alpha_i) - \sum_{i=1}^{d_0(r)} m_i \log(t - \gamma_i)\right)$$

$$= \sum_{i=1}^{d_0(p)} \frac{k_i}{t - \alpha_i} - \sum_{i=1}^{d_0(r)} \frac{m_i}{t - \gamma_i}$$

and there is a similar formula for $g'(t)/g(t)$. Let $s(t)$ be the polynomial

$$\prod_{i=1}^{d_0(p)} (t - \alpha_i) \prod_{i=1}^{d_0(q)} (t - \beta_i) \prod_{i=1}^{d_0(r)} (t - \gamma_i)$$

whose degree is $d_0(pqr)$ (because p, q, and r have no common factor). The calculations above show that

$$s(t)f'(t)/f(t)$$

is a polynomial of degree at most $d_0(pqr) - 1$; similarly,

$$s(t)g'(t)/g(t)$$

is a polynomial of degree at most $d_0(pqr) - 1$. But then

$$\frac{p(t)}{q(t)} = \frac{-s(t)f'(t)/f(t)}{s(t)g'(t)/g(t)}$$

is written as the quotient of two polynomials of degree at most $d_0(pqr) - 1$. Since $p(t)$ and $q(t)$ have no common factor, the rational function p/q is 'in lowest terms', so the degrees $d(p)$ and $d(q)$ must both be at most $d_0(pqr) - 1$. Since $r = p + q$ it follows that $d(r) \le d_0(pqr) - 1$ also. □

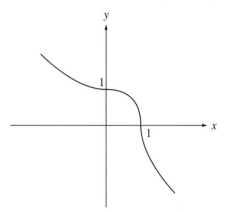

Fig. 7.9. The Fermat curve $x^3 + y^3 = 1$.

Some classical examples of curves of higher degree are the *Fermat curves*

$$x^n + y^n = 1$$

for $n \geq 3$. We saw in 5.2.7 that we can find Pythagorean triads by looking for rational points on the circle $x^2 + y^2 = 1$. In the same kind of way, we could try to prove (or disprove) 'Fermat's last theorem' by looking for rational points on the higher Fermat curves. The next result shows that the method we used for the circle will not work.

(7.5.2)   **Proposition:** *The Fermat curve*

$$x^n + y^n = 1$$

is not rational for $n \geq 3$.

Proof: Suppose that the curve were rational. There would then be rational functions $f(t)$ and $g(t)$ (of degree greater than zero) such that

$$f(t)^n + g(t)^n = 1.$$

Multiplying by a lowest common denominator, we could then find polynomials p, q, and r without common factor such that

$$p(t)^n + q(t)^n = r(t)^n.$$

Now we apply Mason's theorem, noticing that

$$d(p^n) = nd(p), \quad d_0(p^n) = d_0(p).$$

This gives us

$$n \max\{d(p), d(q), d(r)\} \leq d_0(pqr) - 1 \leq d(p) + d(q) + d(r) - 1.$$

Adding the three inequalities for $d(p)$, $d(q)$, and $d(r)$, we find

$$n(d(p) + d(q) + d(r)) \leq 3(d(p) + d(q) + d(r)) - 3.$$

If $n \geq 3$, this is impossible. □

Although this calculation shows that not *all* higher-degree curves are rational, the possibility remains that *some* of them are. And indeed this proves to be the case. Consider, for example, the folium of Descartes, given by the equation $x^3 + y^3 - 3xy = 0$. Suppose that we try to find a rational parameterization in the same way that we did for the circle: consider the intersection of the folium with the line $y = tx$ (Figure 7.10). Substituting, we obtain

$$x^3(1 + t^3) - 3tx^2 = 0$$

and so

(7.5.3)
$$x = \frac{3t}{1 + t^3}, \quad y = \frac{3t^2}{1 + t^3}$$

giving a rational parameterization.

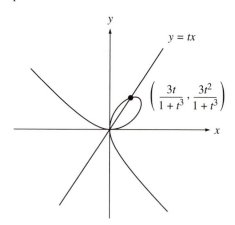

Fig. 7.10. A rational parameterization for the folium of Descartes.

Why does this work for the folium of Descartes but not for the Fermat cubic? The answer can be found by thinking about the intersections of a line and a cubic. In general, a line will meet a cubic in three points (just as it will meet a conic in two points) and these three points can be obtained by solving a certain third-degree equation of the form

$$at^3 + bt^2 + ct + d = 0.$$

In general such a solution involves extracting roots, not just rational operations. But if two of the roots are known, then the third root can be found rationally from the fact that the three roots add up to $-b/a$ (or from the fact that their product is $-d/a$). This tells us that if we know two points (x_1, y_1) and (x_2, y_2) on a cubic curve, then the coordinates of the third point (x_3, y_3) where the line through the two given points meets the curve again can be calculated as rational functions of x_1, y_1, x_2, and y_2.

This does not immediately seem relevant to the folium of Descartes; after all, we obtained our parameterization by intersecting the curve with a *variable* line through the point $(0,0)$. But $(0,0)$ is a *double point* of the folium of Descartes, and therefore it counts as a *double root* of the equation giving the points of intersection of the curve with a line through it. (This corresponds to the step in the argument above where we cancelled the factor of x^2.) The point of intersection that we want is then the third root of a third-degree equation two of whose roots are known, which explains why it can be expressed as a rational function. The argument is quite general, and shows that any cubic with a double point (or any quartic with a triple point, and so on) is a rational curve.

This calculation shows the importance of the operation which assigns to two points of a cubic the third point of intersection of the cubic with the line through the two given points. If some fixed cubic C is given, and A and B are points on the cubic, let $\langle A, B \rangle$ denote the third point of intersection of $\overrightarrow{AB}$ with C. If $A = B$, $\langle A, A \rangle$ will denote the second point of intersection of the tangent line at A with C; obviously this is in some sense a limiting case. Our calculations above show:

(7.5.4)   **Proposition:** *If the coefficients of the cubic C are rational numbers, and the coordinates of the points P_1 and P_2 are rational numbers, then the coordinates of the point $P_3 = \langle P_1, P_2 \rangle$ are rational numbers also.*

(7.5.5)   **Example:** Diophantus of Alexandria gave the solution $x = \frac{21}{4}$, $y = \frac{71}{8}$ to the problem of finding rational numbers x and y satisfying the equation

$$x^3 - 3x^2 + 3x + 1 = y^2.$$

It is easy enough to check that this solution is correct, but how could Diophantus have thought of it in the first place? One obvious rational solution is

$x = 0, y = 1$. Let us work out the tangent to the cubic at this point. The cubic is defined by $f(x, y) = 0$, where $f(x, y) = x^3 - 3x^2 + 3x + 1 - y^2$. Therefore

$$\nabla f(x, y) = (3x^2 - 6x + 3)\mathbf{i} - 2y\mathbf{j},$$

and so the tangent line at $(0, 1)$ is perpendicular to the vector $3\mathbf{i} - 2\mathbf{j}$. The tangent line can therefore be given by a parametric equation

$$x = 2t, \quad y = 3t + 1.$$

Substituting this into the original cubic to find the third point of intersection, we get

$$8t^3 - 12t^2 + 6t + 1 = 9t^2 + 6t + 1$$

and so

$$t^2(8t - 21) = 0.$$

The double root at $t = 0$ corresponds to the original point $(0, 1)$, and the other root $t = \frac{21}{8}$ gives us Diophantus' rational solution $x = \frac{21}{4}, y = \frac{71}{8}$.

Diophantus stands out among classical mathematicians for his interest in number theory; problems involving the solution of equations in rational numbers or integers are now called *Diophantine problems* in his honour. Little is known of his life, and even his dates are uncertain (250 AD, plus or minus a hundred years). It was Fermat's investigations of Diophantus' work that led to modern number theory. Fittingly, it was in the margin of his copy of Diophantus that Fermat inscribed his 'last theorem'.

We see, therefore, that results in number theory can be obtained by studying the geometry of cubic curves. One of the most beautiful results in this area is the *Mordell–Weil theorem*, proved in the early 1930s, which says that on any cubic curve with rational coefficients there are *finitely many* rational points $P_1, \ldots, P_N$ on the curve such that *all* rational points on the curve can be obtained from them by repeated application of the $\langle, \rangle$ operation. Unfortunately, the proof does not provide any means of estimating the number N for any particular given curve.

When we consider curves of degree higher than 3, there are still more possibilities. Many of the properties of a curve of degree d are determined by its *genus g*, which is a non-negative integer which can be defined by a formula like

(7.5.6)
$$g = \frac{(d - 1)(d - 2)}{2} - \text{Contribution from singular points}$$

where of course one has to explain precisely how the 'contribution from singular points' is worked out; for instance, the double point of the folium of

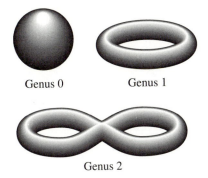

Genus 0   Genus 1

Genus 2

Fig. 7.11. Topology of surfaces of differing genus.

Descartes makes 'contribution' 1 and so the folium has genus 0. But if we consider only *nonsingular* curves for simplicity, we find that curves of degree 1 or 2 (that is, lines and conics) have genus 0, curves of degree 3 (cubics) have genus 1, and curves of higher degree have genus ≥ 2.

The best way of interpreting the genus, however, involves *complex* numbers. If we consider x and y as being complex numbers, then the equation of the curve will describe a two-dimensional surface in the four-dimensional space $\mathbf{C}^2$. It turns out that g is just the 'number of holes' in the surface, so that $g = 0$ corresponds to a sphere, $g = 1$ to a doughnut-shaped surface or *torus*, and $g \geq 2$ to a multiple torus (see Figure 7.11). Each such surface has associated to it a natural geometry; for $g = 0$ it is the spherical geometry that we will discuss in section 12.2, for $g = 1$ it is ordinary Euclidean plane geometry, and for $g \geq 2$ it is the 'hyperbolic' geometry of Lobatchewsky and Bolyai. The interplay between geometry, complex analysis, topology, and number theory brings delight to many mathematicians working in the theory of curves. At the time of writing there is also considerable involvement with theoretical physics. In *string theory*, the 'elementary particles' are regarded not as particles but as one-dimensionally extended loops or *strings*. As a string moves, it sweeps out a two-dimensional surface or *world sheet* in space-time, and this world sheet can be studied by the methods of complex curve theory.

The natural analogue of the Mordell–Weil theorem turns out to be that there are only finitely many rational points on a curve of genus ≥ 2. This statement was for many years a conjecture (the *Mordell conjecture*), until Faltings gave a proof in 1983. Since the Fermat curve $x^n + y^n = 1$ has genus ≥ 2 for $n \geq 4$, Faltings' result implies that for each n there can be at most finitely many essentially different counterexamples to Fermat's 'last theorem'. This is the most general result that is presently known about the Fermat problem.

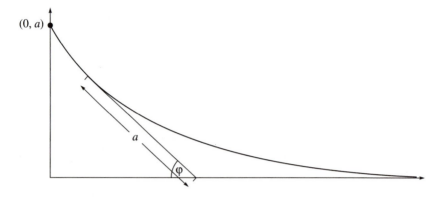

Fig. 7.12. The tractrix.

7.6 Exercises

1. Find the equation of the tangent line at (x_0, y_0) to

    (i) the *cissoid of Diocles*, with equation

$$y^2 = \frac{x^3}{2a - x};$$

    (ii) the *lemniscate*, with equation

$$(x^2 + y^2)^2 = a^2(x^2 - y^2), \quad \text{where } a \neq 0 \text{ is constant.}$$

Find also the singular points (if any) of these curves.

2. Find the equation of the tangent line at the point with parameter value t_0 to the folium of Descartes given parametrically by

$$x = \frac{3t}{1 + t^3}, \quad y = \frac{3t^2}{1 + t^3}.$$

3. A heavy particle lies on the plane at the point $(0, a)$. It is connected to a string of length a, the other end of which is initially at the origin. The free end of the string is then moved along the x-axis in the positive direction, dragging the particle along a curve C called the *tractrix*. The defining property of C is therefore that the length of the segment of the tangent line between a given point of the curve and the x-axis is always equal to a (see Figure 7.12).

Let φ be the angle between the tangent line and the x-axis. Show that if $(x = x(\varphi), y = y(\varphi))$ is a point on the curve, then $y = a \sin \varphi$ and

$$\frac{dy}{d\varphi} = -\tan \varphi \frac{dx}{d\varphi}.$$

Hence show that the tractrix has parametric equation

$$x = a(\log \cot \tfrac{1}{2}\varphi - \cos \varphi), \quad y = a \sin \varphi.$$

4. Find the equations (in Cartesian coordinates) of

   (i) the ellipse with foci $(0, \pm 2)$ that passes through $(1, 0)$;

   (ii) the hyperbola with asymptotes $y = \pm 2x$ and directrices $x = \pm 1$;

   (iii) the ellipse consisting of all points P such that $|PA| + |PB| = 7$, where $A = (0, 3)$ and $B = (0, -3)$.

5. $\mathcal{C}$ is a nondegenerate conic in a plane, with focus F. A straight line $\mathcal{L}$ through F intersects $\mathcal{C}$ in two points A and B. Prove that the quantity

$$\frac{1}{|AF|} + \frac{1}{|BF|}$$

is independent of the choice of $\mathcal{L}$.

6. Show that the line $lx + my + n = 0$ is tangent to the ellipse $\dfrac{x^2}{a^2} + \dfrac{y^2}{b^2} = 1$ at the point (x_0, y_0) on the ellipse if and only if

$$\begin{aligned} lx_0 + my_0 &= -n \\ mb^2 x_0 - la^2 y_0 &= 0. \end{aligned}$$

By solving these equations for (x_0, y_0) and then substituting into the equation of the ellipse, or otherwise, show that the line is tangent to the ellipse *somewhere* if and only if $a^2 l^2 + b^2 m^2 = n^2$.

7. What sorts of conics are described by the following equations?

   (i) $5x^2 - 6xy + 5y^2 = 9$;

   (ii) $157x^2 + 270xy + 13y^2 + 34 = 0$;

   (iii) $39x^2 - 6xy + 31y^2 + 30\sqrt{10}x + 10\sqrt{10}y = 0$;

   (iv) $3x^2 - 2\sqrt{3}xy + y^2 + 2x + 2\sqrt{3}y = 0$;

   (v) $4x^2 - 4xy + y^2 - 12x + 6y + 9 = 0$;

   (vi) $x^2 - xy + y^2 - x + y = 0$;

   (vii) $x^2 - y^2 + 2y = 1$.

In each case find a coordinate change that will simplify the equations as far as possible, and find the eccentricities if they are defined.

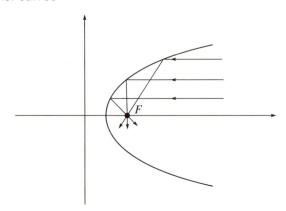

Fig. 7.13. The parabolic mirror.

8. Show that a general point on the parabola with equation $y^2 = 4ax$ may be expressed in the form $(x, y) = (at^2, 2at)$. Find the equation of the tangent line to the parabola at this point, and show that the tangent lines at two points with parameters t_1 and t_2 intersect at the point $(at_1t_2, a(t_1 + t_2))$.

Consider the triangle formed by three tangent lines to the parabola. Prove that the *orthocentre* (Exercise 4.6.7) of this triangle lies on the directrix, and that its circumcircle passes through the focus.

9. Find all the common tangents to the two parabolas $y^2 = 4ax$ and $x^2 = 4by$.

(A *common tangent* is a line that is tangent to both parabolas.)

10. Show that any ray of light parallel to the axis of a parabolic mirror will be reflected so as to pass through the focus. See Figure 7.13. This is why parabolic reflectors are used in car headlights, satellite dishes, radio telescopes and so on; such a reflector makes a parallel beam from a distant source converge at a point, or used the other way round it transforms light from a point source into a parallel beam. To analyse this problem, you need to know that the laws of reflection state that the incident and reflected rays make equal angles with the tangent line to the parabola.

11. Prove that the orthocentre of the triangle formed by three points on a rectangular hyperbola also lies on the hyperbola.

12. (For those with some knowledge of topology.) Prove that the three kinds of nondegenerate conic can be distinguished by their topological properties, as follows: an ellipse is connected and compact, a parabola is connected but not compact, and a hyperbola is neither connected nor compact.

13. Let C be a nondegenerate conic given by the equation $Ax^2 + 2Bxy + Cy^2 + Dx + Ey = 0$. Notice that $(0, 0)$ is a point on C. Find the coordinates of the other point of intersection of C with the line $y = tx$. Hence show that C is rational.

14. Find some pairs of rational numbers (x, y) that satisfy the equation

$$x^3 - 3x^2y + 5y^2 - 2y = 1.$$

15. Consider the quartic (fourth-degree) curve given by

$$5x^2y^2 + x^2y + y^2x - 3x^2 + y^2 = 0.$$

Show that if (x, y) lies on the curve, then $(1/x, 1/y)$ lies on a certain conic. Deduce that this quartic curve is rational.

8

Solid geometry

8.1 Orientation and the vector product

In this chapter we will use vector methods to study the geometry of three dimensions, traditionally known as *solid geometry*. As we did for plane geometry, we will begin by discussing the concept of *orientation*. In two dimensions, an orientation specifies a 'positive direction of rotation' (6.1.2), and thereby divides the possible coordinate systems into right-handed and left-handed. In three dimensions, we will find that an orientation makes possible a similar distinction between right-handed and left-handed coordinate systems. For a lively discussion of the physical significance (if any) of this distinction, see Gardner [17].

An orientation of a plane gives a coherent choice of one of the two possible perpendicular unit vectors to a given unit vector. We want to define something analogous in three dimensions, but clearly it is not much use asking for a standard choice of perpendicular to *one* vector in three-dimensional space. There are infinitely many possibilities! But there are only two possibilities for a unit vector perpendicular to *two* (independent) vectors in space. An orientation in three dimensions, therefore, ought to select one of these.

This motivates the following definition.

(8.1.1)   **Definition:** *An* orientation *for a three-dimensional Euclidean space S is an operation which assigns to each pair of vectors $\mathbf{u}$ and $\mathbf{v}$ in $\vec{S}$ a new vector $\mathbf{u} \wedge \mathbf{v}$ in $\vec{S}$ (called their* vector product*), having the following properties:*

(i) *The mapping $(\mathbf{u}, \mathbf{v}) \mapsto \mathbf{u} \wedge \mathbf{v}$ is bilinear, in other words*

$$(\lambda_1 \mathbf{u}_1 + \lambda_2 \mathbf{u}_2) \wedge \mathbf{v} \;=\; \lambda_1 \mathbf{u}_1 \wedge \mathbf{v} + \lambda_2 \mathbf{u}_2 \wedge \mathbf{v}$$
$$\mathbf{u} \wedge (\mu_1 \mathbf{v}_1 + \mu_2 \mathbf{v}_2) \;=\; \mu_1 \mathbf{u} \wedge \mathbf{v}_1 + \mu_2 \mathbf{u} \wedge \mathbf{v}_2.$$

(ii) *The vector $\mathbf{u} \wedge \mathbf{v}$ is perpendicular to both $\mathbf{u}$ and $\mathbf{v}$.*

(iii) *For any vector* **u**, **u** ∧ **u** = **0**.

(iv) *If* **u** *and* **v** *are perpendicular unit vectors, then* **u** ∧ **v** *is also a unit vector.*

The first of these requirements is the natural analogue of our requirement in the plane case that $\mathbf{v} \mapsto \mathbf{v}^{\perp}$ should be a linear map. The third requirement is there because there can be no standard choice of a vector perpendicular to just one vector **u** — so the only sensible answer is zero.

From the axioms we can prove that the vector product is *antisymmetric*: for all vectors **u** and **v**,

$$\mathbf{u} \wedge \mathbf{v} = -\mathbf{v} \wedge \mathbf{u}.$$

To prove this consider the identity

$$(\mathbf{u} + \mathbf{v}) \wedge (\mathbf{u} + \mathbf{v}) = \mathbf{u} \wedge \mathbf{u} + \mathbf{u} \wedge \mathbf{v} + \mathbf{v} \wedge \mathbf{u} + \mathbf{v} \wedge \mathbf{v}$$

which follows from bilinearity. Since $(\mathbf{u} + \mathbf{v}) \wedge (\mathbf{u} + \mathbf{v}) = \mathbf{u} \wedge \mathbf{u} = \mathbf{v} \wedge \mathbf{v} = \mathbf{0}$ by 8.1.1(iii), we get

$$\mathbf{u} \wedge \mathbf{v} + \mathbf{v} \wedge \mathbf{u} = \mathbf{0}.$$

Suppose that an orientation has been chosen, and let **i**, **j**, **k** be an orthonormal basis. Then **i** ∧ **j** must equal either **k** or −**k**: we call the basis *right-handed* in the first case and *left-handed* in the second case. If the basis is right-handed, so that $\mathbf{i} \wedge \mathbf{j} = \mathbf{k}$, then $\mathbf{i} \wedge \mathbf{k}$ must be either **j** or −**j**: it can't be **j**, for then $\mathbf{i} \wedge (\mathbf{j} + \mathbf{k})$ would equal $\mathbf{j} + \mathbf{k}$, so it must be −**j**. Therefore by antisymmetry, $\mathbf{k} \wedge \mathbf{i} = \mathbf{j}$. Continuing in this way, we can obtain the six identities for a right-handed basis:

$$\mathbf{i} \wedge \mathbf{j} = \mathbf{k}, \quad \mathbf{j} \wedge \mathbf{k} = \mathbf{i}, \quad \mathbf{k} \wedge \mathbf{i} = \mathbf{j}, \quad \mathbf{j} \wedge \mathbf{i} = -\mathbf{k}, \quad \mathbf{i} \wedge \mathbf{k} = -\mathbf{j}, \quad \mathbf{k} \wedge \mathbf{j} = -\mathbf{i}.$$

Notice that there is a positive sign if the three letters **i**, **j**, and **k** appear in cyclic order, and a negative sign otherwise. For a left-handed basis all the signs would be reversed.

Since any vector can be written as a linear combination of the basis vectors, we can use these calculations to work out the vector product in coordinate form. Specifically, suppose that **v** and **w** are vectors, written in component form as

$$\mathbf{v} = v_1\mathbf{i} + v_2\mathbf{j} + v_3\mathbf{k}, \quad \mathbf{w} = w_1\mathbf{i} + w_2\mathbf{j} + w_3\mathbf{k}$$

relative to a right-handed orthonormal basis **i**, **j**, **k**. Then

$$\mathbf{v} \wedge \mathbf{w} = (v_2w_3 - v_3w_2)\mathbf{i} + (v_3w_1 - v_1w_3)\mathbf{j} + (v_1w_2 - v_2w_1)\mathbf{k}$$

is the coordinate form of the vector product. Conversely, starting with an orthonormal basis we may *define* a vector product by this formula, and it is not hard to verify that this definition will satisfy the four axioms 8.1.1. The hardest to check is 8.1.1(iv), and this follows from the identity

$$(v_2w_3 - v_3w_2)^2 + (v_3w_1 - v_1w_3)^2 + (v_1w_2 - v_2w_1)^2$$
$$= (v_1^2 + v_2^2 + v_3^2)(w_1^2 + w_2^2 + w_3^2) - (v_1w_1 + v_2w_2 + v_3w_3)^2.$$

It is often convenient to abbreviate the coordinate expression for the vector product by using determinant notation

(8.1.2)
$$\mathbf{v} \wedge \mathbf{w} = \begin{vmatrix} \mathbf{i} & v_1 & w_1 \\ \mathbf{j} & v_2 & w_2 \\ \mathbf{k} & v_3 & w_3 \end{vmatrix} ;$$

this should just be regarded as a handy way of remembering the formula, rather than as the honest determinant of some mysterious 'matrix with vectors in it'. One advantage of this notation is that it provides a very memorable formula for the so-called *scalar triple product*, which is a real number obtained by taking the dot product of $\mathbf{v} \wedge \mathbf{w}$ with another vector, $\mathbf{u}$. We find that

(8.1.3)
$$\mathbf{u} \cdot (\mathbf{v} \wedge \mathbf{w}) = \begin{vmatrix} u_1 & v_1 & w_1 \\ u_2 & v_2 & w_2 \\ u_3 & v_3 & w_3 \end{vmatrix}$$

where $\mathbf{u} = u_1\mathbf{i} + u_2\mathbf{j} + u_3\mathbf{k}$. The scalar triple product $\mathbf{u} \cdot (\mathbf{v} \wedge \mathbf{w})$ is often written $[\mathbf{u}, \mathbf{v}, \mathbf{w}]$.

The determinant formula for the scalar triple product can be expressed more concisely by means of the notation we introduced in 6.2.5 for a matrix made up of column vectors. Let u, v, and w be the column vectors representing the components of $\mathbf{u}$, $\mathbf{v}$, and $\mathbf{w}$ with respect to the right-handed orthonormal basis $\mathbf{i}, \mathbf{j}, \mathbf{k}$; and let $(\mathsf{u}|\mathsf{v}|\mathsf{w})$ be the 3×3 matrix whose columns are u, v, and w. Then our formula can be written

$$[\mathbf{u}, \mathbf{v}, \mathbf{w}] = \det(\mathsf{u}|\mathsf{v}|\mathsf{w}).$$

(8.1.4)   **Proposition:**

(i) *The scalar triple product $[\mathbf{u}, \mathbf{v}, \mathbf{w}]$ depends linearly on each of $\mathbf{u}$, $\mathbf{v}$, and $\mathbf{w}$.*

(ii) *The scalar triple product $[\mathbf{u}, \mathbf{v}, \mathbf{w}]$ changes sign whenever any two of the vectors $\mathbf{u}$, $\mathbf{v}$, and $\mathbf{w}$ are interchanged.*

(iii) *If $\mathbf{u}$, $\mathbf{v}$, and $\mathbf{w}$ are linearly dependent then $[\mathbf{u}, \mathbf{v}, \mathbf{w}] = 0$. Conversely, if $[\mathbf{u}, \mathbf{v}, \mathbf{w}] = 0$, then $\mathbf{u}$, $\mathbf{v}$, and $\mathbf{w}$ are linearly dependent.*

Proof: The first fact comes from the linearity of the scalar and vector products; the other two come from the corresponding properties of determinants — a determinant changes sign if two of its rows are interchanged, and vanishes if and only if its rows are linearly dependent. □

The next result, which expresses the effect of a linear transformation on a scalar triple product, is a three-dimensional analogue of 6.2.6.

(8.1.5)   **Proposition:** *Let L be any linear transformation acting on vectors. Then for all vectors* **u**, **v**, *and* **w**,

$$[L\mathbf{u}, L\mathbf{v}, L\mathbf{w}] = \det(L)[\mathbf{u}, \mathbf{v}, \mathbf{w}].$$

Proof: The proof is just the same as that of 6.2.6; apply the product rule for determinants to the matrix identity

$$(L\mathbf{u}|L\mathbf{v}|L\mathbf{w}) = L(\mathbf{u}|\mathbf{v}|\mathbf{w}). \qquad \square$$

As well as the scalar triple product, we may consider the *vector triple product* $\mathbf{u} \wedge (\mathbf{v} \wedge \mathbf{w})$. This is a vector which is perpendicular to $\mathbf{v} \wedge \mathbf{w}$, which is itself perpendicular to the plane of **v** and **w**: we therefore expect $\mathbf{u} \wedge (\mathbf{v} \wedge \mathbf{w})$ to be in this plane, in other words to be a linear combination of **v** and **w**. The next result gives the exact formula.

(8.1.6)   **Proposition:** *For any three vectors* **u**, **v**, *and* **w**,

$$\mathbf{u} \wedge (\mathbf{v} \wedge \mathbf{w}) = (\mathbf{u} \cdot \mathbf{w})\mathbf{v} - (\mathbf{u} \cdot \mathbf{v})\mathbf{w}.$$

Proof: To simplify the calculation, choose a right-handed orthonormal basis $\mathbf{i}, \mathbf{j}, \mathbf{k}$ in which $\mathbf{u} = u\mathbf{i}$, and write $\mathbf{v} = v_1\mathbf{i} + v_2\mathbf{j} + v_3\mathbf{k}$ and $\mathbf{w} = w_1\mathbf{i} + w_2\mathbf{j} + w_3\mathbf{k}$. Then

$$\mathbf{u} \wedge (\mathbf{v} \wedge \mathbf{w})$$
$$\begin{aligned}
&= u\mathbf{i} \wedge ((v_2w_3 - v_3w_2)\mathbf{i} + (v_3w_1 - v_1w_3)\mathbf{j} + (v_1w_2 - v_2w_1)\mathbf{k}) \\
&= u(v_3w_1 - v_1w_3)\mathbf{k} + (v_2w_1 - v_1w_2)\mathbf{j} \\
&= uw_1(v_1\mathbf{i} + v_2\mathbf{j} + v_3\mathbf{k}) - uv_1(w_1\mathbf{i} + w_2\mathbf{j} + w_3\mathbf{k}) \\
&= (\mathbf{u} \cdot \mathbf{w})\mathbf{v} - (\mathbf{u} \cdot \mathbf{v})\mathbf{w}. \qquad \square
\end{aligned}$$

There are also two types of *quadruple products* with interesting properties, the *scalar quadruple product*

$$(\mathbf{u}_1 \wedge \mathbf{v}_1) \cdot (\mathbf{u}_2 \wedge \mathbf{v}_2)$$

and the *vector quadruple product*

$$(\mathbf{u}_1 \wedge \mathbf{v}_1) \wedge (\mathbf{u}_2 \wedge \mathbf{v}_2).$$

We consider first the scalar quadruple product.

(8.1.7)   **Proposition:** *For any four vectors* $\mathbf{u}_1$, $\mathbf{v}_1$, $\mathbf{u}_2$, *and* $\mathbf{v}_2$,

$$(\mathbf{u}_1 \wedge \mathbf{v}_1) \cdot (\mathbf{u}_2 \wedge \mathbf{v}_2) = (\mathbf{u}_1 \cdot \mathbf{u}_2)(\mathbf{v}_1 \cdot \mathbf{v}_2) - (\mathbf{u}_1 \cdot \mathbf{v}_2)(\mathbf{v}_1 \cdot \mathbf{u}_2).$$

Proof: We may regard the quadruple product as a scalar triple product $[\mathbf{u}_1 \wedge \mathbf{v}_1, \mathbf{u}_2, \mathbf{v}_2]$. Rearranging this scalar triple product by 8.1.4, we find that

$$
\begin{aligned}
(\mathbf{u}_1 \wedge \mathbf{v}_1) \cdot (\mathbf{u}_2 \wedge \mathbf{v}_2) &= [\mathbf{u}_2, \mathbf{v}_2, \mathbf{u}_1 \wedge \mathbf{v}_1] \\
&= \mathbf{u}_2 \cdot (\mathbf{v}_2 \wedge (\mathbf{u}_1 \wedge \mathbf{v}_1)) \\
&= \mathbf{u}_2 \cdot ((\mathbf{v}_2 \cdot \mathbf{v}_1)\mathbf{u}_1 - (\mathbf{v}_2 \cdot \mathbf{u}_1)\mathbf{v}_1)
\end{aligned}
$$

where in the last line we used the vector triple product expansion (8.1.6) for $\mathbf{v}_2 \wedge (\mathbf{u}_1 \wedge \mathbf{v}_1)$. $\square$

(8.1.8)   **Corollary:** *Let* $\mathbf{u}$ *and* $\mathbf{v}$ *be vectors. Then*

$$|\mathbf{u} \wedge \mathbf{v}| = |\mathbf{u}||\mathbf{v}| \sin \theta$$

where θ *is the (unoriented) angle between* $\mathbf{u}$ *and* $\mathbf{v}$.

Proof: By the previous result

$$|\mathbf{u} \wedge \mathbf{v}|^2 = (\mathbf{u} \wedge \mathbf{v}) \cdot (\mathbf{u} \wedge \mathbf{v}) = |\mathbf{u}|^2 |\mathbf{v}|^2 - (\mathbf{u} \cdot \mathbf{v})^2.$$

Since the definition of θ is that $0 \le \theta \le \pi$ and

$$\mathbf{u} \cdot \mathbf{v} = |\mathbf{u}||\mathbf{v}| \cos \theta$$

this shows that $|\mathbf{u} \wedge \mathbf{v}|^2 = |\mathbf{u}|^2 |\mathbf{v}|^2 (1 - \cos^2 \theta) = |\mathbf{u}|^2 |\mathbf{v}|^2 \sin^2 \theta$. As $\sin \theta \ge 0$, we can take the positive square root of both sides to get the answer. $\square$

This corollary justifies the school-textbook definition of $\mathbf{u} \wedge \mathbf{v}$ as a vector of magnitude $|\mathbf{u}||\mathbf{v}| \sin \theta$ in the right-handed normal direction to $\mathbf{u}$ and $\mathbf{v}$.

Now we will consider the vector quadruple product.

(8.1.9)   **Proposition:** *For any four vectors* $\mathbf{u}_1$, $\mathbf{v}_1$, $\mathbf{u}_2$, *and* $\mathbf{v}_2$,

$$
\begin{aligned}
(\mathbf{u}_1 \wedge \mathbf{v}_1) &\wedge (\mathbf{u}_2 \wedge \mathbf{v}_2) \\
&= [\mathbf{u}_2, \mathbf{v}_2, \mathbf{u}_1]\mathbf{v}_1 - [\mathbf{u}_2, \mathbf{v}_2, \mathbf{v}_1]\mathbf{u}_1 = [\mathbf{u}_1, \mathbf{v}_1, \mathbf{v}_2]\mathbf{u}_2 - [\mathbf{u}_1, \mathbf{v}_1, \mathbf{u}_2]\mathbf{v}_2.
\end{aligned}
$$

Proof: We consider $(\mathbf{u}_1 \wedge \mathbf{v}_1) \wedge (\mathbf{u}_2 \wedge \mathbf{v}_2)$ as the vector triple product of $\mathbf{u}_1 \wedge \mathbf{v}_1$, $\mathbf{u}_2$, and $\mathbf{v}_2$. Using the vector triple product expansion (8.1.6), this is equal to

$$((\mathbf{u}_1 \wedge \mathbf{v}_1) \cdot \mathbf{v}_2)\mathbf{u}_2 - ((\mathbf{u}_1 \wedge \mathbf{v}_1) \cdot \mathbf{u}_2)\mathbf{v}_2.$$

This gives the second formula. The first one is obtained by writing

$$(\mathbf{u}_1 \wedge \mathbf{v}_1) \wedge (\mathbf{u}_2 \wedge \mathbf{v}_2) = -(\mathbf{u}_2 \wedge \mathbf{v}_2) \wedge (\mathbf{u}_1 \wedge \mathbf{v}_1)$$

and using the same argument. $\square$

Forgetting for the moment its origin in the vector quadruple product, the formula

$$[\mathbf{u}_2, \mathbf{v}_2, \mathbf{u}_1]\mathbf{v}_1 - [\mathbf{u}_2, \mathbf{v}_2, \mathbf{v}_1]\mathbf{u}_1 = [\mathbf{u}_1, \mathbf{v}_1, \mathbf{v}_2]\mathbf{u}_2 - [\mathbf{u}_1, \mathbf{v}_1, \mathbf{u}_2]\mathbf{v}_2$$

is of some interest in its own right. What it shows is that there is always a *linear relation* between four arbitrarily chosen vectors $\mathbf{u}_1$, $\mathbf{v}_1$, $\mathbf{u}_2$, and $\mathbf{v}_2$. This is a special case of a general result from linear algebra, that $n+1$ vectors in an n-dimensional vector space cannot be linearly independent.

From linear algebra we also know that three linearly independent vectors (in a three-dimensional space) always form a *basis* (perhaps not orthonormal) — any other vector can be written as a linear combination of them in a unique way. Let $\mathbf{u}$, $\mathbf{v}$, and $\mathbf{w}$ be independent vectors; then their scalar triple product $[\mathbf{u}, \mathbf{v}, \mathbf{w}]$ is nonzero (see 8.1.4). The quadruple product identity shows explicitly how a fourth vector $\mathbf{x}$ may be written as a linear combination of them:

$$\mathbf{x} = \frac{[\mathbf{x}, \mathbf{v}, \mathbf{w}]}{[\mathbf{u}, \mathbf{v}, \mathbf{w}]}\mathbf{u} + \frac{[\mathbf{u}, \mathbf{x}, \mathbf{w}]}{[\mathbf{u}, \mathbf{v}, \mathbf{w}]}\mathbf{v} + \frac{[\mathbf{u}, \mathbf{v}, \mathbf{x}]}{[\mathbf{u}, \mathbf{v}, \mathbf{w}]}\mathbf{w}.$$

If you know about Cramer's rule for solving simultaneous equations, you might like to think how this formula is related to it.

8.2   Vector equations of lines and planes

Using the scalar and vector products, we can give equations for lines and planes in an (oriented) three-dimensional space $\mathcal{S}$. We begin with the equation of a line.

(8.2.1)   **Lemma:** *Let $\mathbf{u}$ and $\mathbf{v}$ be vectors in $\vec{\mathcal{S}}$. Then the vector product $\mathbf{u} \wedge \mathbf{v}$ is $\mathbf{0}$ if and only if $\mathbf{u}$ and $\mathbf{v}$ are linearly dependent (i.e. one is a multiple of the other).*

        Proof: It is clear that if $\mathbf{u}$ is a multiple of $\mathbf{v}$, then $\mathbf{u} \wedge \mathbf{v} = \mathbf{0}$. Conversely, suppose that $\mathbf{u} \neq \mathbf{0}$ and $\mathbf{u} \wedge \mathbf{v} = \mathbf{0}$. Then

$$\mathbf{0} = \mathbf{u} \wedge (\mathbf{u} \wedge \mathbf{v}) = (\mathbf{u} \cdot \mathbf{v})\mathbf{u} - |\mathbf{u}|^2\mathbf{v}$$

by the vector triple product expansion (8.1.6). Since $|\mathbf{u}| \neq 0$, this shows that $\mathbf{u}$ and $\mathbf{v}$ are linearly dependent. $\square$

(8.2.2)   **Proposition:** *Let O be an origin in $\mathcal{S}$, and let $\mathbf{u}$ and $\mathbf{v}$ be fixed vectors with $\mathbf{u} \neq \mathbf{0}$ and $\mathbf{u} \cdot \mathbf{v} = 0$. Let $\mathbf{r} = \overrightarrow{OP}$ be the position vector of a variable point P. Then the equation*

$$\mathbf{r} \wedge \mathbf{u} = \mathbf{v}$$

says that P lies on some line $\mathcal{L}$ in $\mathcal{S}$; and every line can be described by an equation of this form.

Proof: Given the equation $\mathbf{r} \wedge \mathbf{u} = \mathbf{v}$, let P_0 be the point with position vector

$$\mathbf{r}_0 = \frac{\mathbf{u} \wedge \mathbf{v}}{|\mathbf{u}|^2}.$$

Then

$$\mathbf{r}_0 \wedge \mathbf{u} = \frac{|\mathbf{u}|^2 \mathbf{v} - (\mathbf{u} \cdot \mathbf{v})\mathbf{u}}{|\mathbf{u}|^2} = \mathbf{v}$$

by the vector triple product expansion (8.1.6). So if $\mathbf{r} \wedge \mathbf{u} = \mathbf{v}$, then $(\mathbf{r} - \mathbf{r}_0) \wedge \mathbf{u} = \mathbf{0}$. By lemma 8.2.1, $\mathbf{r} - \mathbf{r}_0$ must be a multiple of $\mathbf{u}$. Thus P lies on the line through P_0 in the direction of the vector $\mathbf{u}$. In the notation of 2.1.3, this is the line $U[P_0]$, where U is the one-dimensional subspace of $\vec{S}$ spanned by $\mathbf{u}$.

Conversely, given a line $\mathcal{L} = U[P_0]$, let $\mathbf{u}$ be a vector parallel to it (that is, a nonzero element of U) and let $\mathbf{r}_0$ be the position vector of P_0. If $\mathbf{r}$ is the position vector of a general point on $\mathcal{L}$, then $\mathbf{r} - \mathbf{r}_0 \in U$, and so

$$\mathbf{r} \wedge \mathbf{u} = \mathbf{r}_0 \wedge \mathbf{u} = \mathbf{v}, \text{ say}.$$

So $\mathbf{r}$ satisfies an equation of the required form. $\square$

If we introduce a right-handed orthonormal basis $\mathbf{i}, \mathbf{j}, \mathbf{k}$, let $\mathbf{r} = x\mathbf{i} + y\mathbf{j} + z\mathbf{k}$, and write $\mathbf{u}$ and $\mathbf{v}$ in terms of components in the usual way, we obtain three simultaneous equations describing a straight line:

$$
\begin{aligned}
yu_3 - zu_2 &= v_1 \\
zu_1 - xu_3 &= v_2 \\
xu_2 - yu_1 &= v_3
\end{aligned}
$$

These three equations are not independent: if you multiply the first by u_1, the second by u_2, the third by u_3, and add them all, you get $0 = 0$. Thus one of them can always be obtained from the other two. A line in S is therefore given by two independent linear equations: this is not surprising since, as we shall see, one linear equation describes a plane and a line is the intersection of two planes.

We now turn to the equation of a plane $\mathcal{P}$. According to 2.1.3, $\mathcal{P}$ is simply a two-dimensional affine subspace, of the form $W[P_0]$ where $P_0 \in S$ is a fixed point and W is a two-dimensional vector subspace of $\vec{S}$. So, if we choose an origin O in S, then the plane $\mathcal{P}$ can be described as the set of all points P with position vectors

$$\mathbf{r} = \overrightarrow{OP} = \mathbf{r}_0 + \lambda\mathbf{u} + \mu\mathbf{v}, \quad \lambda, \mu \in \mathbf{R}$$

where $\mathbf{r}_0$ is the position vector of the fixed point P_0 and $\mathbf{u}$ and $\mathbf{v}$ are two linearly independent vectors forming a basis for W.

(8.2.3)   **Proposition:** *Let O be an origin in S. Let* **n** *be a fixed nonzero vector, and let k be a constant. Let* **r** *be the position vector of a variable point P. Then the equation*

$$\mathbf{r} \cdot \mathbf{n} = k$$

says that P lies on some plane $\mathcal{P}$ in S; and every plane can be described by an equation of this form.

Proof: Given the equation $\mathbf{r} \cdot \mathbf{n} = k$, choose an orthonormal basis $\mathbf{i}, \mathbf{j}, \mathbf{k}$ such that $\mathbf{n} = n\mathbf{i}$. Let P_0 be the point with position vector $\mathbf{r}_0 = k\mathbf{i}/n$; then $\mathbf{r}_0 \cdot \mathbf{n} = k$, and so $\mathbf{r} \cdot \mathbf{n} = k$ if and only if $(\mathbf{r} - \mathbf{r}_0) \cdot \mathbf{i} = 0$. This equation tells us that $\mathbf{r} - \mathbf{r}_0$ has no component in the $\mathbf{i}$-direction, so it is a linear combination of $\mathbf{j}$ and $\mathbf{k}$; in other words

$$\mathbf{r} = \mathbf{r}_0 + \lambda \mathbf{j} + \mu \mathbf{k}$$

and so belongs to the plane $W[P_0]$, where W is the two-dimensional subspace spanned by $\mathbf{j}$ and $\mathbf{k}$.

Conversely, given a plane $\mathcal{P} = W[P_0]$, let $\mathbf{u}$ and $\mathbf{v}$ form a basis for W and let $\mathbf{n} = \mathbf{u} \wedge \mathbf{v}$. Let $\mathbf{r}_0$ be the position vector of P_0. If $\mathbf{r}$ is the position vector of a general point on $\mathcal{P}$, then $\mathbf{r} - \mathbf{r}_o \in W$, and so

$$\mathbf{r} \cdot \mathbf{n} = \mathbf{r}_0 \cdot \mathbf{n} = k, \text{ say,}$$

because $\mathbf{n}$ is orthogonal to all vectors in W. $\square$

The vector $\mathbf{n}$ is called a *normal vector* to the plane.

(8.2.4)   **Orientation of a plane:** The plane described by the equation $\mathbf{r} \cdot \mathbf{n} = k$ is unchanged if we multiply both $\mathbf{n}$ and k by some nonzero constant. We can choose the constant so that $\mathbf{n}$ is a unit vector, a *unit normal* to the plane. There are still two possible choices of unit normal to a given plane, though. Can we pick out one of them? As we have come to expect, this 'one out of two' choice is once again a question of orientation. Let $\mathbf{n}$ be a unit normal vector to the plane $\mathcal{P}$, and consider the operation that maps a vector $\mathbf{v} \in \vec{\mathcal{P}}$ to $\mathbf{n} \wedge \mathbf{v}$. This vector is also in $\vec{\mathcal{P}}$ because it is perpendicular to $\mathbf{n}$. From properties of the vector product, $\mathbf{n} \wedge \mathbf{v}$ is perpendicular to $\mathbf{v}$ and has the same length as $\mathbf{v}$. Therefore, the operation of taking the vector product with $\mathbf{n}$ defines an *orientation* (in the two-dimensional sense, 6.1.2) on the plane $\mathcal{P}$.

Thus the two possible choices of unit normal vector correspond to the two possible choices of orientation of the plane $\mathcal{P}$. Conversely, if we have already chosen an orientation for $\mathcal{P}$, then this choice corresponds to just one of the two possibilities for a unit normal vector. For an oriented plane $\mathcal{P}$ there is therefore a canonical choice of *oriented unit normal vector* $\mathbf{n}$.

If we write the equation $\mathbf{r} \cdot \mathbf{n} = k$ in coordinate form it becomes

(8.2.5)
$$n_1 x + n_2 y + n_3 z = k$$

with the usual notation $\mathbf{r} = x\mathbf{i} + y\mathbf{j} + z\mathbf{k}$, $\mathbf{n} = n_1\mathbf{i} + n_2\mathbf{j} + n_3\mathbf{k}$. Thus, a single linear equation in three dimensions describes a plane.

In general, two planes in three-dimensional space will intersect in a line. We can find the equation of the line by vector methods:

(8.2.6)   **Proposition:** *Let O be an origin, and let*

$$\mathbf{r} \cdot \mathbf{n}_1 = k_1, \quad \mathbf{r} \cdot \mathbf{n}_2 = k_2$$

be the equations of two planes $\mathcal{P}_1$ and $\mathcal{P}_2$. Then, provided that $\mathbf{n}_1$ and $\mathbf{n}_2$ are linearly independent, the two planes intersect in the line $\mathcal{L}$ with equation

$$\mathbf{r} \wedge \mathbf{u} = \mathbf{v}$$

where
$$\mathbf{u} = \mathbf{n}_1 \wedge \mathbf{n}_2, \quad \mathbf{v} = k_2\mathbf{n}_1 - k_1\mathbf{n}_2.$$

Proof: We must show two things: that every point on $\mathcal{L}$ is on both $\mathcal{P}_1$ and $\mathcal{P}_2$, and also that every point on $\mathcal{P}_1 \cap \mathcal{P}_2$ is on $\mathcal{L}$.

For the first, suppose that P is on $\mathcal{L}$, so that $\mathbf{r} = \overrightarrow{OP}$ satisfies $\mathbf{r} \wedge \mathbf{u} = \mathbf{v}$. Substituting the definitions of $\mathbf{u}$ and $\mathbf{v}$ into this equation, and using the expansion of the vector triple product (8.1.6), we find that

$$(\mathbf{r} \cdot \mathbf{n}_2)\mathbf{n}_1 - (\mathbf{r} \cdot \mathbf{n}_1)\mathbf{n}_2 = k_2\mathbf{n}_1 - k_1\mathbf{n}_2.$$

Because $\mathbf{n}_1$ and $\mathbf{n}_2$ are linearly independent vectors, this implies that $\mathbf{r} \cdot \mathbf{n}_1 = k_1$ and $\mathbf{r} \cdot \mathbf{n}_2 = k_2$.

To prove the converse, simply note that if P is on both $\mathcal{P}_1$ and $\mathcal{P}_2$, then

$$\mathbf{r} \wedge \mathbf{u} = \mathbf{r} \wedge (\mathbf{n}_1 \wedge \mathbf{n}_2) = (\mathbf{r} \cdot \mathbf{n}_2)\mathbf{n}_1 - (\mathbf{r} \cdot \mathbf{n}_1)\mathbf{n}_2 = \mathbf{v}$$

by the vector triple product expansion again. □

If $\mathbf{n}_1$ and $\mathbf{n}_2$ are not linearly independent, then the two planes are either parallel or coincident.

8.3 Isometries

We began our study of plane geometry with a classification of all possible isometries of a plane. This was important because isometries arise in many

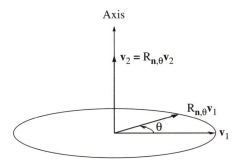

Fig. 8.1. A rotation in three dimensions.

different guises, both as congruences (the alibi interpretation) and as coordinate changes (the alias interpretation). For the same reasons, we now want to classify all the isometries of a three-dimensional oriented Euclidean space S.

The vector definitions of a *translation* $T_{\mathbf{v}}$ (4.5.2) and of a *reflection* $S_{\mathbf{n}}$ (6.1.1) still make sense in three dimensions, and the proofs that they are isometries are exactly the same. But there is a difference regarding the definition of a *rotation*. To specify a rotation in three dimensions one needs to give not only the angle of rotation θ but also the *axis* (see Figure 8.1). The axis of rotation can be specified by a unit vector $\mathbf{n}$: the idea is that points along the axis are left unchanged, whereas the plane $\mathcal{P}_{\mathbf{n}}$ normal to $\mathbf{n}$ is rotated through θ. (Recall from 8.2.4 that the choice of unit normal vector $\mathbf{n}$ specifies an *orientation* for the normal plane; it is in terms of this orientation that we define what we mean by 'rotation of the plane through θ'.)

Let us choose an origin O on the axis of rotation, and represent points by their position vectors relative to O. To give a formula for a rotation we need first of all to have some way of resolving a general vector $\mathbf{v}$ into its components parallel and perpendicular to $\mathbf{n}$:

(8.3.1)   **Lemma:** *Let $\mathbf{n}$ be a unit vector and let $\mathbf{v}$ be any vector. Then*

$$\mathbf{v} = (\mathbf{v} \cdot \mathbf{n})\mathbf{n} + \mathbf{n} \wedge (\mathbf{v} \wedge \mathbf{n}).$$

The first term on the right-hand side is parallel to $\mathbf{n}$ and the second is perpendicular to $\mathbf{n}$.

Proof: This follows from the vector triple product identity (8.1.6) and the fact that $\mathbf{n} \cdot \mathbf{n} = 1$. $\square$

To rotate $\mathbf{v}$ through θ around the axis $\mathbf{n}$, we must leave the term $(\mathbf{v} \cdot \mathbf{n})\mathbf{n}$ alone and replace $\mathbf{n} \wedge (\mathbf{v} \wedge \mathbf{n})$ by the result of rotating it through θ in the plane $\mathcal{P}_{\mathbf{n}}$, which is

$$\cos \theta \, \mathbf{n} \wedge (\mathbf{v} \wedge \mathbf{n}) + \sin \theta \, \mathbf{n} \wedge (\mathbf{n} \wedge (\mathbf{v} \wedge \mathbf{n}))$$

— remember (8.2.4) that the orientation operation in $\mathcal{P}_{\mathbf{n}}$ is given by taking the vector product with $\mathbf{n}$. Fortunately this can be simplified a bit: it is easy to show (using the vector triple product expansion) that

$$\mathbf{n} \wedge (\mathbf{n} \wedge (\mathbf{v} \wedge \mathbf{n})) = \mathbf{n} \wedge \mathbf{v},$$

so we arrive at the following definition of a rotation:

(8.3.2)   **Definition:** *Let $\mathcal{S}$ be a three-dimensional oriented Euclidean space, $O \in \mathcal{S}$ an origin. Let $\mathbf{n} \in \vec{S}$ be a unit vector and θ an angle. The* rotation through angle θ with axis through O in direction $\mathbf{n}$, $R_{\mathbf{n},\theta}$, *is the mapping $\mathcal{S} \to \mathcal{S}$ defined in terms of position vectors by*

$$R_{\mathbf{n},\theta}(\mathbf{v}) = (\mathbf{v} \cdot \mathbf{n})\mathbf{n} + \sin\theta\,\mathbf{n} \wedge \mathbf{v} + \cos\theta\,\mathbf{n} \wedge (\mathbf{v} \wedge \mathbf{n}).$$

Clearly, $R_{\mathbf{n},\theta}$ is a linear transformation. Suppose that we take a right-handed orthonormal basis $\mathbf{i}, \mathbf{j}, \mathbf{k}$ whose first element $\mathbf{i}$ is the given unit vector $\mathbf{n}$. Then it is easy to calculate

$$
\begin{aligned}
R_{\mathbf{n},\theta}(\mathbf{i}) &= \mathbf{i} \\
R_{\mathbf{n},\theta}(\mathbf{j}) &= \cos\theta\,\mathbf{j} + \sin\theta\,\mathbf{k} \\
R_{\mathbf{n},\theta}(\mathbf{k}) &= -\sin\theta\,\mathbf{j} + \cos\theta\,\mathbf{k}.
\end{aligned}
$$

Therefore $R_{\mathbf{n},\theta}$ is represented by the matrix

(8.3.3)
$$
\begin{pmatrix}
1 & 0 & 0 \\
0 & \cos\theta & -\sin\theta \\
0 & \sin\theta & \cos\theta
\end{pmatrix}.
$$

This matrix is orthogonal, which verifies that $R_{\mathbf{n},\theta}$ is an isometry.

In two dimensions we proved that any isometry with a fixed point is either a rotation or a reflection. This statement is no longer true in three dimensions. A simple example is the 'central inversion' ι defined in terms of position vectors by $\iota(\mathbf{r}) = -\mathbf{r}$. In three dimensions this is neither a rotation nor a reflection. However, we will be able to prove that every direct isometry that has a fixed point is a rotation. The proof depends on the concept of the *eigenvectors* of a matrix.

Recall (7.3.4) that we defined the *eigenvalues* of a square matrix M to be the roots of the characteristic equation

$$\det(\mathsf{M} - \lambda\mathsf{I}) = 0.$$

So if λ is an eigenvalue, the matrix $\mathsf{M} - \lambda\mathsf{I}$ is *singular*, and this means that there is a column vector $\mathsf{v} \neq 0$ such that $(\mathsf{M} - \lambda\mathsf{I})\mathsf{v} = 0$, or in other words $\mathsf{Mv} = \lambda\mathsf{v}$.

(8.3.4) **Definition:** *A column vector* v *satisfying* $\mathsf{Mv} = \lambda\mathsf{v}$ *is called an* eigenvector *of* M *belonging to the eigenvalue* λ.

An eigenvector with eigenvalue 1 is a *fixed vector* (one that is not changed by the matrix).

(8.3.5) **Proposition:** *Let* M *be an orthogonal* 3×3 *matrix with determinant* $+1$. *Then* $+1$ *is an eigenvalue of* M.

Proof: We need to show that the matrix $\mathsf{M} - \mathsf{I}$ is singular. We can compute

$$
\begin{aligned}
\det(\mathsf{M} - \mathsf{I}) &= \det((\mathsf{M} - \mathsf{I})\mathsf{M}^t) &&\text{since } \det \mathsf{M}^t = \det \mathsf{M} = 1 \\
&= \det(\mathsf{I} - \mathsf{M}^t) &&\text{since } \mathsf{M}\mathsf{M}^t = \mathsf{I} \\
&= \det((\mathsf{I} - \mathsf{M})^t) \\
&= \det(\mathsf{I} - \mathsf{M}) &&\text{since } \det \mathsf{A}^t = \det \mathsf{A} \\
&= -\det(\mathsf{M} - \mathsf{I}) &&\text{since for an } n \times n \text{ matrix } \mathsf{A}, \\
& && \det(-\mathsf{A}) = (-1)^n \det \mathsf{A}.
\end{aligned}
$$

Therefore $\det(\mathsf{M} - \mathsf{I}) = 0$. Notice that the argument would work in any odd dimension. $\square$

Now let T be a direct isometry of $\mathcal{S}$. By definition (5.3.4), the vectorialization $\vec{T}$ of T can be represented by an orthogonal matrix with determinant $+1$. So by the preceding proposition, there is a nonzero vector $\mathbf{n}$ such that $\vec{T}\mathbf{n} = \mathbf{n}$. Without loss of generality, we may assume that $\mathbf{n}$ is a *unit* vector. If the isometry T fixes a point O, then it must also fix every point on the line through O parallel to $\mathbf{n}$. This line is called an *axis* for the isometry.

(8.3.6) **Theorem:** *Let* $T \colon \mathcal{S} \to \mathcal{S}$ *be a direct isometry that leaves a point* O *fixed. Then* T *is a rotation.*

Proof: We can consider T as acting on position vectors relative to the origin O. This amounts to replacing T by its vectorialization $\vec{T}$, which is an orthogonal linear transformation. So by the discussion above, there is a unit vector $\mathbf{n}$ such that $\vec{T}(\mathbf{n}) = \mathbf{n}$.

Now form a right-handed orthonormal basis $\mathbf{i}, \mathbf{j}, \mathbf{k}$, where $\mathbf{i} = \mathbf{n}$. Then $\vec{T}$ maps $\mathbf{i}$ to itself. Now T is an isometry and so preserves angles; hence $\vec{T}$ maps any vector perpendicular to $\mathbf{i}$ to another vector perpendicular to $\mathbf{i}$. In particular, $\vec{T}(\mathbf{j}) = a\mathbf{j} + c\mathbf{k}$ and $\vec{T}(\mathbf{k}) = b\mathbf{j} + d\mathbf{k}$. So the matrix of $\vec{T}$ relative to the basis $\mathbf{i}, \mathbf{j}, \mathbf{k}$ must be of the form

$$
\begin{pmatrix} 1 & 0 & 0 \\ 0 & a & b \\ 0 & c & d \end{pmatrix}.
$$

The submatrix $\begin{pmatrix} a & b \\ c & d \end{pmatrix}$ is a 2×2 orthogonal matrix of determinant $+1$. By our classification of such matrices (5.3.7), this 2×2 matrix must be of the form

$$
\begin{pmatrix}
\cos\theta & -\sin\theta \\
\sin\theta & \cos\theta
\end{pmatrix}
$$

for some angle θ. So the 3×3 matrix of $\vec{T}$ represents a rotation through θ with axis in direction **n**. $\square$

Remark: Some writers *define* a rotation to be an orthogonal matrix with determinant $+1$; they then state this theorem in the form 'every rotation has an axis'.

(8.3.7)   **Corollary:** *Any direct isometry of S can be written as the composite of a rotation and a translation.*

When we defined isometries (4.5.1), we mentioned the idea that some isometries can be brought about 'in a continuous movement', whereas others cannot. To formalize this, let us say that an isometry T can be *performed continuously* if there is a continuous[1] mapping $t \mapsto T_t$ of $[0, 1]$ to the isometry group $\mathrm{Iso}(S)$, with T_0 equal to the identity and $T_1 = T$. (In topological language, this says that T is in the same *path component* of $\mathrm{Iso}(S)$ as the identity.) This idea is illustrated in Figure 8.2, which shows the successive images $T_t(F)$ of a certain subset $F \subseteq S$.

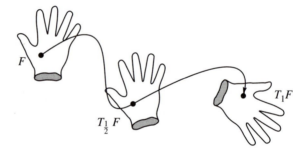

Fig. 8.2. An isometry that can be performed continuously.

Which isometries can be performed continuously? The answer is very simple.

(8.3.8)   **Proposition:** *An isometry of S can be performed continuously if and only if it is direct.*

Proof: A rotation $R_{\mathbf{n},\theta}$ can certainly be performed continuously, because it can be joined to the identity by the continuous family of isometries $t \mapsto R_{\mathbf{n},t\theta}$.

---

[1]We have not said precisely what we mean by a continuity for a family of isometries, or in fancier language what is the topology on $\mathrm{Iso}(S)$. One possible definition is that $t \mapsto T_t$ is continuous if and only if all the maps $[0, 1] \to S$ given by $t \mapsto T_t(P)$ for fixed points P are continuous. If you know enough to worry about such things, then you should not find it too hard to check that this concept of continuity has all the properties that we need.

Similarly, a translation $T_{\mathbf{v}}$ can be performed continuously, using the family $t \mapsto T_{t\mathbf{v}}$. So a composite of a translation and a rotation can be performed continuously, and such composites give all direct isometries.

Conversely, the determinant $\det(\vec{T}_t)$ must vary continuously along any continuous family T_t of isometries. Since the only possible values for it are ± 1, it must stay constant — the intermediate value theorem says that it can't jump from $+1$ to -1 without going through 0 in between. So any isometry that can be joined to the identity by a continuous path is direct. $\square$

Thus one should think of the collection $\text{Iso}(\mathcal{S})$ of all isometries as being *disconnected*, like the hyperbola; it has two *connected components* given by the two possible values of the determinant. Of course $\text{Iso}(\mathcal{S})$ is also a *group*, and the determinant gives a *homomorphism* of this group onto the two-element group $\{\pm 1\}$; the direct isometries form the *kernel* of this homomorphism. Thus the isometries have two interacting structures: topological and group-theoretic. Groups of this kind are now called *Lie groups*, after the Norwegian mathematician Sophus Lie (1842–1899) who first systematically investigated them. They play a fundamental rôle in many studies in physics and chemistry where symmetry is involved — for instance, the idea that some 'fundamental particles' (baryons and mesons) might be made up of still tinier particles (quarks) with fractional electric charge and other odd properties was arrived at from the theory of the Lie group known as $SU(3)$.

(8.3.9)   **Example:** Suppose that we are given an orthogonal matrix with determinant $+1$, such as

$$M = \begin{pmatrix} 0 & 1 & 0 \\ 0 & 0 & 1 \\ 1 & 0 & 0 \end{pmatrix}.$$

We know from our theorems that this must represent a rotation $R_{\mathbf{a},\theta}$, but how can we explicitly calculate the axis and the angle?

To find the *axis* we need to find a unit vector $\mathbf{a}$ that is fixed under the matrix. In other words, we must solve the equations

$$\begin{pmatrix} 0 & 1 & 0 \\ 0 & 0 & 1 \\ 1 & 0 & 0 \end{pmatrix} \begin{pmatrix} a_1 \\ a_2 \\ a_3 \end{pmatrix} = \begin{pmatrix} a_1 \\ a_2 \\ a_3 \end{pmatrix}.$$

These are easily solved to give $a_1 = a_2 = a_3$; so a unit vector along the axis is

$$\mathbf{a} = \frac{1}{\sqrt{3}}(\mathbf{i} + \mathbf{j} + \mathbf{k}).$$

Of course, $-\mathbf{a}$ would also be a possible choice for unit vector along the axis.

To find the angle of rotation we need to find a right-handed orthonormal basis of which $\mathbf{a}$ is the first element. Picking any unit vector $\mathbf{b}$ perpendicular to **a** as the second element — for example, $\mathbf{b} = \dfrac{1}{\sqrt{2}}(\mathbf{i} - \mathbf{j})$ — we can compute the third element as $\mathbf{c} = \mathbf{a} \wedge \mathbf{b}$. With our choice of **b** we find that $\mathbf{c} = \dfrac{1}{\sqrt{6}}(\mathbf{i} + \mathbf{j} - 2\mathbf{k})$.

Now we must write the transformation in terms of the new basis **a, b, c**. In other words, we have to express the transformation in terms of a new coordinate system x′ related to the original system x by x = Ux′, where U = (a|b|c) is the 'change of basis matrix',

$$U = \begin{pmatrix} 1/\sqrt{3} & 1/\sqrt{2} & 1/\sqrt{6} \\ 1/\sqrt{3} & -1/\sqrt{2} & 1/\sqrt{6} \\ 1/\sqrt{3} & 0 & -2/\sqrt{6} \end{pmatrix}.$$

If x and y in the old coordinate system are related by y = Mx, then the corresponding relationship in the new coordinate system is $\mathbf{y}' = U^{-1}\mathbf{y} = U^{-1}MU\mathbf{x}$. Thus the matrix representing the transformation relative to the new coordinate system is $U^{-1}MU$. Because the basis **a, b, c** is *orthonormal*, U is orthogonal: $U' = U^{-1}$. Thus $U^{-1}MU = U'MU$ and we can calculate this as

$$\begin{pmatrix} 1 & 0 & 0 \\ 0 & -1/2 & \sqrt{3}/2 \\ 0 & -\sqrt{3}/2 & -1/2 \end{pmatrix}.$$

Comparing this with the standard form of a rotation matrix (8.3.3), we see that this represents a rotation through $-2\pi/3$ about the axis **a**.

There is a quicker way to find the angle θ. Recall that the *trace* tr A of a matrix A is defined to be the sum of its diagonal entries. It can be proved that $\mathrm{tr}(AB) = \mathrm{tr}(BA)$ for any matrices A and B. Therefore,

$$\mathrm{tr}(U^{-1}MU) = \mathrm{tr}(MUU^{-1}) = \mathrm{tr}\,M.$$

Since, however, $U^{-1}MU$ is a rotation matrix of the standard form given in 8.3.3, its trace is $1 + 2\cos\theta$. So we find

(8.3.10)   **Proposition:** *Let* M *be an orthogonal* 3 × 3 *matrix with determinant* +1. *Then* M *represents a rotation through an angle* θ, *where*

$$\mathrm{tr}\,M = 1 + 2\cos\theta.$$

In the example above, tr M = 0 so this shows that $\cos\theta = -\tfrac{1}{2}$ which implies that $\theta = \pm 2\pi/3$. Which sign we take is a question of orientation, and depends on which of the two possible unit vectors was selected as the axis; see Exercise 8.5.14.

We have classified all the direct isometries having a fixed point; what about the indirect ones? If T is an indirect isometry having a fixed point O, and ι is the central inversion in O, then ιT is a direct isometry fixing O, and is therefore a rotation. Therefore T is the composite of a rotation and a central inversion. Such an isometry T need *not* be a reflection; see Exercise 8.5.15.

8.4 Quaternions

The theory of vectors in three dimensions that we have been using in this chapter has been built up in several stages. First we defined the abstract concept of a vector in an affine space (Chapter 2). Later, we introduced a dot product to take account of the Euclidean structure of lengths and angles (Chapter 4). Finally in this chapter we introduced the vector product, which is related to orientation and is special to dimension 3. By proceeding in this way we were following a characteristic method of modern mathematics, which is to start with very general concepts and gradually specialize towards particular examples.

The first discoverers of vectors did not take this point of view at all. They did not have our idea of an 'abstract algebraic system' (such as a *group* or a *vector space*) which has only those properties which we lay down by our initial axioms. The only algebraic system with which they were familiar was the system of real numbers, with its addition and multiplication and its commutative, associative and distributive laws, and they searched for 'generalized number systems' satisfying the same laws, which could be used to describe the geometry of more than one dimension. For two dimensions, this point of view had spectacular success; it led to Argand and Wessel's discovery (6.5) of the complex numbers. But in dimension 3, things seemed to get stuck.

The problem was that there seemed to be no sensible way to multiply elements of $\mathbf{R}^3$ (or 'number-triplets' as they were called). The great Irish mathematician William Rowan Hamilton (1805–1865) spent many years in a fruitless attempt to devise such a multiplication law. Later he wrote (in a letter to his son)

Every morning, on my coming down to breakfast, you used to ask me: 'Well, Papa, can you multiply triplets?' Whereto I was always obliged to reply, with a sad shake of the head: 'No, I can only add and subtract them'.

The breakthrough came in 1843, when he realized that by going to *four* dimensions rather than three it was possible to define an algebraic structure having all the required properties *except* the commutative law of multiplication. He called his new 'numbers' *quaternions* and spent the rest of his life working out how to apply them to geometrical and other problems. Towards the end of the nineteenth century, though, it gradually became clear that for nearly all purposes it was much simpler to decompose a four-dimensional quaternion into a one-dimensional scalar and a three-dimensional vector, and to decompose the quaternion product into the scalar and vector products. Thus vector algebra in its modern form arose from Hamilton's theory. But we will reverse the historical order and *define* quaternions in terms of vector algebra.

(8.4.1) **Definition:** *Let S be an oriented three-dimensional Euclidean space. A quaternion for S is an ordered pair $(\alpha, \mathbf{a})$, where $\alpha \in \mathbf{R}$ and $\mathbf{a} \in \vec{S}$.*
   The sum *of two quaternions is defined by*

$$(\alpha, \mathbf{a}) + (\beta, \mathbf{b}) = (\alpha + \beta, \mathbf{a} + \mathbf{b}).$$

The product *of two quaternions is defined by*

$$(\alpha, \mathbf{a})(\beta, \mathbf{b}) = (\alpha\beta - \mathbf{a} \cdot \mathbf{b}, \alpha\mathbf{b} + \beta\mathbf{a} + \mathbf{a} \wedge \mathbf{b}).$$

A quaternion may be denoted by a single letter such as q. If $q = (\alpha, \mathbf{a})$, then the number α is called the *real part* $\Re q$ of the quaternion q, and the vector $\mathbf{a}$ is called the *imaginary part* $\Im q$ of q. It is convenient to identify each real number α with the quaternion $(\alpha, \mathbf{0})$, just as we normally identify a real number with a complex number that has zero imaginary part. Similarly, it is convenient to identify each vector $\mathbf{a} \in \vec{S}$ with the corresponding 'pure imaginary' quaternion $(0, \mathbf{a})$.

   In honour of Hamilton, the set of all quaternions is denoted by **H**. Hamilton discovered quaternions while on a walk in Dublin on Monday, October 16, 1843, and he was so excited that he immediately carved with a penknife into the stone of the Brougham Bridge the fundamental equations

$$\mathbf{i}^2 = \mathbf{j}^2 = \mathbf{k}^2 = \mathbf{ijk} = -1$$

from which the whole multiplication law can be recovered.

   It is clear that quaternion addition is commutative and associative, and that quaternion addition is distributive over quaternion multiplication. A less obvious fact is

(8.4.2) **Proposition:** *Quaternion multiplication is associative.*

Proof: This can be proved by a laborious computation using properties of the scalar and vector triple products, but we will use a more indirect approach. If $q = a + b\mathbf{i} + c\mathbf{j} + d\mathbf{k}$ is a quaternion, then let M_q denote the 2×2 matrix of complex numbers

$$M_q = \begin{pmatrix} a + bi & c + di \\ -c + di & a - bi \end{pmatrix}.$$

One introduces these matrices because it can be shown by direct calculation (see Exercise 8.5.18) that the laws for quaternion and matrix multiplication correspond: $M_{q_1 q_2} = M_{q_1} M_{q_2}$. Thus quaternion multiplication is associative because matrix multiplication is associative! □

Quaternion multiplication is not commutative; for instance, if **i**, **j**, **k** are the usual basis vectors (identified with pure imaginary quaternions), then **ij** = **k**, whereas **ji** = −**k**. However, if the quaternion q is *real* (has zero imaginary part), then it commutes with every other quaternion.

(8.4.3) **Definition:** *Let* $q = (\alpha, \mathbf{a})$ *be a quaternion. Then the* conjugate quaternion *$\bar{q}$ is defined to be* $(\alpha, -\mathbf{a})$.

Quaternion conjugation is analogous to complex conjugation, which reverses the sign of the imaginary part of a complex number. We can calculate that

$$q\bar{q} = \bar{q}q = \alpha^2 + |\mathbf{a}|^2.$$

This is a non-negative real number, and is zero if and only if $q = 0$. It is denoted by $|q|^2$, and the non-negative real number $|q| = \sqrt{q\bar{q}}$ is called the *modulus* of the quaternion q.

(8.4.4) **Proposition:** *For every nonzero quaternion q there is an inverse q^{-1} such that $qq^{-1} = q^{-1}q = 1$.*

Proof: We can define $q^{-1} = \dfrac{1}{|q|^2}\bar{q}$. □

If q_1 and q_2 are two quaternions, then $\bar{q}_2\bar{q}_1 = \overline{q_1 q_2}$; quaternion conjugation reverses the order of multiplication. If we now consider the expression $q_1 q_2 \bar{q}_2 \bar{q}_1$ it is equal on the one hand to $|q_1 q_2|^2$, and on the other it is equal to $q_1 |q_2|^2 \bar{q}_1 = q_1 \bar{q}_1 |q_2|^2 = |q_1|^2 |q_2|^2$ using the fact that the real number $|q_2|^2$ commutes with $\bar{q}_1$. So we find that

$$|q_1 q_2| = |q_1||q_2|$$

as for complex numbers. A quaternion q with $|q| = 1$ is called a *unit quaternion*; any unit quaternion is of the form $q = \cos\varphi + \sin\varphi\mathbf{n}$ for some angle φ and unit vector **n**. This can be proved by applying A.1.7 to the two-dimensional subspace of the quaternions spanned by $\Re q$ and $\Im q$.

Hamilton discovered a close link between his quaternions and *rotations*. We will now study this.

(8.4.5) **Definition:** *Let q be a unit quaternion. The* conjugation map C_q *is the linear transformation* **H** → **H** *defined by*

$$C_q(x) = qx\bar{q}$$

*for all quaternions $x \in$ **H**.*

The transformation C_q is an isometry: $|C_q(x)| = |x|$. To prove this, write

$$|C_q(x)|^2 = qx\bar{q}\overline{qx\bar{q}} = qx\bar{q}\bar{q}x\bar{q} = q|x|^2\bar{q} = |x|^2$$

using the fact that $|q| = 1$. Moreover, $C_q(x) = x$ if x is real (because then x commutes with q). The interesting question, then, is what C_q does to purely imaginary quaternions, which we recall can be identified with vectors in $\vec{S}$.

(8.4.6)   **Proposition:** *Let q and C_q be as above. Then*

(i) *If x is a purely imaginary quaternion, then $C_q(x)$ is also a purely imaginary quaternion.*

(ii) *The transformation C_q on the space $\vec{S}$ of purely imaginary quaternions represents a rotation. In fact, if $q = \cos\varphi + \sin\varphi\mathbf{n}$, then C_q represents a rotation through angle 2φ with axis $\mathbf{n}$.*

(iii) *Any rotation can be represented by a conjugation map C_q, and two conjugation maps C_{q_1} and C_{q_2} represent the same rotation if and only if $q_1 = \pm q_2$.*

(iv) *The rotation represented by $C_{q_1 q_2}$ is the composite of the rotations represented by C_{q_1} and C_{q_2}.*

Proof:

(i) This follows from the identity $C_q(\bar{x}) = \overline{C_q(x)}$.

(ii) Let $q = \cos\varphi + \sin\varphi\mathbf{n}$, and choose an orthonormal basis $\mathbf{i}, \mathbf{j}, \mathbf{k}$ for $\vec{S}$ with $\mathbf{i} = \mathbf{n}$. Now compute

$$\begin{aligned}
C_q(\mathbf{i}) &= (\cos\varphi + \sin\varphi\mathbf{i})\mathbf{i}(\cos\varphi - \sin\varphi\mathbf{i}) \\
&= (\cos\varphi\mathbf{i} - \sin\varphi)(\cos\varphi - \sin\varphi\mathbf{i}) &&= \mathbf{i} \\
C_q(\mathbf{j}) &= (\cos\varphi + \sin\varphi\mathbf{i})\mathbf{j}(\cos\varphi - \sin\varphi\mathbf{i}) \\
&= (\cos\varphi\mathbf{j} + \sin\varphi\mathbf{k})(\cos\varphi - \sin\varphi\mathbf{i}) &&= \cos 2\varphi\mathbf{j} + \sin 2\varphi\mathbf{k} \\
C_q(\mathbf{k}) &= (\cos\varphi + \sin\varphi\mathbf{i})\mathbf{k}(\cos\varphi - \sin\varphi\mathbf{i}) \\
&= (\cos\varphi\mathbf{k} - \sin\varphi\mathbf{j})(\cos\varphi - \sin\varphi\mathbf{i}) &&= \cos 2\varphi\mathbf{k} - \sin 2\varphi\mathbf{j}.
\end{aligned}$$

This represents a rotation through 2φ about $\mathbf{i}$ (see 8.3.3).

(iii) Since any rotation has an axis $\mathbf{n}$ and an angle θ, any rotation can be represented by a suitable C_q, $q = \cos\frac{1}{2}\theta + \sin\frac{1}{2}\theta\mathbf{n}$. The sign ambiguity comes from the half angles; because θ is defined only up to a multiple of 2π, $\frac{1}{2}\theta$ is defined only up to a multiple of π, and so $-q$ and q represent the same rotation.

(iv) We calculate

$$C_{q_1 q_2}(x) = q_1 q_2 x \overline{q_1 q_2} = q_1 q_2 x \bar{q}_2 \bar{q}_1 = q_1 C_{q_2}(x) \bar{q}_1 = C_{q_1}(C_{q_2}(x)). \quad \square$$

This result can be expressed in fancier language as follows. The unit quaternions form a *group* under multiplication. For rather complicated historical reasons, this group is denoted $Sp(1)$. The rotations also form a group, denoted $SO(3)$ (for less complicated reasons; S stands for 'special' (determinant $+1$), O stands for 'orthogonal', and 3 gives the dimension). The proposition says that the map $q \mapsto C_q$ defines a *homomorphism* of groups $Sp(1) \to SO(3)$, and that this homomorphism is *surjective* (onto) and has *kernel* the two-element subgroup $\{\pm 1\}$ of $Sp(1)$.

The ambiguity of sign (the kernel of the homomorphism described above) may seem like a blemish, but in fact it is telling us something rather interesting. To see what, try the following experiment.

Take a bowl of soup[2] and hold it in your right hand. Now rotate the bowl continuously in an anticlockwise direction until you have rotated it through angle 2π and it has returned more or less to where it started. You should notice that your arm has become somewhat twisted. Although the bowl has returned to its original position after the rotation, it is in a different 'orientation–entanglement relationship[3]' with its surroundings.

What has this got to do with quaternions? A rotation through angle θ about the $\mathbf{k}$-axis is represented by the unit quaternion $\cos \frac{1}{2}\theta + \sin \frac{1}{2}\theta \mathbf{k}$. As θ moves continuously from 0 to 2π, the corresponding unit quaternion moves continuously from $+1$ to -1. The final position of the soup bowl is therefore described by the quaternion -1 rather than $+1$, and this reflects the fact that its 'orientation–entanglement relationship' has changed.

This analysis has a surprising consequence. Suppose you continue rotating the bowl in the same direction until it has been moved through a total angle of 4π. The corresponding quaternion has now changed back to $+1$ again, and this suggests that the original orientation–entanglement relationship has been restored. At first sight this does not seem plausible; surely two rotations will twist your arm up twice as badly as one rotation? But in fact you can make the second rotation in the same direction and return your arm to its original untwisted state. The second rotation is made by raising your arm above your head, but you always keep turning the bowl in the same direction. Try it and see!

To explain precisely what is going on here requires the language of algebraic topology. But without going into details, we can say that we have used the quaternions to examine the structure of *paths* in $SO(3)$, and we have shown

[2]If you are nervous, just take an empty bowl.
[3]This phrase comes from Misner *et al.* [30].

that even if a path begins and ends at the same point it can be *topologically nontrivial* in a certain sense. An analogous situation is that of a path which goes 'round the hole' of a torus; even though it begins and ends at the same point it is nontrivial because it cannot be shrunk to a point while remaining on the torus.

Does the orientation–entanglement relationship have physical significance? It is tempting to answer, 'Of course not!'. Surely rotating a physical object through 360 degrees cannot change it it in any way! But in fact the answer given by quantum mechanics is 'Yes'. Quantum mechanics divides fundamental particles into two classes: *bosons* and *fermions*. Bosons are unaffected by the orientation–entanglement relationship, but fermions are affected; for instance, electrons, which are examples of fermions, are thought of as existing in two states, 'spin up' and 'spin down', corresponding to the two possible orientation–entanglement relationships with their surroundings. Moreover, only fermions satisfy the *Pauli exclusion principle*, which prohibits two particles from being in exactly the same state. It is the Pauli principle which is responsible for the gradual filling of electron 'shells' which gives rise to the periodic table, and thereby to all of chemistry. So that ± 1 is quite important after all.

8.5 Exercises

1. Three points R, S, and T have position vectors $\mathbf{r}$, $\mathbf{s}$, and $\mathbf{t}$ respectively. Show that R, S, and T are collinear if and only if

$$\mathbf{r} \wedge \mathbf{s} + \mathbf{s} \wedge \mathbf{t} + \mathbf{t} \wedge \mathbf{r} = \mathbf{0}.$$

2. Prove the identity

$$[\mathbf{a}, \mathbf{b}, \mathbf{c}]^2 = \begin{vmatrix} \mathbf{a} \cdot \mathbf{a} & \mathbf{a} \cdot \mathbf{b} & \mathbf{a} \cdot \mathbf{c} \\ \mathbf{b} \cdot \mathbf{a} & \mathbf{b} \cdot \mathbf{b} & \mathbf{b} \cdot \mathbf{c} \\ \mathbf{c} \cdot \mathbf{a} & \mathbf{c} \cdot \mathbf{b} & \mathbf{c} \cdot \mathbf{c} \end{vmatrix}.$$

3. Prove that the vector product satisfies the *Jacobi identity*

$$\mathbf{a} \wedge (\mathbf{b} \wedge \mathbf{c}) + \mathbf{b} \wedge (\mathbf{c} \wedge \mathbf{a}) + \mathbf{c} \wedge (\mathbf{a} \wedge \mathbf{b}) = \mathbf{0}.$$

4. Let T be an isometry of an oriented Euclidean space. Show that for any two vectors $\mathbf{u}$ and $\mathbf{v}$,

$$\vec{T}(\mathbf{u} \wedge \mathbf{v}) = \pm \vec{T}(\mathbf{u}) \wedge \vec{T}(\mathbf{v}),$$

where the sign is $+$ if the isometry T is direct, and $-$ if it is indirect.

5. Let $\mathbf{a}, \mathbf{b}$, and $\mathbf{c}$ be three linearly independent vectors. Prove that the vector equations

$$\mathbf{v}\cdot\mathbf{a}=\alpha, \quad \mathbf{v}\cdot\mathbf{b}=\beta, \quad \mathbf{v}\cdot\mathbf{c}=\gamma$$

have the unique solution

$$\mathbf{v}=\frac{\gamma(\mathbf{a}\wedge\mathbf{b})+\alpha(\mathbf{b}\wedge\mathbf{c})+\beta(\mathbf{c}\wedge\mathbf{a})}{[\mathbf{a},\mathbf{b},\mathbf{c}]}.$$

6. Prove that the three vectors $\mathbf{a}, \mathbf{b}$, and $\mathbf{c}$ are linearly independent if and only if the three vectors $\mathbf{a}\wedge\mathbf{b}, \mathbf{b}\wedge\mathbf{c}$, and $\mathbf{c}\wedge\mathbf{a}$ are linearly independent.

7. Prove that for any six vectors $\mathbf{u},\mathbf{v},\mathbf{w},\mathbf{u}',\mathbf{v}',\mathbf{w}'$ in three-dimensional oriented Euclidean space,

$$[\mathbf{u}\wedge\mathbf{u}',\mathbf{v}\wedge\mathbf{v}',\mathbf{w}\wedge\mathbf{w}']=[\mathbf{w},\mathbf{u},\mathbf{u}'][\mathbf{w}',\mathbf{v},\mathbf{v}']-[\mathbf{w}',\mathbf{u},\mathbf{u}'][\mathbf{w},\mathbf{v},\mathbf{v}'].$$

Deduce that if *both* $\mathbf{u}\wedge\mathbf{v}'$, $\mathbf{v}\wedge\mathbf{w}'$, and $\mathbf{w}\wedge\mathbf{u}'$ are linearly dependent *and* $\mathbf{u}'\wedge\mathbf{v}$, $\mathbf{v}'\wedge\mathbf{w}$, and $\mathbf{w}'\wedge\mathbf{u}$ are linearly dependent, then $\mathbf{u}\wedge\mathbf{u}'$, $\mathbf{v}\wedge\mathbf{v}'$, and $\mathbf{w}\wedge\mathbf{w}'$ are linearly dependent.

8. Prove that there is always a unique point equidistant from four given noncoplanar points in three-dimensional space. Illustrate your answer by finding the centre and radius of the unique sphere passing through the four points $(1,-3,3)$, $(1,6,6)$, $(6,2,3)$, and $(5,5,3)$.

9. Two lines $\mathcal{L}_1$ and $\mathcal{L}_2$ in three-dimensional space have equations $\mathbf{r}\wedge\mathbf{a}_1=\mathbf{b}_1$ and $\mathbf{r}\wedge\mathbf{a}_2=\mathbf{b}_2$. The two lines are not parallel. Show that the least distance between them is equal to

$$\pm\frac{\mathbf{b}_1\cdot\mathbf{a}_2+\mathbf{a}_1\cdot\mathbf{b}_2}{|\mathbf{a}_1\wedge\mathbf{a}_2|}.$$

10. Two skew lines $\mathcal{L}_1$ and $\mathcal{L}_2$ in three-dimensional space have equations $\mathbf{r}\wedge\mathbf{a}_1=\mathbf{b}_1$ and $\mathbf{r}\wedge\mathbf{a}_2=\mathbf{b}_2$. A third line $\mathbf{r}\wedge\mathbf{a}=\mathbf{b}$ meets $\mathcal{L}_1$ and $\mathcal{L}_2$ and passes through the point with position vector $\mathbf{r}_0$. Find $\mathbf{a}$.

11. Find the axis and angle of a rotation which maps the three points $(0,0,0)$, $(1,1,0)$, and $(2,0,1)$ to $(0,0,0)$, $(1,0,1)$, and $(0,-1,2)$ respectively.

12. Explain why the composite of two rotations of a three-dimensional oriented Euclidean space is another rotation. Illustrate your answer by finding the axis and the angle of the rotation

$$R_{\mathbf{i},\frac{\pi}{2}}\circ R_{\mathbf{j},\frac{\pi}{2}}$$

where $R_{\mathbf{a},\theta}$ denotes the rotation with axis $\mathbf{a}$ and angle θ.

13. One of the two orthogonal matrices below represents a rotation and one represents a reflection. Find out which is which, and find the axis and angle of the rotation and the plane of the reflection.

$$\frac{1}{25}\begin{pmatrix} 20 & 15 & 0 \\ -12 & 16 & 15 \\ 9 & -12 & 20 \end{pmatrix}, \quad \frac{1}{25}\begin{pmatrix} -7 & 0 & -24 \\ 0 & 25 & 0 \\ -24 & 0 & 7 \end{pmatrix}.$$

14. Let T be a rotation (fixing the origin). Let $\mathbf{n}$ be a unit vector along the axis of T and let $\mathbf{v}$ be any vector not a multiple of $\mathbf{n}$. Show that if the scalar triple product $[\mathbf{n}, \mathbf{v}, T\mathbf{v}]$ is positive then the angle of rotation lies between 0 and π, whereas if the scalar triple product is negative then the angle of rotation lies between $-\pi$ and 0.

15. Let M be a 3×3 orthogonal matrix with determinant -1. Show that M represents a reflection if and only if $\operatorname{tr} M = +1$.

16. Show that any 3×3 real matrix has at least one real eigenvalue. Show also that any real eigenvalue of an orthogonal matrix must be ± 1. What kinds of transformations can be represented by orthogonal matrices all of whose eigenvalues are real?

17. A *real unital division algebra* is a real vector space A equipped with a bilinear 'multiplication' map $A \times A \rightarrow A$ and a unit element $e \in A$ such that $ex = xe = x$ for all $x \in A$ and such that for each nonzero $x \in A$ there is an 'inverse' $y \in A$ with $xy = yx = e$. Thus **R**, **C**, and **H** are examples of real division algebras.

Hamilton, in his search for a method of multiplying 'triplets', was trying to find a three-dimensional division algebra. Show that his search was hopeless, by proving that *there is no real unital division algebra of odd dimension* > 1.

18. Check that quaternion multiplication and matrix multiplication coincide under the correspondence of Proposition 8.4.2.

19. Use quaternions to prove the identity

$$
\begin{aligned}
(w_1^2 + x_1^2 + y_1^2 + z_1^2)(w_2^2 + x_2^2 + y_2^2 + z_2^2) = \; & (w_1 w_2 - x_1 x_2 - y_1 y_2 - z_1 z_2)^2 \\
& + (w_1 x_2 + x_1 w_2 + y_1 z_2 - z_1 y_2)^2 \\
& + (w_1 y_2 - x_1 z_2 + y_1 w_2 + z_1 x_2)^2 \\
& + (w_1 z_2 + x_1 y_2 - y_1 x_2 + z_1 w_2)^2.
\end{aligned}
$$

There are similar identities for one square (trivial!), two squares, and eight squares. It is natural to guess that such an identity will exist for 2^n squares; but in fact it has been proved that such an identity can exist only for one, two, four, or eight squares and for no other number.

20. Let x_θ denote the unit quaternion $\cos\theta + \sin\theta\,\mathbf{i}$ and let y_θ denote the unit quaternion $\cos\theta + \sin\theta\,\mathbf{j}$. Show that any unit quaternion can be represented as a product $x_\alpha y_\beta x_\gamma$ for appropriate α, β, and γ. Deduce that any rotation can be represented as a product of three rotations about the $\mathbf{i}$ and $\mathbf{j}$ axes, in the form

$$
R_{\mathbf{i},\varphi} R_{\mathbf{j},\chi} R_{\mathbf{i},\psi}.
$$

(The angles φ, χ, and ψ are called *Euler angles*. This representation of rotations is very useful in rigid body mechanics.)

9

Area and volume

9.1 Polygonal regions

The concept of *area* is sometimes introduced by talking about the 'amount of paint' required to cover a certain figure in the plane. Thus a square of side 2 feet would need four times as much paint to cover it as a square of side 1 foot, a circular disc of radius 1 foot would need a little over three times as much, and so on. This simple idea of the 'amount of paint' suggests to us several properties that area should have, which we could use as axioms; for example, the property that the area of a figure made up of two separate pieces is the sum of the areas of the individual pieces. However, if our theory is to be precise, we need to start one stage further back, and to say what kinds of 'figures' we intend to find the area of.

In school geometry we apply the concept of area to simple figures like squares, triangles, and circles. These are all examples of *subsets* of a Euclidean plane $\mathcal{P}$, and our first thought might well be that an area should be assigned to every subset of $\mathcal{P}$. But this approach runs into difficulties. The problem is that a general subset of $\mathcal{P}$ might be very unlike the simple ones from which we build up our intuition about areas. Consider for example the following curious subset of the plane $\mathbf{R}^2$:

(9.1.1) $K = \{(x, y) : 0 \leq x, y \leq 1 \text{ and } x \text{ and } y \text{ are } \textit{rational} \text{ numbers}\}.$

(Remember that a *rational number* is just a fraction p/q, where p and q are integers.) Now for any rational number, there are irrational numbers arbitrarily close to it; and for any irrational number, there are rational numbers arbitrarily close to it. How then am I to work out how much paint is needed to cover the set K? If I put a spot of paint on each point of K, then however small the spots are they will overlap and fill up the whole unit square. If I decide to be more careful and only to put down paint completely inside K, then I will leave everything unpainted[1]. So it looks as though I do not have a clear idea what should be the area of so strange a set as K.

To avoid this problem, it is necessary to abandon the assumption that *every* subset of the plane can be given a definite area. The subsets to which the area axioms will apply are *bounded polygonal regions*, which are defined in terms of *half-planes*.

(9.1.2)   **Definition:** *A closed half-plane H in a Euclidean plane $\mathcal{P}$ is a subset represented in a Cartesian coordinate system by*

$$H = \{(x,y) : ax + by + c \geq 0\}$$

where a, b, and c are constants, and not both of a and b are zero.
    An open half-plane *is a subset represented in a Cartesian coordinate system by*

$$H = \{(x,y) : ax + by + c > 0\}$$

where a, b, and c are as above.

A line in the plane $\mathcal{P}$ separates it into two open half-planes; if the line has equation $ax+by+c = 0$, then the two half-planes are given by $ax+by+c > 0$ and $-ax - by - c > 0$. Closed half-planes can be obtained by taking the union of the open half-planes with the bounding line.

(9.1.3)   **Definition:** *A polygonal region in a Euclidean plane $\mathcal{P}$ is a subset P of $\mathcal{P}$ which can be obtained from finitely many half-planes (closed or open) by the operations of union ($\cup$) and intersection ($\cap$).*

For example, if $H_1, \ldots, H_4$ are half-planes, then the sets

$$(H_1 \cap H_2) \cup (H_3 \cap H_4), \quad H_1 \cap H_3 \cap H_4, \quad H_4 \cup (H_2 \cap (H_1 \cup H_3))$$

are all polygonal regions. It is clear from the definition that the union or intersection of two polygonal regions is again a polygonal region.

If X and Y are two subsets of the plane, their *set-theoretic difference $X \setminus Y$* is the intersection of X with the complement of Y, that is

$$X \setminus Y = \{x \in X : x \notin Y\}.$$

Using the fact that the complement of an open half-plane is a closed half-plane and vice versa, it is easy to give a proof that the set-theoretic difference of two polygonal regions is again a polygonal region.

---

[1] In more formal language, which will be introduced later, the *outer content* of K is 1, while the *inner content* of K is 0.

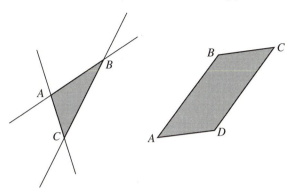

Fig. 9.1. Interior of a triangle, and of a parallelogram.

(9.1.4)   **Example:** Let ABC be a proper triangle in the plane $\mathcal{P}$. The line $\overleftrightarrow{AB}$ then divides $\mathcal{P}$ into two open half-planes, one of which contains the point C; call this half-plane H_C. Similarly let H_A be an open half-plane containing A and bounded by the line $\overleftrightarrow{BC}$ and let H_B be an open half-plane containing B and bounded by the line $\overleftrightarrow{AC}$. The intersection $H_A \cap H_B \cap H_C$ is then a polygonal region called the *interior of the triangle ABC* (see Figure 9.1).

It is a little pedantic to keep up the distinction between the triangle ABC and the polygonal region which is its interior, and we will not worry too much about this. In particular, we will refer to the 'area of the triangle ABC' rather than the strictly accurate 'area of the interior of triangle ABC'.

(9.1.5)   **Example:** Let $ABCD$ be a parallelogram. There is then an open half-plane H_{AB} bounded by $\overleftrightarrow{CD}$ and containing A and B, and there are similarly defined open half-planes H_{BC}, H_{CD}, and H_{DA}. The intersection of these four half-planes, $H_{AB} \cap H_{BC} \cap H_{CD} \cap H_{DA}$, is called the *interior of the parallelogram ABCD*. It is tempting to say that the interior of the parallelogram $ABCD$ is the union of the interiors of the triangles ABC and CDA; this is not strictly accurate, because the diagonal AC does not belong to the interior of either triangle. We will see, however, that the inaccuracy makes no difference from the point of view of area.

The area of a rectangle will be of particular interest to us. A *rectangle* is a parallelogram whose adjoining sides are perpendicular. Its interior may also be defined as a set given by $\{(x, y) : a < x < b, \; c < y < d\}$ in an appropriate Cartesian coordinate system.

(9.1.6)   **Definition:** *Let S be a subset of a Euclidean plane $\mathcal{P}$. The* diameter *of S is the least upper bound of the distances between points of S; in symbols,*

$$\mathrm{Diam}(S) = \sup\{|PQ| : P, Q \in S\}.$$

The diameter is a positive number, possibly $+\infty$. If the diameter is finite the set S is called *bounded*. It is clear that the union of two bounded sets is bounded, and that the intersection of a bounded set with any set is bounded.

(9.1.7)   **Proposition:** *The interior of a triangle, or of a parallelogram, is a bounded set.*

Proof: It is enough to give the proof for a parallelogram, since any triangle is contained in a parallelogram. Choose one vertex of the parallelogram as an origin, and let $\mathbf{a}$ and $\mathbf{b}$ be vectors along the sides, so that the interior consists of all points with position vectors

$$\mathbf{r} = \lambda\mathbf{a} + \mu\mathbf{b} : 0 < \lambda, \mu < 1.$$

Then
$$|\mathbf{r}|^2 \le \lambda^2|\mathbf{a}|^2 + 2\lambda\mu|\mathbf{a}||\mathbf{b}| + \mu^2|\mathbf{b}|^2 \le (|\mathbf{a}| + |\mathbf{b}|)^2$$

by the Cauchy–Schwarz inequality (4.3.1). So, if $\mathbf{r}_1$ and $\mathbf{r}_2$ are the position vectors of two points in the interior, then by the triangle inequality (4.3.2), $|\mathbf{r}_1 - \mathbf{r}_2| \le 2(|\mathbf{a}| + |\mathbf{b}|)$. So the diameter is finite. $\square$

9.2 Axioms for area

Formally speaking, *area* is a new undefined term in our axiomatic system for geometry; the *area* $\mathcal{A}(P)$ of a bounded polygonal region P is a non-negative real number. The properties of area are given by the three axioms below.

AXIOM 13 (AREA INVARIANCE AXIOM): Let P be a bounded polygonal region. Let T be an isometry of the plane $\mathcal{P}$, mapping P to a new polygonal region $T(P)$. Then $\mathcal{A}(P) = \mathcal{A}(T(P))$.

AXIOM 14 (AREA ADDITION AXIOM): Let P_1 and P_2 be bounded polygonal regions that are *disjoint* (that is, $P_1 \cap P_2 = \emptyset$). Then

$$\mathcal{A}(P_1 \cup P_2) = \mathcal{A}(P_1) + \mathcal{A}(P_2).$$

AXIOM 15 (AREA NORMALIZATION AXIOM): The area of a rectangle of sides a and b is ab.

From the addition axiom we can immediately deduce the *monotonicity property*: if P_1 and P_2 are bounded polygonal regions, and $P_1 \subseteq P_2$, then $\mathcal{A}(P_1) \leq \mathcal{A}(P_2)$. For since $P_1 \subseteq P_2$, we can write P_2 as the disjoint union of P_1 and the difference set $P_2 \setminus P_1$. From the addition axiom, we deduce that

$$\mathcal{A}(P_2) = \mathcal{A}(P_1) + \mathcal{A}(P_2 \setminus P_1) \geq \mathcal{A}(P_1)$$

using the fact that the area of $P_2 \setminus P_1$ is non-negative.

We can also obtain a more general form of the addition axiom that applies when the sets involved are not necessarily disjoint.

(9.2.1)   **Proposition:** *Let P_1 and P_2 be bounded polygonal regions. Then*

$$\mathcal{A}(P_1 \cup P_2) = \mathcal{A}(P_1) + \mathcal{A}(P_2) - \mathcal{A}(P_1 \cap P_2).$$

Proof: Write $P_1 \cup P_2$ as the disjoint union of P_1 and $P_2 \setminus P_1$ to get

$$\mathcal{A}(P_1 \cup P_2) = \mathcal{A}(P_1) + \mathcal{A}(P_2 \setminus P_1).$$

Write P_2 as the disjoint union of $P_1 \cap P_2$ and $P_2 \setminus P_1$ to get

$$\mathcal{A}(P_2) = \mathcal{A}(P_1 \cap P_2) + \mathcal{A}(P_2 \setminus P_1).$$

Combine these two equations to obtain the result. □

We frequently will want to apply the addition axiom to a polygonal region made up of simpler regions such as triangles. When we put triangles together there tend to be some line segments left over at the edge; the next result assures us that this doesn't matter from the point of view of area.

(9.2.2)   **Lemma:** *A (bounded) line segment has zero area.*

Proof: A line segment of length l can be enclosed in a rectangle which can be as thin as we like; for instance, in a rectangle of sides $l + \varepsilon$ and ε, where ε can be any small positive quantity. So the area of the line segment must be less than or equal to the area of the rectangle, which is $\varepsilon(l + \varepsilon)$. This can be made as small as we like by making ε small, so the area must be zero. □

(9.2.3)   **Proposition:** *The area of a right-angled triangle ABC is $\frac{1}{2}|AB||BC|$.*

Proof: Let X be the midpoint of the hypotenuse AC, and let CDA be the triangle obtained from ABC by central inversion in X. Then $\overrightarrow{AB} = \overrightarrow{DC}$ and $\overrightarrow{AD} = \overrightarrow{BC}$, so $ABCD$ is a parallelogram; in fact it is a rectangle, because AB is perpendicular to BC. (See Figure 9.2.)

Now the interior of $ABCD$ is made up of three disjoint pieces: the interior of ABC, the interior of CDA, and the line segment AC. The line segment has

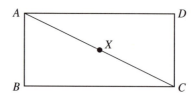

Fig. 9.2. Area of a right-angled triangle.

zero area by 9.2.2, and the two triangles have the same area because they are congruent by central inversion. So by the addition axiom,

$$A(ABC) = \tfrac{1}{2}A(ABCD) = \tfrac{1}{2}|AB||BC|. \qquad \square$$

We can calculate the area of a general triangle from that of a right-angled triangle:

(9.2.4)   **Proposition:** *The area of a triangle ABC is equal to $\tfrac{1}{2}|AB||CX|$, where CX is the* altitude *(3.3.5) from C to $\overset{\frown}{AB}$.*

Remark: This proposition really gives us three different ways of calculating the area of a triangle because any one of the three vertices may be chosen as the one from which we measure the altitude. The proposition is often stated in words as 'the area of a triangle is equal to half the base times the altitude'.

Proof: There are two cases to consider (Figure 9.3); either X is between A and B, or it is not.

If X is between A and B, then the right-angled triangles ACX and CXB are disjoint. So

$$A(ABC) = A(ACX) + A(CXB).$$

Use the previous result to work out the areas of the right-angled triangles ACX and CXB, getting

$$A(ABC) = \tfrac{1}{2}|AX||CX| + \tfrac{1}{2}|XB||CX| = \tfrac{1}{2}|AB||CX|$$

because $|AB| = |AX| + |XB|$.

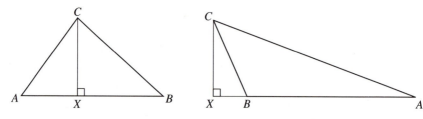

Fig. 9.3. The area of a triangle.

If X is not between A and B, then one of A and B is between the other one and X. Assume without loss of generality that B is between A and X. Then the triangles ABC and BCX are disjoint, and so

$$\mathcal{A}(ABC) = \mathcal{A}(ACX) - \mathcal{A}(BCX) = \tfrac{1}{2}|AX||CX| - \tfrac{1}{2}|BX||CX| = \tfrac{1}{2}|AB||CX|$$

because now $|AB| = |AX| - |BX|$. □

(9.2.5)   **Corollary:** *The area of a parallelogram $ABCD$ is $|AB||CX|$, where CX is the altitude from C to $\overleftrightarrow{AB}$.*

               Proof: This is clear on dividing up the parallelogram into two triangles which are congruent by central inversion in the midpoint M. □

The formula for the area of a triangle can be reinterpreted in several different ways. One interpretation makes use of an orientation for the plane $\mathcal{P}$. Recall that an *orientation* for $\mathcal{P}$ is an operation $\mathbf{u} \mapsto \mathbf{u}^{\perp}$ which rotates a vector $\mathbf{u}$ through $\frac{\pi}{2}$ radians.

(9.2.6)   **Proposition:** *Let A, B, and C be points in an oriented Euclidean plane $\mathcal{P}$, with $\overrightarrow{AB} = \mathbf{u}$ and $\overrightarrow{AC} = \mathbf{v}$. Then*

$$\mathcal{A}(ABC) = \tfrac{1}{2}|\mathbf{u}^{\perp} \cdot \mathbf{v}|,$$

and if D is the unique fourth point that makes $ABCD$ a parallelogram, then

$$\mathcal{A}(ABCD) = |\mathbf{u}^{\perp} \cdot \mathbf{v}|.$$

Proof: We have already seen that the area of a parallelogram is twice the area of the corresponding triangle, so we need only prove the formula for triangles. To do this, introduce a Cartesian coordinate system with origin A and x-axis along the line AB. Then B will have coordinates $(x_2, 0)$ and C will have coordinates (x_3, y_3) (see Figure 9.4). The altitude $|CX|$ is equal to $|y_3|$, so that the area of the triangle is $\frac{1}{2}|x_2 y_3|$. On the other hand, the dot product $\mathbf{u}^{\perp} \cdot \mathbf{v}$ can be worked out as a determinant (6.2.5),

$$\mathbf{u}^{\perp} \cdot \mathbf{v} = \begin{vmatrix} x_2 & x_3 \\ 0 & y_3 \end{vmatrix} = x_2 y_3.$$

So $\mathcal{A}(ABC) = \frac{1}{2}|\mathbf{u}^{\perp} \cdot \mathbf{v}|$, as required. □

Remark: If $\mathcal{P}$ is a plane in a three-dimensional oriented Euclidean space $\mathcal{S}$, and $\mathbf{n}$ is a unit normal vector to $\mathcal{P}$, then an orientation can be defined on $\mathcal{P}$ by $\mathbf{u}^{\perp} = \mathbf{n} \wedge \mathbf{u}$ (see 8.2.4). So the area formula becomes

(9.2.7) $\mathcal{A}(ABC) = \tfrac{1}{2}\mathcal{A}(ABCD) = \tfrac{1}{2}|[\mathbf{n}, \mathbf{u}, \mathbf{v}]| = \tfrac{1}{2}|\mathbf{u} \wedge \mathbf{v}|.$

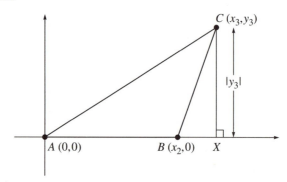

Fig. 9.4. Vector formula for the area of a triangle.

Suppose now that the three points A, B, and C have Cartesian coordinates (x_1, y_1), (x_2, y_2), and (x_3, y_3) respectively. Then the area formula gives

(9.2.8)
$$A(ABC) = \pm\frac{1}{2}\begin{vmatrix} x_2 - x_1 & x_3 - x_1 \\ y_2 - y_1 & y_3 - y_1 \end{vmatrix} = \pm\frac{1}{2}\begin{vmatrix} 1 & 1 & 1 \\ x_1 & x_2 & x_3 \\ y_1 & y_2 & y_3 \end{vmatrix}.$$

We have already seen (5.2.1) that this determinant vanishes if and only if the three points are collinear. The sign ($+$ or $-$) can be explained in terms of orientation (see Exercise 9.7.6).

Yet another expression for the area of a triangle is in terms of the lengths of its sides. If we use the standard notation a, b, c for the sides and A, B, C for the angles of the triangle ABC, then the altitude $|CX|$ is equal to $b \sin A$, by trigonometry in the right-angled triangle CXA. Using the formula for $\sin A$ in 4.4.2, this gives us *Héron's formula:*

(9.2.9)
$$A(ABC) = \sqrt{s(s-a)(s-b)(s-c)},$$

where $s = \frac{1}{2}(a+b+c)$ is the semiperimeter.

(9.2.10)   **Example:** Suppose that you have a loop of string which you want to stretch out into a triangle enclosing as large an area as possible. A few experiments will quickly convince you that long thin triangles are not a good idea; you want as short and fat a triangle as possible, and in fact an equilateral one seems to be best. This is a simple example of an *isoperimetric* problem, to enclose the greatest area within a specified perimeter. We can use Héron's formula to check that the solution to this problem is indeed an equilateral triangle.

We will need the celebrated *arithmetic–geometric mean inequality*, which states that for any n positive numbers $x_1, \ldots, x_n$ their geometric mean is always less than or equal to their arithmetic mean[2], that is $(x_1 \ldots x_n)^{1/n} \le \frac{1}{n}(x_1 + \cdots + x_n)$; equality occurs only if $x_1 = \cdots = x_n$. Applying this inequality to the three numbers $(s-a)$, $(s-b)$, and $(s-c)$, we find that

$$(s-a)(s-b)(s-c) \le \frac{1}{27}\left(3s - a - b - c\right)^3 = \frac{s^3}{27}.$$

(Remember that $a + b + c = 2s$.) Therefore by Héron's formula,

$$\mathcal{A}(ABC) \le \frac{s^2}{3\sqrt{3}},$$

and equality occurs if and only if $s - a = s - b = s - c$, which is to say that $a = b = c$ and the triangle is equilateral.

9.3 The method of exhaustion

Although we have seen that there are difficulties in the way of defining area for *all* subsets of the plane, it is too restrictive to confine our attention to *polygonal regions*. We all 'know' that the area of a circle of radius r is πr^2, and a system of geometry that stopped short of this result would rightly be regarded as defective. But a circle is not a polygonal region, so our axioms do not guarantee that it has an area; and even assuming that it does, how can we calculate the area?

A method for answering such questions was developed by the Greeks Eudoxus and Archimedes. Although little is known about the life of Eudoxus, he is credited with two of the most profound ideas in Euclid's *Elements*: the *theory of proportion*, which anticipated modern constructions of the real numbers, and the *method of exhaustion* for calculating areas. By contrast, we have a lot of information on the life of Archimedes, who put Eudoxus' method to work in a series of remarkable calculations of area and volume. Archimedes was born in Syracuse in Sicily around 287 BC, and worked there on mathematics, physics, and mechanics. Legend tells of him hitting on the laws of hydrostatics while floating in his bath, and running naked through the streets shouting 'Eureka!' ('I have found it!'). It is significant, though, that his discovery was related to a practical problem: whether a crown given to Hieron, ruler of Syracuse, was indeed made of pure gold. Archimedes was famous for his inventions in Hieron's service, and when Syracuse was besieged by the Romans in 212 BC he built giant catapults for its defence. Nevertheless, the city eventually fell, and Archimedes was killed. According to tradition, his last words were 'Don't disturb my diagrams'.

---

[2]For a proof, see Binmore [5, page 23]; or try Exercise 9.7.4.

We can illustrate the method of exhaustion by considering the example of a circle. Suppose that the circle has unit radius. We can inscribe in it a regular hexagon made up of six disjoint equilateral triangles of side 1. The area of each such triangle is $3/(4\sqrt{3})$ by Héron's formula, so the area of the hexagon is $9/(2\sqrt{3}) \approx 2.59808$. The area of the circle should be greater than this. On the other hand, we can circumscribe a regular hexagon of side $2/\sqrt{3}$ around the unit circle, and (again by Héron's formula) the area of this hexagon is $6/\sqrt{3} \approx 3.46410$. The area of the circle should be less than this. Thus the area is sandwiched between 2.59808 and 3.46410. We would expect that by considering polygons with more and more sides, we could arrive at more and more accurate approximations to the area π of the circle. The method of exhaustion takes its name from this idea of 'exhausting' the circle by increasingly complicated polygonal regions.

To develop the method formally, we begin by defining two 'contents' or areas for a general subset of a plane, one obtained by 'exhaustion from inside' and one obtained by 'exhaustion from outside'. For a nice geometrical subset like a circle the two 'contents' will coincide, and we will then define their common value to be the area.

(9.3.1) **Definition:** *Let X be a bounded subset of a Euclidean plane. We define the* inner content $\underline{A}(X)$ *of X to be the least upper bound of the areas of all polygonal regions contained in X, that is*

$$\underline{A}(X) = \sup\{\mathcal{A}(P) : P \text{ polygonal, } P \subseteq X\}.$$

Similarly we define the outer content $\overline{A}(X)$ *of X to be the greatest lower bound of the areas of all bounded polygonal regions containing X, that is*

$$\overline{A}(X) = \inf\{\mathcal{A}(P) : P \text{ bounded polygonal, } X \subseteq P\}.$$

Thus $\underline{A}(X)$ is the smallest number that is greater than the areas of all 'inscribed polygons', and $\overline{A}(X)$ is the largest number that is less than the areas of all 'circumscribed polygons'.

Some properties of the inner and outer contents follow easily from this definition. They are invariant under isometries, because the areas of polygonal regions are invariant under isometries. They are also *monotonic*; if $X_1 \subseteq X_2$, then any polygonal region contained in X_1 is also contained in X_2, and any polygonal region containing X_2 also contains X_1, so $\underline{A}(X_1) \leq \underline{A}(X_2)$ and $\overline{A}(X_1) \leq \overline{A}(X_2)$. Finally, since any 'inscribed' polygonal region is contained in any 'circumscribed' polygonal region, $\underline{A}(X) \leq \overline{A}(X)$.

If X itself is a polygonal region, then $\underline{A}(X) = \mathcal{A}(X) = \overline{A}(X)$. To see this, notice that any inscribed polygonal region must have area less than or equal to $\mathcal{A}(X)$, by the monotonicity of area for polygonal regions; so $\underline{A}(X) \leq \mathcal{A}(X)$. On the other hand, X itself satisfies the definition of an inscribed polygonal

Fig. 9.5. A contented set.

region, so $\underline{\mathcal{A}}(X) \geq \mathcal{A}(X)$. Thus $\underline{\mathcal{A}}(X) = \mathcal{A}(X)$, and there is a similar proof for outer content.

In general, the two numbers $\underline{\mathcal{A}}(X)$ and $\overline{\mathcal{A}}(X)$ need not be equal. The set K defined in 9.1.1, for example, has outer content 1 (any circumscribed polygonal region must contain the unit square) and inner content 0 (no polygonal region at all can be inscribed in K). But if the inner and outer contents are equal, their common value is the only possible definition for the area of X.

(9.3.2)   **Definition:** *A subset X of a Euclidean plane will be called a* contented set[3] *if its inner and outer contents are equal. If this is so, the common value $\underline{\mathcal{A}}(X) = \overline{\mathcal{A}}(X)$ will just be called the* area *of X, and denoted by $\mathcal{A}(X)$.*

(9.3.3)   **Examples:** Any bounded polygonal region is contented, and the new definition of the area agrees with the old one.

A circular disc D of radius r is contented, and its area is equal to πr^2. To prove this we will calculate the areas I_n and C_n of inscribed and circumscribed regular polygons of n sides. The inscribed polygon is made up of n isosceles triangles each of which has base $2r\sin(\pi/n)$ and altitude $r\cos(\pi/n)$. The circumscribed polygon is made up of n isosceles triangles each of which has base $2r\tan(\pi/n)$ and altitude r. (See Figure 9.6.) Therefore

$$I_n = nr^2 \sin(\pi/n)\cos(\pi/n), \quad C_n = nr^2 \frac{\sin(\pi/n)}{\cos(\pi/n)}.$$

As $n \to \infty$, $n\sin(\pi/n) \to \pi$ and $\cos(\pi/n) \to 1$, so I_n and C_n both tend to πr^2. Therefore

$$\underline{\mathcal{A}}(D) \geq \sup I_n \geq \pi r^2, \quad \overline{\mathcal{A}}(D) \leq \inf C_n \leq \pi r^2.$$

Since $\underline{\mathcal{A}}(D) \leq \overline{\mathcal{A}}(D)$, they must both be equal to πr^2. If X is a contented set, and T is an isometry, then $T(X)$ is contented too and has the same area, because both inner and outer content are invariant under isometry.

---

[3]See Figure 9.5. The reader is warned that this terminology is not standard.

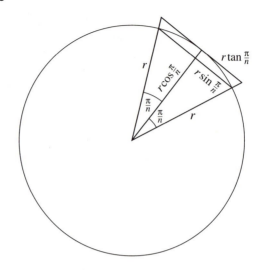

Fig. 9.6. Calculation of the area of a regular polygon.

(9.3.4) **Proposition:** *Let X_1 and X_2 be contented sets. Then $X_1 \cup X_2$ and $X_1 \cap X_2$ are contented also, and*

$$A(X_1 \cup X_2) = A(X_1) + A(X_2) - A(X_1 \cap X_2).$$

Proof: Let P_1 and P_2 be polygonal regions contained in X_1 and X_2 respectively. Then $P_1 \cup P_2 \subseteq X_1 \cup X_2$ and $P_1 \cap P_2 \subseteq X_1 \cap X_2$. Therefore

$$\underline{A}(X_1 \cup X_2) + \underline{A}(X_1 \cap X_2) \geq A(P_1 \cup P_2) + A(P_1 \cap P_2) = A(P_1) + A(P_2)$$

using the addition formula 9.2.1. Now take the supremum over all polygonal regions P_1 and P_2 contained in X_1 and X_2, to find that

$$\underline{A}(X_1 \cup X_2) + \underline{A}(X_1 \cap X_2) \geq \underline{A}(X_1) + \underline{A}(X_2).$$

By a similar argument for polygonal regions containing X_1 and X_2,

$$\overline{A}(X_1 \cup X_2) + \overline{A}(X_1 \cap X_2) \leq \overline{A}(X_1) + \overline{A}(X_2).$$

But the inner and outer contents of X_1 and X_2 are equal. So we may use the fact that the inner content is less than or equal to the outer content to form a chain of inequalities with $A(X_1) + A(X_2)$ at each end. The conclusion is that all the inequalities must be equalities, so $X_1 \cup X_2$ and $X_1 \cap X_2$ are contented, and their areas sum to $A(X_1) + A(X_2)$ as required. □

How can one *calculate* the areas defined by the method of exhaustion? The Greek pioneers had to tailor a specific sequence of exhausting regions to fit each new problem that they wanted to solve. Some of their constructions, particularly those relating to volume rather than area, were very difficult. Their approach, too, was suited more to rigorous verification of answers already guessed by other means than to calculation of areas and volumes from scratch. Both of these deficiencies were remedied in the seventeenth century with the development of the systematic version of the method of exhaustion that we know as the *integral calculus*.

Informally, one can say that the main idea of the integral calculus is to divide up a region (say the area under the graph of a function) into 'infinitely many infinitely thin rectangular strips'. One needs to use the concept of *limit* to give this phrase a rigorous sense. The idea, though, was already known to Archimedes. In 1906 the German scholar Heiberg discovered in the library of a monastery in Constantinople the text of a previously unknown work by Archimedes, explaining how he arrived at the results which he later rigorously proved by exhaustion. He used the idea of dividing up a region into infinitely thin strips, together with mechanical ideas about balance. Archimedes was generous enough to reveal his informal reasoning as well as his formal results. Not all mathematicians have followed his example!

Recall some basic ideas about integration, for which we refer to Binmore [5]. Let $[a, b]$ be an interval of real numbers, and let $f: [a, b] \to \mathbf{R}$ be a continuous function. A *partition* of $[a, b]$ is a finite sequence $a = c_0 < c_1 < \cdots < c_N = b$ of real numbers. Define

$$M_i = \sup\{f(x) : c_{i-1} \leq x \leq c_i\}, \quad m_i = \inf\{f(x) : c_{i-1} \leq x \leq c_i\}.$$

Then the *upper and lower Riemann sums* associated to the given partition are, respectively,

$$\sum_{i=1}^{N} M_i(c_i - c_{i-1}) \quad \text{and} \quad \sum_{i=1}^{N} m_i(c_i - c_{i-1}).$$

This is illustrated in Figure 9.7. It can be proved using the continuity of f that the least upper bound of all possible lower Riemann sums and the greatest lower bound of all possible upper Riemann sums are equal[4]. Their common value is called the *integral* of f over $[a, b]$ and denoted $\int_a^b f(x)\, dx$. Moreover, this integral can be evaluated by means of the *fundamental theorem of calculus*, which says that if F is a function whose derivative is f, then $\int_a^b f(x)\, dx = F(b) - F(a)$.

---

[4]This property, which is analogous to the equality of inner and outer content, is expressed by saying that f is *Riemann integrable*. All continuous functions are Riemann integrable, but there are other examples such as functions continuous except at finitely many points.

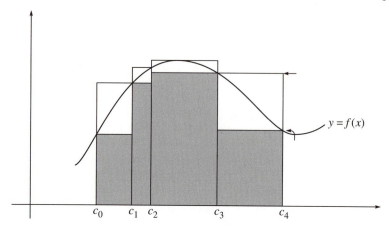

Fig. 9.7. Upper and lower Riemann sums associated to a partition.

(9.3.5) **Proposition:** *Let f be a continuous, non-negative function on $[a, b]$. Let X be the region under the graph of f between the limits $x = a$ and $x = b$; that is, the subset of the standard Euclidean plane $\mathbf{R}^2$ defined by*

$$X = \{(x, y) : a < x < b,\ 0 < y < f(x)\}.$$

Then X is contented, and $\mathcal{A}(X) = \int_a^b f(x)\,\mathrm{d}x$.

In other words, the integral equals the area under the curve.

Proof: Any lower Riemann sum for $\int_a^b f(x)\,\mathrm{d}x$ is equal to the area of a polygonal region, made up of disjoint rectangles, that is contained in X. Therefore the inner content $\underline{\mathcal{A}}(X)$ is greater than or equal to any lower Riemann sum, and so it is greater than or equal to the least upper bound of all such lower Riemann sums. But this least upper bound is the integral, so

$$\underline{\mathcal{A}}(X) \geq \int_a^b f(x)\,\mathrm{d}x.$$

By similar reasoning with upper Riemann sums

$$\overline{\mathcal{A}}(X) \leq \int_a^b f(x)\,\mathrm{d}x.$$

Since $\underline{\mathcal{A}}(X) \leq \overline{\mathcal{A}}(X)$, we have equality everywhere. $\square$

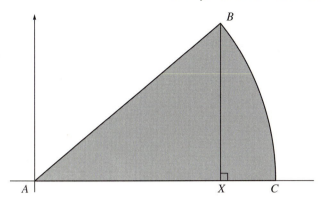

Fig. 9.8. A circular sector.

(9.3.6)   **Example:** Let us calculate the area of a circular sector or 'pie piece', of radius
r and angle $\theta < \pi/2$ (Figure 9.8). Let BX be perpendicular to AC; then the area
of the circular sector ABC is the sum of the areas of the right-angled triangle
AXB and of the piece BXC. The area of ABX is $\frac{1}{2}r^2 \sin\theta\cos\theta = \frac{1}{4}r^2 \sin 2\theta$,
and the area of BXC can be calculated by integration as

$$\int_{r\cos\theta}^{r} \sqrt{r^2 - x^2}\, dx.$$

The substitution $x = r\cos\psi$ transforms this integral into

$$r^2 \int_0^\theta \sin^2\psi\, d\psi = r^2 \left[\frac{\psi}{2} - \frac{\sin 2\psi}{4}\right]_0^\theta = r^2 \left(\frac{\theta}{2} - \frac{\sin 2\theta}{4}\right).$$

Therefore the area of the sector is $\frac{1}{2}r^2\theta$.

9.4   The question of consistency

In Chapters 1 and 3, when the axioms for geometry were introduced, we
mentioned the question of whether these axioms are *consistent*. They can be
shown to be consistent by exhibiting a model. If you are not satisfied that the
'intuitive picture' of geometry is sufficiently precise to qualify as a model,
then to prove consistency you need to *construct* a model in terms of other
mathematical objects. In Chapters 2 and 4 we showed, as a by-product of
our study of vectors and dot products, how this can be done for the axioms of
geometry that we had previously introduced.

     The same question of consistency can be asked about the axioms of area,
and again there are two possible answers. One can say that the concept of

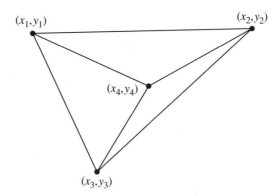

Fig. 9.9. A triangle decomposed into three smaller triangles.

area is such a clear part of 'intuitive geometry' that no further justification for the axioms is required, or one can try to construct a model for them. In this case, to construct a model means to define (by means of some kind of formula) the area of a bounded polygonal region, in such a way that the area so defined satisfies the addition, invariance, and normalization axioms.

At first sight this does not seem too hard. After all, we already have a formula (9.2.8) whereby we can define the area of a *triangle*, and it is easy to see that any bounded polygonal region can be decomposed (give or take a few line segments) into the union of finitely many disjoint triangles. So why can we not simply define the area of a bounded polygonal region as the sum of the areas of the triangles making it up, and thereby prove the consistency of the axioms of area?

The problem is that a bounded polygonal region can be written as the union of nonoverlapping triangles in many different ways. Even a triangle itself can be written as the union of three smaller triangles (Figure 9.9). How do we know that we will arrive at the same answer by adding up the areas of the triangles in all these different decompositions? Remember that we are not allowed to use the addition axiom to say that the answer will always be the area of the polygonal region in question, because the consistency of this axiom is precisely what we are trying to prove! We have to show *directly from the formulae* that the answer is always the same, and this is obviously going to be a messy business. Here is how we might go about it for the triangle illustrated in Figure 9.9. Consider the 4×4 determinant

$$
\begin{vmatrix}
1 & 1 & 1 & 1 \\
1 & 1 & 1 & 1 \\
x_1 & x_2 & x_3 & x_4 \\
y_1 & y_2 & y_3 & y_4
\end{vmatrix}.
$$

This determinant is zero because the first two rows are the same. But expanding by the first row, we get

$$\begin{vmatrix} 1 & 1 & 1 \\ x_2 & x_3 & x_4 \\ y_2 & y_3 & y_4 \end{vmatrix} - \begin{vmatrix} 1 & 1 & 1 \\ x_1 & x_3 & x_4 \\ y_1 & y_3 & y_4 \end{vmatrix} + \begin{vmatrix} 1 & 1 & 1 \\ x_1 & x_2 & x_4 \\ y_1 & y_2 & y_4 \end{vmatrix} - \begin{vmatrix} 1 & 1 & 1 \\ x_1 & x_2 & x_3 \\ y_1 & y_2 & y_3 \end{vmatrix} = 0.$$

Taking account of the orientation of the triangles, we see that this tells us that the area of the large triangle is equal to the sum of the areas of the three smaller ones.

A general proof of the consistency of the area axioms can be given along these lines, but it is rather long winded. I have decided to omit it, taking the view that most readers will find other things more interesting. If you would like to see the details, you can find them in Pogorelov [33].

Remark: A different, although related, question is whether our axioms for area can be *strengthened* in any way while still preserving consistency. Around 1900 the French mathematician Lebesgue discovered that one can obtain a consistent theory of area satisfying the axiom of 'countable additivity', which is a strengthening of the addition axiom: if $(X_n)_{n=1}^{\infty}$ is a sequence of mutually disjoint sets, then

$$\mathcal{A}\left(\bigcup_{n=1}^{\infty} X_n\right) = \sum_{n=1}^{\infty} \mathcal{A}(X_n).$$

Lebesgue's theory allows one to assign areas to many more sets than the theory that we have developed; in particular it assigns the area zero to the noncontented set K introduced in 9.1.1. But its chief importance is not in geometry but in analysis, where it allows one to develop a theory of integration which is 'complete' in a sense in which Riemann integration is not. Very roughly, the relationship between Riemann integration and Lebesgue integration is the same as that between the rational numbers and the real numbers.

9.5 Volume

The theory of *volume* in three dimensions runs parallel to the theory of area in two dimensions, and can be based on analogous axioms. Thus we can define the concepts of *half-space* in a three-dimensional Euclidean space S, and of *polyhedral region* in S, by analogy with the concepts of half-plane and of polygonal region in two dimensions: a half-space is given by an equation of the form $ax + by + cz + d > 0$ (open half-space) or $ax + by + cz + d \geq 0$ (closed half-space), and a polyhedral region is obtained from half-spaces by

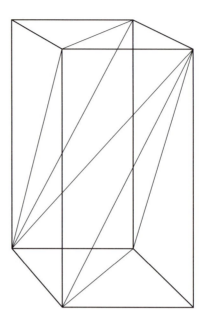

Fig. 9.10. A cuboid dissected into tetrahedra; they are not all congruent.

intersection and union. A particular example of a polyhedral region is a *cuboid* of sides a, b, and c, most easily defined as the region

$$\{(x, y, z) : 0 < x < a,\ 0 < y < b,\ 0 < z < c\}$$

in an appropriate Cartesian coordinate system. This is the three-dimensional generalization of a rectangle. Using these concepts one can introduce three axioms for volume which are analogous to the three axioms for area, saying that volume is invariant under isometries, that it is additive on disjoint unions, and that the volume of a cuboid of sides a, b, and c is abc. From these axioms the whole theory can be built up.

There are, however, some problems in the theory of volume that have no counterparts for area. The first of these appears when we try to calculate the volume of a tetrahedron (the three-dimensional analogue of a triangle). In two dimensions it was easy to find the area of a right-angled triangle, by putting two congruent ones together to form a rectangle. But in three dimensions it is not possible to put six congruent right-angled tetrahedra together to form a cuboid; although a cuboid can indeed be divided into six tetrahedra, the six tetrahedra are *not* all congruent to one another (Figure 9.10)! To find the volume of a tetrahedron it is therefore necessary to resort to more powerful methods; the answer (one-third the base area times the altitude) was first found by Eudoxus and is one of the earliest triumphs of the method of exhaustion.

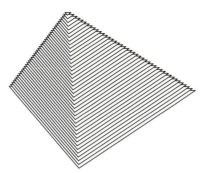

Fig. 9.11. Slicing a tetrahedron.

We will obtain the result by using integration, thinking of the tetrahedron as decomposed into infinitely many plane 'slices'. To illustrate the idea, imagine building an approximate model of a tetrahedron from a stack of thin cards (Figure 9.11). We expect that the volume of the tetrahedron will be approximately equal to the total volume of all the cards, which is the sum of their areas multiplied by their thickness. In the limit we anticipate that the sum will be replaced by an integral.

To do this formally, let P be a bounded polyhedral region in space S. Let x, y, z be Cartesian coordinates in S and for each fixed w let $\Pi(w)$ denote the plane $\{(x, y, z) : z = w\}$. Then $\Pi(w)$ is a Euclidean plane and the 'slice' $P \cap \Pi(w)$ is a bounded polygonal region in that plane, and so has an area. We then have:

(9.5.1)   **Proposition:** (CAVALIERI'S PRINCIPLE) *With notation as above, the function* $w \mapsto \mathcal{A}(P \cap \Pi(w))$ *is Riemann integrable, and*

(9.5.2)
$$\mathcal{V}(P) = \int_{-\infty}^{\infty} \mathcal{A}(P \cap \Pi(w)) \, \mathrm{d}w.$$

The range of integration is from $-\infty$ to ∞, but this causes no problems because the integrand is zero outside some finite range.

Remark: It is not quite historically accurate to call this 'Cavalieri's principle'. Strictly speaking, Cavalieri's principle is the statement that if P_1 and P_2 are polyhedral regions, and $\mathcal{A}(P_1 \cap \Pi(w)) = \mathcal{A}(P_2 \cap \Pi(w))$ for all w, then $\mathcal{V}(P_1) = \mathcal{V}(P_2)$. This is an easy corollary of our result.

Proof: We won't give all the details of the proof, which belong more to integration theory than to geometry. The general idea is quite straightforward, though, and it is one which is often used in mathematical analysis. It is that one should prove the result first for some restricted class of 'simple' regions, and then extend to all regions by an approximation argument based on the method of exhaustion.

We start the proof by noticing that Cavalieri's principle is true if P is a cuboid whose sides are parallel to the coordinate axes. To check it, let $P = \{(x, y, z) : x_0 < x < x_0 + a,\ y_0 < y < y_0 + b,\ z_0 < z < z_0 + c\}$. The volume of P is abc. The area of $P \cap \Pi(w)$ is ab if $z_0 < w < z_0 + c$ and 0 otherwise, so

$$\int_{-\infty}^{\infty} \mathcal{A}(P \cap \Pi(w))\, dw = \int_{z_0}^{z_0 + c} ab\, dw = abc$$

and Cavalieri's principle is verified.

It is obvious that both the left-hand and the right-hand sides of equation 9.5.2 are additive on disjoint unions; this follows from the additivity axioms for volume and area. So Cavalieri's principle holds, not only for cuboids, but also for regions made up of *disjoint unions* of finitely many cuboids (with sides parallel to the axes).

To complete the proof of Cavalieri's principle we need to know that any polyhedral region can be approximated by a union of finitely many disjoint cuboids. To be precise, we need the following:

(9.5.3)   **Claim:** *Let P be a bounded polyhedral region in* $\mathbf{R}^3$. *Then for any real number* $\varepsilon > 0$ *there is a region* $Q \subseteq P$ *which is a union of finitely many disjoint cuboids with sides parallel to the coordinate axes, such that* $\mathcal{V}(P) - \mathcal{V}(Q) < \varepsilon$.

If you want a formal proof of this claim, work through Exercise 9.7.14. Informally, all it says is that we can find the volume of P to a good approximation by adding up the total volume of very small cuboids contained in it.

To prove Cavalieri's principle, assuming the claim, let $\varepsilon > 0$ and let Q be the corresponding union of cuboids. Then

$$\int_{-\infty}^{\infty} \mathcal{A}(P \cap \Pi(w))\, dw \geq \int_{-\infty}^{\infty} \mathcal{A}(Q \cap \Pi(w))\, dw = \mathcal{V}(Q) > \mathcal{V}(P) - \varepsilon$$

using the fact that Cavalieri's principle holds for Q. The first and the last terms of this inequality do not involve the approximating set Q. Considering only these two terms, we may let $\varepsilon \to 0$ and obtain

$$\int_{-\infty}^{\infty} \mathcal{A}(P \cap \Pi(w))\, dw \geq \mathcal{V}(P).$$

We would like to turn this inequality into an equality, and to do this we enclose P in some large cuboid B and apply the same argument to $B \setminus P$, obtaining

$$\int_{-\infty}^{\infty} \mathcal{A}((B \setminus P) \cap \Pi(w))\, dw \geq \mathcal{V}(B \setminus P).$$

Together these give

$$\int_{-\infty}^{\infty} \Big(\mathcal{A}(P \cap \Pi(w)) + \mathcal{A}((B \setminus P) \cap \Pi(w))\Big)\, dw \geq \mathcal{V}(P) + \mathcal{V}(B \setminus P).$$

But the left-hand side of this inequality is just the integral $\int \mathcal{A}(B \cap \Pi(w))\, dw$, whereas the right-hand side is the volume $\mathcal{V}(B)$. Since Cavalieri's principle does hold for the cuboid B, these two are equal. The inequality is in fact an equality, which means that the two inequalities from which it was derived must be equalities as well. □

We will use Cavalieri's principle to find the volume of a tetrahedron. Formally speaking, a *tetrahedron* is defined by four noncoplanar points A, B, C, and D in space, and the *interior P of the tetrahedron* (which is what we want to find the volume of) is the intersection of four open half-spaces each of which contains one of the points and is bounded by the plane of the other three.

Let us choose coordinates so that the plane of A, B, and C has equation $z = 0$ and D belongs to the half-space $z > 0$. We will refer to triangle ABC as the *base* of the tetrahedron, and the corresponding *altitude h* is the perpendicular distance from D to the plane of the base, in other words the z-coordinate of D. The plane $\Pi(w)$ meets the tetrahedron if $0 < w < h$, and the intersection is a triangle similar to the base but scaled by a factor w/h. (See Exercise 4.6.14 for the facts we are using about similarities.) Therefore $\mathcal{A}(P \cap \Pi(w)) = (w/h)^2 \mathcal{A}(ABC)$. Now we may apply Cavalieri's principle to find that

$$\mathcal{V}(P) = \int_0^h \left(\frac{w}{h}\right)^2 \mathcal{A}(ABC)\, dw = \frac{h}{3}\mathcal{A}(ABC);$$

the volume of a tetrahedron is one-third the base area times the altitude.

We can express this result in vector terms. Let $\mathbf{a}$, $\mathbf{b}$, and $\mathbf{c}$ be the position vectors of A, B, and C relative to D as origin. Then the volume of the tetrahedron can be expressed in terms of the scalar triple product as $\frac{1}{6}|[\mathbf{a}, \mathbf{b}, \mathbf{c}]|$. (This is analogous to the formula 9.2.6 for the area of a triangle.) To prove this, think of the tetrahedron as a pyramid with vertex A and base the triangle BCD. By 9.2.7, the vector $\mathbf{b} \wedge \mathbf{c}$ has magnitude $2\mathcal{A}(BCD)$ and direction perpendicular to the plane of the base. Therefore the scalar triple product $\mathbf{a} \cdot (\mathbf{b} \wedge \mathbf{c})$ has magnitude $2h\mathcal{A}(BCD)$, where h is the length of the projection of **a** in the direction perpendicular to the plane of the base, that is the altitude. So

$$\|[\mathbf{a}, \mathbf{b}, \mathbf{c}]\| = 2h\mathcal{A}(BCD) = 6V(ABCD).$$

Remark: By comparison with the simple calculation of the area of a triangle, this process seems very roundabout. Is there really no way of calculating the volume of a tetrahedron without recourse to the method of exhaustion? This question (in a rather more precise form) was the third in the famous list of 23 problems which David Hilbert presented to the 1900 International Congress of Mathematicians. Hilbert was one of the leaders of the mathematical world, and his problems were intended to suggest major directions of research for the twentieth century; several of them remain unsolved. His third problem was the first to succumb. In the same year (1900), Dehn produced an example of two tetrahedra which have the same volume but nevertheless cannot be transformed into each other by any finite process of dissection (along plane cuts) and rearrangement. This example showed that infinite processes such as exhaustion are essential in the theory of volume even for polyhedra.

So far we have only discussed the volume of polyhedra. However, just as we did for area, we can use the method of exhaustion to *extend* the concept of volume from bounded polyhedral regions to more general 'contented' subsets of $\mathcal{S}$. The arguments are exactly the same; we introduce an 'inner content' $\underline{V}$ and an 'outer content' $\overline{V}$, and define a subset to be *contented* if the two contents are equal. Moreover, Cavalieri's principle can also be extended; if X is a contented subset of $\mathcal{S}$, and if for all w the cross-section $X \cap \Pi(w)$ is a contented subset of $\Pi(w)$, then

(9.5.4)
$$V(X) = \int_{-\infty}^{\infty} \mathcal{A}(X \cap \Pi(w)) \, dw.$$

The proof is easy; simply apply the version of Cavalieri's principle that we have already proved to polyhedral regions approximating X from inside and from outside.

(9.5.5)   **Example:** Let f be a continuous, non-negative function on the interval $[a, b]$. The region R in $\mathbf{R}^3$ defined by

$$R = \{(x, y, z) : x \in [a, b], \ y^2 + z^2 < f(x)^2\}$$

is called a *solid of revolution*, because it is obtained by revolving the area under the graph of $y = f(x)$ about the x-axis (see Figure 9.12).

It can be proved that the solid of revolution R defined in this way is contented (see Exercise 9.7.15). Therefore, the extended form of Cavalieri's principle (9.5.4) applies to show that its volume is given by

$$V(R) = \int_a^b \pi f(x)^2 \, dx.$$

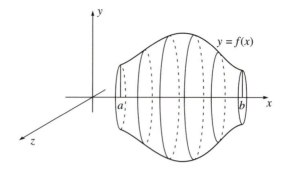

Fig. 9.12. A solid of revolution.

(9.5.6)   **Example:** As a special case we can calculate the volume of a solid sphere, defined as the set of all points at distance $\leq r$ from the origin O. The sphere can be regarded as a solid of revolution, with

$$a = -r, \quad b = r, \quad f(x) = \sqrt{r^2 - x^2}.$$

So the volume of the sphere is

$$\int_{-r}^{r} \pi(r^2 - x^2) \, dx = \pi\left[r^2x - \tfrac{1}{3}x^3\right]_{-r}^{r} = \tfrac{4}{3}\pi r^3.$$

9.6   Area integrals and Green's theorem

In our study of areas and volumes, we have so far only used the simplest form of integration theory — the integral $\int_a^b f(x) \, dx$ of a function of one variable over an interval on the real line. Books on advanced calculus (such as Apostol [3] or Kaplan [25], for example) describe several kinds of generalized integrals, and these too are very useful in calculating areas and volumes. In this section we will see how one of these generalizations, the area or 'double' integral, can be applied to geometry. We won't give the proofs of the basic facts about area integrals, which really belong in a calculus book rather than in this one.

Let X be a contented subset of $\mathbf{R}^2$, and let $f: \mathbf{R}^2 \to \mathbf{R}$ be a function. A *partition* of X is a decomposition of X into the union of finitely many disjoint contented subsets $X_1 \cup \cdots \cup X_N$. Define

$$M_i = \sup\{f(x) : x \in X_i\}, \quad m_i = \inf\{f(x) : x \in X_i\}.$$

Then the *upper and lower Riemann sums* associated to the partition are, respectively,

$$\sum_{i=1}^{N} M_i A(X_i) \quad \text{and} \quad \sum_{i=1}^{N} m_i A(X_i).$$

If the least upper bound of all possible lower Riemann sums and the greatest lower bound of all possible upper Riemann sums are equal, then f is said to be *Riemann integrable*, and their common value is called the *area integral* of f over X and denoted by

$$\int_X f \, dA = \int_X f(x, y) \, dA(x, y).$$

A sufficient condition for the Riemann integrability of f is that it should be *uniformly continuous* on the set X. This condition is automatically satisfied if f is the restriction to X of a continuous function on $\mathbf{R}^2$.

If $f(x, y) \geq 0$ for all $(x, y) \in X$, then it can be proved as in 9.3.5 that the area integral is equal to the *volume* of the region

$$Z = \{(x, y, z) : (x, y) \in X, 0 \leq z \leq f(x, y)\}$$

which is a contented set in $\mathbf{R}^3$.

(9.6.1)   **Example:** What is the integral of the constant function 1? If $X = X_1 \cup \cdots \cup X_N$ is a partition, then $A(X) = \sum_{i=1}^{N} A(X_i)$ by the additivity of area. Therefore *any* Riemann sum (upper or lower) for the integral $\int_X 1 \, dA$ is equal to $A(X)$, and so $\int_X 1 \, dA = A(X)$. This can also be seen by interpreting the integral as the volume of a right cylinder with base X and height 1.

(9.6.2)   **Example:** It is a theorem (related to Cavalieri's principle) that an area integral can be evaluated by composing two ordinary integrals. We will illustrate this in an example. Suppose we want to find the area integral of the function $f(x, y) = x^2 e^{-xy}$ over the triangular region X defined by $X = \{(x, y) : 0 \leq x \leq 1, \ 0 \leq y \leq x\}$. We write the area integral as a composite of two ordinary integrals:

$$\int_X x^2 e^{-xy} \, dA(x, y) = \int_0^1 \left(\int_0^x x^2 e^{-xy} \, dy \right) dx.$$

Evaluating the inner integral first, this is equal to

$$\int_0^1 x(1 - e^{-x^2}) \, dx = \left[\frac{x^2 + e^{-x^2}}{2} \right]_0^1 = 1/(2\,e).$$

The most important result about integrals of functions of one variable is the fundamental theorem of calculus. Is there an analogous theorem for two variables? In answering this question it is helpful to think of the original version of the theorem in the following way. The formula

$$\int_a^b \frac{df}{dx}\, dx = f(b) - f(a)$$

relates the 'one-dimensional integral' of the derivative of f over the interval $[a, b]$ with the 'zero-dimensional integral' (finite sum) of values of f on the *boundary* of $[a, b]$, which just consists of the two points a and b. With the benefit of 20–20 hindsight it is then easy to see that the appropriate generalization must relate a two-dimensional integral of derivatives over some region X with a one-dimensional integral over a curve forming the boundary of X. To make this precise we need to answer two questions. First, what is an integral over a curve? And second, what is meant by a curve being a boundary of a two-dimensional region?

(9.6.3)   **Definition:** *Let* $\gamma(t) = (x(t), y(t))$, $a \leq t \leq b$ *be a (piecewise regular) parameterized curve in the plane* $\mathbf{R}^2$, *and let* f *and* g *be smooth functions defined on* $\mathbf{R}^2$. *Then the* path integral $\int_\gamma f\, dg$ *is defined as*

$$\int_\gamma f\, dg = \int_a^b f(x(t), y(t)) \frac{dg(x(t), y(t))}{dt}\, dt.$$

The parameter t is not explicitly mentioned in the notation. This makes sense, because if u is another parameter related to t by an oriented parameter change map (7.1.10), then

$$\int f(x, y) \frac{dg(x, y)}{dt}\, dt = \int f(x, y) \frac{dg(x, y)}{du}\, du$$

by the chain rule. We do need to restrict to *oriented* parameter changes, though; traversing the curve backwards reverses the sign of the integral. One must therefore choose an orientation of γ in order to define the integral.

A parameterized curve γ in the plane is called *simple* if it does not cross itself anywhere, and *closed* if it begins and ends at the same point. In more formal terms, $\gamma: [a, b] \to \mathbf{R}^2$ is a simple closed curve if $\gamma(a) = \gamma(b)$ and γ is an *injective* (one-to-one) map of the half-open interval $[a, b)$ to $\mathbf{R}^2$. It is intuitively clear that any simple closed curve separates the plane into two regions, an interior and an exterior, of which it is the common boundary. (See Figure 9.13.) The first person to realize that this apparently obvious statement required proof[5] was Camille Jordan at the end of the nineteenth

---

[5] One should not imagine that mathematicians spend all their time trying to prove the obvious. The point is that the concept of 'function', and therefore of 'simple closed curve', had become so general that the 'obvious' no longer *was* obvious.

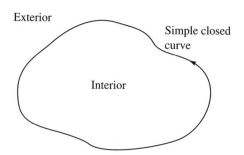

Fig. 9.13. The Jordan curve theorem.

century. Jordan's 'proof' was long and difficult. Worse still, despite its length it was actually incorrect! Although there are now proofs which are shorter, easier, and right, the so-called *Jordan curve theorem* is still too hard a result to include in this book.

Let us take it as read, however, that any piecewise smooth simple closed curve γ has an *interior* $\text{Int}(\gamma)$, which can formally be defined as the unique bounded connected component of the complement of the trace γ^*. (A proof of this result for smooth curves can be found in [12].) The curve γ is said to be *positively oriented* (relative to a given orientation $\perp$ of the plane $\mathbf{R}^2$) if for each parameter value t (except those for which γ is not smooth), there is $\varepsilon > 0$ such that the point $\gamma(t) + s\gamma'(t)^{\perp}$ belongs to the interior if $0 < s < \varepsilon$ and to the exterior if $-\varepsilon < s < 0$. In less formal terms, the curve is positively oriented if, when you traverse it in the direction of increasing parameter value, the interior is on your left.

We can now state the two-dimensional version of the fundamental theorem of calculus.

(9.6.4)   **Theorem:** (GREEN'S THEOREM) *Let γ be a positively oriented piecewise regular simple closed curve in $\mathbf{R}^2$, with interior $\text{Int}(\gamma)$. Let f and g be functions smooth in a neighbourhood of $\text{Int}(\gamma)$. Then*

$$\int_{\text{Int}(\gamma)} \left(\frac{\partial g}{\partial x} - \frac{\partial f}{\partial y} \right) \, \mathrm{d}\mathcal{A}(x,y) = \int_{\gamma} f \, \mathrm{d}x + g \, \mathrm{d}y.$$

We will not give the proof of Green's theorem here, but it may be helpful to work out a special case. Suppose that γ is a *rectangular* curve as shown in Figure 9.14, with $\text{Int}(\gamma) = \{(x,y) : 0 < x < a, \ 0 < y < b\}$. Then

$$\int_{\text{Int}(\gamma)} \frac{\partial g}{\partial x} \, \mathrm{d}\mathcal{A} = \int_{y=0}^{b} \int_{x=0}^{a} \frac{\partial g}{\partial x} \, \mathrm{d}x \, \mathrm{d}y = \int_{y=0}^{b} \Big(g(a,y) - g(0,y) \Big) \, \mathrm{d}y$$

by the fundamental theorem of calculus in one variable. But directly from the definition of a path integral

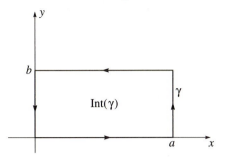

Fig. 9.14. An example of Green's theorem.

$$\int_\gamma g\,\mathrm{d}y = \int_{y=0}^b \Big(g(a,y) - g(0,y)\Big)\,\mathrm{d}y.$$

We can argue similarly for the function f, and putting the two results together we get Green's theorem.

One important application of Green's theorem is to find a formula for the area enclosed by a simple closed curve γ.

(9.6.5)   **Proposition:** *Let γ be a simple closed piecewise regular curve in the plane, with interior $\mathrm{Int}(\gamma)$. Then one has the following formulae for the area of $\mathrm{Int}(\gamma)$:*

$$\mathcal{A}(\mathrm{Int}(\gamma)) = \frac{1}{2}\int_\gamma r^2\,\mathrm{d}\theta \quad \text{in polar coordinates,}$$

$$\mathcal{A}(\mathrm{Int}(\gamma)) = \int_\gamma x\,\mathrm{d}y = -\int_\gamma y\,\mathrm{d}x \quad \text{in Cartesian coordinates.}$$

Proof: Recall (9.6.1) that

$$\mathcal{A}(\mathrm{Int}(\gamma)) = \int_{\mathrm{Int}(\gamma)} 1\,\mathrm{d}\mathcal{A}.$$

The two Cartesian expressions can therefore be derived simply by applying Green's theorem to $f = 0$, $g = x$ and to $f = -y$, $g = 0$.

To check that the polar expression agrees with the Cartesian ones, write $x = r\cos\theta$, $y = r\sin\theta$. Then, by the chain rule,

$$\frac{\mathrm{d}x}{\mathrm{d}t} = \cos\theta\frac{\mathrm{d}r}{\mathrm{d}t} - r\sin\theta\frac{\mathrm{d}\theta}{\mathrm{d}t}, \quad \frac{\mathrm{d}y}{\mathrm{d}t} = \sin\theta\frac{\mathrm{d}r}{\mathrm{d}t} + r\cos\theta\frac{\mathrm{d}\theta}{\mathrm{d}t},$$

and so

$$x\frac{\mathrm{d}y}{\mathrm{d}t} - y\frac{\mathrm{d}x}{\mathrm{d}t} = r^2\frac{\mathrm{d}\theta}{\mathrm{d}t}.$$

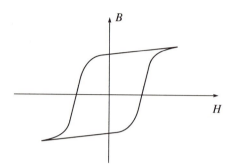

Fig. 9.15. A hysteresis loop.

Therefore

$$\frac{1}{2}\int_\gamma r^2\,d\theta = \frac{1}{2}\int_\gamma x\,dy - y\,dx = \int_\gamma x\,dy = -\int_\gamma y\,dx$$

because we already know that the last two expressions are equal. □

(9.6.6)   **Example:** We can check our formula by working out the area enclosed by a circle once more. This time we take polar coordinates with origin a point on the circle itself, so that the circle has equation $r = d\sin\theta$, where d is the diameter and θ ranges from 0 to π (see 6.3.3). So our formula says that the area is

$$\frac{1}{2}\int_0^\pi r^2\,d\theta = \frac{d^2}{2}\int_0^\pi \sin^2\theta\,d\theta = \frac{\pi d^2}{4}$$

as we expect.

(9.6.7)   **Example:** Let H be the external magnetic field applied to a piece of iron, B the corresponding magnetic induction. According to electromagnetic theory, the integral $\int H\,dB$ represents the work done in magnetizing the iron. A typical graph of B against H is shown in Figure 9.15. This is called a *hysteresis loop*.

  If the iron is part of the core of a power transformer in an electricity substation, then it is being cycled around this loop 50 times a second (60 times in the United States). At each cycle an amount of energy equal to the area of the loop, $\int H\,dB$, is expended in magnetizing and demagnetizing the iron, and is eventually dissipated as heat. The iron becomes quite warm, which is why power transformers are filled with oil and have large cooling fins. Power engineers strive to make the area of the hysteresis loop as small as possible.

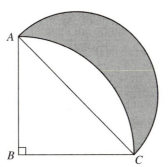

Fig. 9.16. Hippocrates' quadrature of the lune

9.7 Exercises

1. Find the area under the graph of $y = \cosh x$ between $x = -a$ and $x = a$.

2. A bowl of height h is formed by revolving the curve $y = x^2/h$ (between $x = 0$ and $x = h$) about the y-axis. Find the volume of the bowl. How does it compare with the volume of a hemispherical bowl of radius h?

3. ABC is a right-angled isosceles triangle. Two circular arcs are drawn joining A and C, one with centre B and the other with centre the midpoint of the hypotenuse AC (see Figure 9.16). The region enclosed between the two arcs is called a *lune*. Show that the area of the lune is equal to the area of the isosceles triangle ABC.

    (This is called the *quadrature of the lune*, and is due to Hippocrates of Chios (about 430 BC). As the Greek geometers were able to 'square the lune', it was perhaps not so unreasonable for them to hope to 'square the circle' (see Section 6.6) as well.)

4. Let $f: (a, b) \rightarrow \mathbf{R}$ be a smooth real-valued function on some interval (a, b), and assume that $f''(x) \leq 0$ for all $x \in (a, b)$. Prove that f is a *convex function*, which is to say that for all $x, y \in (a, b)$ and constants λ, $0 \leq \lambda \leq 1$,

$$f(\lambda x + (1 - \lambda)y) \geq \lambda f(x) + (1 - \lambda)f(y).$$

By induction, extend this property to prove that if $x_1, \ldots, x_n \in (a, b)$ and $\alpha_1, \ldots, \alpha_n \geq 0$ with $\sum_{i=1}^{n} \alpha_i = 1$, then

$$f\left(\sum_{i=1}^{n} \alpha_i x_i\right) \geq \sum_{i=1}^{n} \alpha_i f(x_i).$$

Apply this inequality to the function $f(x) = \log x$, and hence prove the arithmetic–geometric mean inequality.

5. Let ABC be a triangle in the plane $\mathbf{R}^2$. If L is a linear transformation of the plane, with $L(A) = A'$ and so on, show that

$$A(A'B'C') = |\det L|A(ABC).$$

Thus the determinant of L can be interpreted geometrically as the ratio in which it changes areas.

6. Prove that in formula 9.2.8, the $+$ sign should be taken if the three points (x_1, y_1), (x_2, y_2), and (x_3, y_3) are in anticlockwise order around the triangle, and the $-$ sign should be taken if they are in clockwise order, where the concepts of 'clockwise' and 'anticlockwise' are defined in terms of the given orientation of $\mathcal{P}$.

7. Use 9.6.5 to obtain yet another proof of the formula for the area of a triangle with vertices (x_1, y_1), (x_2, y_2), and (x_3, y_3).

8. Find the area of the central loop of the folium of Descartes with equation $x^3 + y^3 - 3xy = 0$.

9. A *cyclic quadrilateral* is a quadrilateral whose four vertices lie on a circle. If a, b, c, and d are the lengths of the sides of a cyclic quadrilateral, prove that its area is

$$\sqrt{(s-a)(s-b)(s-c)(s-d)}$$

where $s = \frac{1}{2}(a + b + c + d)$.

10. Let P, Q, and R be three points on a parabola. Prove that the area of the triangle PQR is twice the area of the triangle whose sides are the tangents to the parabola at P, Q, and R.

11. This question shows how Euclid proved the similarity axiom from properties of areas. For the purposes of this question, therefore, the only properties of areas that you should use are those proved in the first two sections of this chapter.
    Consider Figure 9.17, in which the lines BC and DE are parallel. Show that

$$A(ABC) : A(BCD) = AB : BD, \quad A(ABC) : A(CBE) = AC : CE.$$

Show also that $A(CBE) = A(BCD)$, and deduce that $AB : BD = AC : CE$, which is the similarity axiom.

12. A *lattice point* in the plane $\mathbf{R}^2$ is a point with integer coordinates. If ABC is a triangle whose vertices are lattice points and not all in the same straight line, prove that the area of ABC is at least $\frac{1}{2}$.
    Now suppose that A, B, and C (in that order) lie on a semicircle of radius r. By comparing areas, prove that $r^2\theta^3 > 6$, where θ is the angle subtended by AC at the centre. (You may find it helpful to use the inequality $\theta - \sin\theta < \theta^3/6$, valid for $\theta > 0$.)
    Hence prove that the number of lattice points on the circumference of a circle of radius r cannot be more than $2k$, where k is the least integer greater than $2\pi(r^2/6)^{\frac{1}{3}}$.

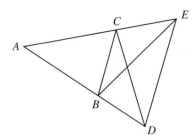

Fig. 9.17. Euclid's proof of the similarity axiom.

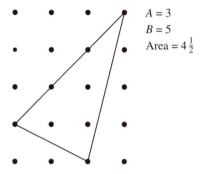

$A = 3$
$B = 5$
Area $= 4\frac{1}{2}$

Fig. 9.18. Pick's theorem.

13. Consider a convex polygon whose vertices are lattice points. Let A denote the number of lattice points completely contained in the polygon, and B the number of lattice points on the boundary (including the vertices). *Pick's theorem* states that the area of the polygon is equal to $A + \frac{1}{2}B - 1$; see Figure 9.18.

To prove Pick's theorem, proceed as follows. First, check the formula for a rectangle with sides parallel to the axes. Next, prove that the formula satisfies the analogue of the area addition axiom when a convex polygon is dissected into two smaller convex polygons. By dissecting a rectangle, prove that the formula is correct for a right-angled triangle, then for any triangle, and then for any polygon.

14. Let X be a polyhedral region in $\mathbf{R}^3$ (not necessarily bounded). Given $\delta > 0$, we define a δ-*cube* to be a cube

$$\{(x, y, z) : l\delta < x < (l+1)\delta, m\delta < y < (m+1)\delta, n\delta < z < (n+1)\delta\}$$

where l, m, n are integers, and we define the δ-*interior* X_δ of X to be the union of all δ-cubes contained completely within X. We say that X is *good* if for any *bounded* polyhedral region B,

$$\mathcal{V}((X \setminus X_\delta) \cap B) \to 0 \quad \text{as } \delta \to 0.$$

Prove that a half-space is good, and that the union or intersection of two good regions is again good. Deduce that, in fact, all polyhedral regions are good; and hence prove Claim 9.5.3.

15. Prove from first principles that a right circular cylinder (that is, a solid of revolution obtained by rotating the area under the graph $y = $ constant about the x-axis) is a contented set.

Consider now a solid of revolution R obtained by revolving the graph of a continuous function $y = f(x)$, $a < x < b$, about the x-axis. Show that any lower Riemann sum for the integral $\int_a^b \pi f(x)^2 \, dx$ represents the volume of a contented set made up of a union of right circular cylinders contained within R, and deduce that

$$\underline{V}(R) \geq \int_a^b \pi f(x)^2 \, dx.$$

Make a similar argument for outer content using upper Riemann sums, and deduce that R is in fact a contented set.

10

Quadric surfaces

10.1 Definition and examples

In this chapter we will study the simplest examples of curved *surfaces* in three-dimensional space. Just as a curve can be defined by a locus equation $f(x, y) = 0$ relating two variables, so a surface can be defined by a locus equation $f(x, y, z) = 0$ relating three variables. The surfaces we will study are analogous to conics, in that the function $f(x, y, z)$ is a second-degree polynomial in x, y, and z. They are called *quadric* surfaces.

Using matrices, we can give the formal definition as follows:

(10.1.1) **Definition:** *A* quadric *is a surface in a three-dimensional space given in terms of coordinates x, y, z by the equation*

$$\mathbf{x}^t \mathbf{M} \mathbf{x} + \mathbf{N} \mathbf{x} + k = 0$$

where k is a constant, $\mathbf{M}$ is a nonzero 3×3 symmetric matrix, $\mathbf{N}$ is a 1×3 row matrix, and $\mathbf{x}$ is the column vector $\begin{pmatrix} x \\ y \\ z \end{pmatrix}$.

We can write this out at length if we wish: putting

$$\mathbf{M} = \begin{pmatrix} A & D & F \\ D & B & E \\ F & E & C \end{pmatrix}, \qquad \mathbf{N} = \begin{pmatrix} P & Q & R \end{pmatrix}$$

the equation becomes

$$Ax^2 + By^2 + Cz^2 + 2Dxy + 2Eyz + 2Fzx + Px + Qy + Rz + k = 0.$$

For theoretical work it is generally easier to use the matrix form. The quadric is called *central* if for a suitable choice of coordinates $P = Q = R = 0$, so that there are no linear terms.

There are many different kinds of quadric. Here are some examples:

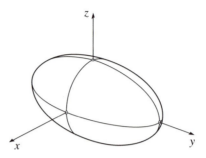

Fig. 10.1. An ellipsoid.

(10.1.2) **Example:** An *ellipsoid* (Figure 10.1) is a quadric with equation

$$\frac{x^2}{a^2} + \frac{y^2}{b^2} + \frac{z^2}{c^2} = 1$$

in an appropriate coordinate system. An ellipsoid is a bounded figure in three dimensions, just as an ellipse is a bounded figure in two dimensions. It can be obtained by squashing a sphere along two perpendicular axes, just as an ellipse can be obtained by squashing a circle along one axis. The numbers a, b, and c are called the *lengths of the principal axes* of the ellipsoid.

A special case is an *ellipsoid of revolution* or *spheroid*, where two of the three numbers a, b, and c are equal. If for instance we assume that $b = c$, then the ellipsoid can be obtained by rotating the ellipse $x^2/a^2 + y^2/b^2 = 1$ (in the *xy*-plane) about the *x*-axis.

The surface of the earth is approximately an ellipsoid of revolution, slightly flattened at the poles because of the rotation of the earth. In fact, the equatorial radius is about 6,378 kilometres, whereas the polar radius is about 6,356 kilometres. This means that a cross-section of the earth through the poles is an ellipse of eccentricity about 8 percent.

(10.1.3) **Example:** A *hyperboloid of one sheet* (Figure 10.2) is a quadric with equation

$$\frac{x^2}{a^2} + \frac{y^2}{b^2} - \frac{z^2}{c^2} = 1$$

in an appropriate coordinate system. If $a = b$ it is a *hyperboloid of revolution*, because it can be obtained by rotating a hyperbola in the *xz*-plane about the *z*-axis.

A *hyperboloid of two sheets* (Figure 10.3) is given by the equation

$$\frac{x^2}{a^2} + \frac{y^2}{b^2} - \frac{z^2}{c^2} = -1$$

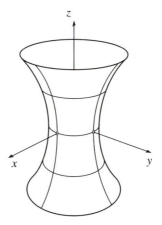

Fig. 10.2. A hyperboloid of one sheet.

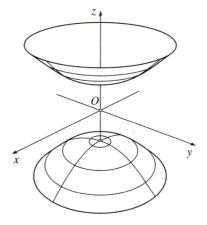

Fig. 10.3. A hyperboloid of two sheets.

in an appropriate coordinate system. If $a = b$ this is also called a *hyperboloid of revolution*. Notice that the hyperboloid of one sheet is *connected*, whereas the hyperboloid of two sheets is not.

The ellipsoid and the two types of hyperboloid are the most usual kinds of quadric, corresponding to the ellipse and hyperbola among conics. Any equation of the form $Ax^2 + By^2 + Cz^2 = 1$ with A, B, and C all nonzero defines one of these three or else has no solutions. The possibilities can be summed up in a table as follows:

Number of A, B, C that are > 0	Type of quadric
3	Ellipsoid
2	Hyperboloid of one sheet
1	Hyperboloid of two sheets
0	Empty set $\emptyset$

(10.1.4) **Example:** The two types of *paraboloid* (Figure 10.4) occur when one variable (for example, z) only appears to the first power in the equation of the quadric. An *elliptic paraboloid* is a quadric of the form

$$z = \frac{x^2}{a^2} + \frac{y^2}{b^2}$$

in suitable coordinates; and a *hyperbolic paraboloid* is one of the form

$$z = \frac{x^2}{a^2} - \frac{y^2}{b^2}.$$

Either of the two types of paraboloid intersects all the planes $x = $ constant and $y = $ constant in two families of congruent parabolas. Because of its shape, the hyperbolic paraboloid is also sometimes called a *saddle surface*.

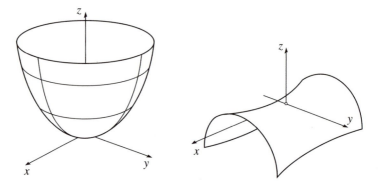

Fig. 10.4. Elliptic paraboloid.   Hyperbolic paraboloid.

(10.1.5) **Example:** A *cone* (Figure 10.5) is a quadric of the form

$$z^2 = \frac{x^2}{a^2} + \frac{y^2}{b^2}$$

in an appropriate coordinate system. The cone is made up of all the straight lines between $(0,0,0)$ and points of the ellipse $x^2/a^2 + y^2/b^2 = 1$ in the plane $z = 1$. If $a = b$ we have a *circular cone*.

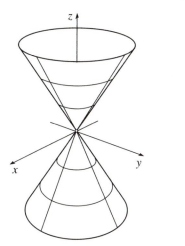
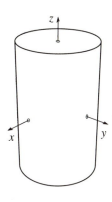

Fig. 10.5. A cone.   An elliptic cylinder.

(10.1.6)   **Example:** A *cylinder* (Figure 10.5) is a quadric whose equation can be expressed in a form in which one of the variables is absent. For example, the *elliptic cylinder* is represented by the equation

$$\frac{x^2}{a^2} + \frac{y^2}{b^2} = 1$$

in an appropriate coordinate system. This equation defines an ellipse in the plane $z = 0$; the elliptic cylinder is made up of all the lines through this ellipse and parallel to the z-axis. Similarly one can define *hyperbolic* and *parabolic* cylinders.

10.2 Reduction to canonical form

Given a general quadric

$$\mathbf{x}^t \mathbf{M} \mathbf{x} + \mathbf{N} \mathbf{x} + k = 0,$$

can we change coordinates so as to bring its equation to one of the simpler forms given in the previous section? In Section 7.3 we studied this problem for conics. We found in that case that one could simplify the equation by a two-stage process. The first stage was to rotate the coordinates so as to remove the term in xy, and the second stage was to translate the coordinates so as to remove, if possible, the terms in x and in y. Moreover the first stage (rotation) could be expressed by finding an orthogonal matrix $\mathbf{U}$ such that $\mathbf{U}^t \mathbf{M} \mathbf{U}$ is diagonal.

The same two-stage process can be carried out for quadrics. The second stage is exactly the same as before, but to generalize the first stage to three or indeed to n dimensions, we need the following theorem, which replaces our explicit discussion of rotation angles in the two-dimensional case:

(10.2.1)   **Theorem:** (DIAGONALIZATION THEOREM) *Let* M *be an* $n \times n$ *symmetric matrix. Then there is an* $n \times n$ *orthogonal matrix* U *such that* $U^t M U$ *is a diagonal matrix (one that has zeros everywhere except on the main diagonal).*

The proof of this theorem will be given in the next section. But assuming for the moment that it is true, let us investigate how we might go about *calculating* the matrix U. Suppose therefore that $U^t M U = D$ for some diagonal matrix

$$D = \begin{pmatrix} \lambda_0 & \cdots & 0 \\ \vdots & \ddots & \vdots \\ 0 & \cdots & \lambda_{n-1} \end{pmatrix}.$$

Then $MU = UD$. Let $c_0, \ldots, c_{n-1}$ denote the columns of the matrix U, so that (in the notation introduced in 6.2.5) $U = (c_0 | \ldots | c_{n-1})$. Then from the definition of matrix multiplication, one can calculate that

$$MU = (Mc_0 | \ldots | Mc_{n-1}).$$

On the other hand,

$$UD = (\lambda_0 c_0 | \ldots | \lambda_{n-1} c_{n-1}).$$

So the condition $MU = UD$ is equivalent to

$$Mc_j = \lambda_j c_j, \quad \forall j = 0, \ldots, n-1;$$

in other words, the columns c_j of U are *eigenvectors* (8.3.4) of M with eigenvalues λ_j.

By proposition 5.3.2, a square matrix such as U is orthogonal if and only if its columns are orthonormal. Thus theorem 10.2.1 is equivalent to the following statement: *any* $n \times n$ *symmetric matrix has* n *orthonormal eigenvectors.*

(10.2.2)   **Example:** Suppose that we want to simplify the quadric

$$4x^2 + 9y^2 + 5z^2 - 4xy + 8yz + 12xz + 9z = 3.$$

This has matrix equation $x^t M x + N x - 3 = 0$, where

$$M = \begin{pmatrix} 4 & -2 & 6 \\ -2 & 9 & 4 \\ 6 & 4 & 5 \end{pmatrix}, \qquad N = (\, 0 \quad 0 \quad 9 \,).$$

Our first task is to find three orthonormal eigenvectors for the matrix M. We can calculate that its characteristic polynomial is

$$\det(M - \lambda I) = -\lambda^3 + 18\lambda^2 - 45\lambda - 324 = -(\lambda - 12)(\lambda - 9)(\lambda + 3)$$

and that the eigenvalues therefore are 12, 9, and -3. To find the eigenvectors corresponding to these eigenvalues we must solve the equations

$$(M - \lambda I)c = 0$$

for column vectors c. For instance, to find the eigenvector corresponding to the eigenvalue $\lambda = 12$ we must solve the equations

$$\begin{pmatrix} -8 & -2 & 6 \\ -2 & -3 & 4 \\ 6 & 4 & -7 \end{pmatrix} \begin{pmatrix} c_1 \\ c_2 \\ c_3 \end{pmatrix} = \begin{pmatrix} 0 \\ 0 \\ 0 \end{pmatrix}.$$

You can use your favourite method for solving simultaneous equations to get the answer $c_2 = c_3 = 2c_1$. Thus, the unit vector

$$\frac{1}{3} \begin{pmatrix} 1 \\ 2 \\ 2 \end{pmatrix}$$

is an eigenvector for the eigenvalue 12. Similarly, we find that the unit vectors

$$\frac{1}{3} \begin{pmatrix} 2 \\ -2 \\ 1 \end{pmatrix}, \quad \frac{1}{3} \begin{pmatrix} -2 \\ -1 \\ 2 \end{pmatrix}$$

are eigenvectors for the eigenvalues 9 and -3 respectively.

Notice that the three unit eigenvectors we have calculated are orthonormal — their dot products vanish. (It can be proved, in fact, that eigenvectors corresponding to *distinct* eigenvalues are automatically orthogonal — see Exercise 10.5.7.) Therefore, the matrix

$$U = \frac{1}{3} \begin{pmatrix} 1 & 2 & -2 \\ 2 & -2 & -1 \\ 2 & 1 & 2 \end{pmatrix}$$

formed from them is orthogonal. We know, therefore, from the theory above, that the matrix $U'MU$ is diagonal, and that its diagonal entries are the eigenvalues 12, 9, and -3. (You may find it reassuring to check this explicitly.) So if we transform to new coordinates x' related to the old by $x = Ux'$, then the quadric is represented by the equation $x''U'MUx' + NUx' - 3 = 0$, which comes out to

$$4x'^2 + 3y'^2 - z'^2 + 2x' + y' + 2z' - 1 = 0$$

after cancelling a common factor of 3.

We have now finished the first stage of the reduction; this equation has no terms in $x'y'$ or other mixed products. We still have the linear terms to deal with, though, and we handle them in the same way as for conics: by completing the square. In this case, by completing the square we can reduce the equation to the form

$$4x''^2 + 3y''^2 - z''^2 = \frac{1}{3},$$

where $x' = x'' - \frac{1}{4}$, $y' = y'' - \frac{1}{6}$, and $z' = z'' + 1$. The quadric is therefore a hyperboloid of one sheet.

(10.2.3)   **Example:** Things can be trickier if the characteristic polynomial has a repeated root. For instance, consider the quadric whose equation is

$$6x^2 + 5y^2 + 5z^2 + 2\sqrt{2}x(y + z) + 2yz = 1.$$

The corresponding matrix is

$$M = \begin{pmatrix} 6 & \sqrt{2} & \sqrt{2} \\ \sqrt{2} & 5 & 1 \\ \sqrt{2} & 1 & 5 \end{pmatrix}$$

whose characteristic polynomial is

$$\det(M - \lambda I) = -\lambda^3 + 16\lambda^2 - 80\lambda + 128 = -(\lambda - 4)^2(\lambda - 8).$$

Calculating as before, we find that a unit eigenvector for the eigenvalue $\lambda = 8$ is

$$c_0 = \frac{1}{2}\begin{pmatrix} \sqrt{2} \\ 1 \\ 1 \end{pmatrix}.$$

For the eigenvalue $\lambda = 4$, we can find *two* linearly independent unit eigenvectors

$$\frac{1}{\sqrt{6}}\begin{pmatrix} \sqrt{2} \\ -2 \\ 0 \end{pmatrix} \quad \text{and} \quad \frac{1}{\sqrt{6}}\begin{pmatrix} \sqrt{2} \\ 0 \\ -2 \end{pmatrix}.$$

These are orthogonal to the $\lambda = 8$ eigenvector, but they are not orthogonal to each other. To choose a pair that are orthogonal to each other, we may proceed as follows. It is clear that any linear combination of eigenvectors with eigenvalue 4 is once again an eigenvector with eigenvalue 4. Therefore, *any* vector perpendicular to c_0 is in fact an eigenvector with eigenvalue 4. Thus we may choose a unit eigenvector c_1 with eigenvalue 4, and then $c_2 = c_0 \wedge c_1$ will automatically be another such eigenvector orthogonal to the first one[1]. For instance, we might take

$$ c_1 = \frac{1}{\sqrt{6}} \begin{pmatrix} \sqrt{2} \\ -2 \\ 0 \end{pmatrix}, \qquad c_2 = \frac{1}{2\sqrt{6}} \begin{pmatrix} 2 \\ \sqrt{2} \\ -3\sqrt{2} \end{pmatrix}. $$

The coordinate transformation $x = Ux'$, where

$$ U = \frac{1}{2\sqrt{6}} \begin{pmatrix} 2\sqrt{3} & 2\sqrt{2} & 2 \\ \sqrt{6} & -4 & \sqrt{2} \\ \sqrt{6} & 0 & -3\sqrt{2} \end{pmatrix}, $$

will therefore reduce our quadric to the form

$$ 8x'^2 + 4y'^2 + 4z'^2 = 1. $$

This is the equation of an ellipsoid.

Remark: Notice that the coefficients 8, 4, and 4 are just the eigenvalues of the original matrix M. Comparing the equation we have obtained with the standard form of the equation of an ellipsoid, we see that the lengths of the principal axes are $1/2\sqrt{2}$, $1/2$, and $1/2$. This result is a general one; the lengths of the principal axes of an ellipsoid are equal to $1/\sqrt{\lambda_i}$, where λ_i, $i = 1, 2, 3$, are the eigenvalues of the corresponding symmetric matrix.

10.3 Proof of the diagonalization theorem

Theorem 10.2.1 will be proved by induction on the size n of the matrix M. For $n = 1$ the theorem is obvious, since a 1×1 matrix is already diagonal.

We may suppose, then, that the theorem is known to be true for matrices of size $(n-1) \times (n-1)$. Let M be an $n \times n$ symmetric matrix. The first thing we need to do is to be sure that M has got at least one eigenvalue and a corresponding eigenvector. Our proof will make use of some ideas from elementary topology.

---

[1] This trick with the vector product is special to three dimensions; in n dimensions one must use the so-called *Gram–Schmidt* process to reduce a general basis of eigenvectors to an orthonormal one.

Let S^{n-1} denote the collection of all $n \times 1$ column vectors $\mathbf{x}$ such that $\mathbf{x}'\mathbf{x} = 1$. If we think of these as position vectors of points in $\mathbf{R}^n$, then S^{n-1} just becomes the set of all points at unit distance from the origin. The case of greatest interest to us is $n = 3$; then S^2 is just the surface of an ordinary unit sphere. The topological fact that we will need is that *any continuous function on S^{n-1} is bounded and attains its bounds*. More precisely, if $\varphi \colon S^{n-1} \to \mathbf{R}$ is continuous, then there is an $\mathbf{x}_0 \in S^{n-1}$ such that $\varphi(\mathbf{x}) \leq \varphi(\mathbf{x}_0)$ for all $\mathbf{x} \in S^{n-1}$. This fact is usually proved as a consequence of the Heine–Borel theorem[2]. A direct proof is also possible, however, and is sketched in Exercise 10.5.6.

In particular, we can consider the function φ defined on S^{n-1} by

$$\varphi(\mathbf{x}) = \mathbf{x}'\mathbf{M}\mathbf{x}.$$

This function is continuous, so according to the theorem that we have mentioned there is a vector $\mathbf{x}_0$ at which φ attains its maximum value. We will prove that this vector $\mathbf{x}_0$ is an eigenvector.

We can resolve $\mathbf{M}\mathbf{x}_0$ into its components parallel and perpendicular to $\mathbf{x}_0$. To do so, we write

$$\mathbf{M}\mathbf{x}_0 = \lambda\mathbf{x}_0 + \mu\mathbf{y}$$

where $\mathbf{x}_0$ and $\mathbf{y}$ are vectors of length 1 and $\mathbf{x}_0$ is perpendicular to $\mathbf{y}$ (that is, $\mathbf{y}'\mathbf{x}_0 = \mathbf{x}_0'\mathbf{y} = 0$). The vector

$$\mathbf{x}_\theta = \cos\theta\mathbf{x}_0 + \sin\theta\mathbf{y}$$

is therefore also of length 1, and so lies on S^{n-1}.

Recall that the function φ attains its maximum on S^{n-1} at the vector $\mathbf{x}_0$. In particular, the function $\theta \mapsto \varphi(\mathbf{x}_\theta)$ must have a maximum at $\theta = 0$. Therefore, we must have

$$\frac{d\varphi(\mathbf{x}_\theta)}{d\theta}\bigg|_{\theta=0} = 0.$$

But we can calculate

$$\begin{aligned}
\varphi(\mathbf{x}_\theta) &= \cos^2\theta\mathbf{x}_0'\mathbf{M}\mathbf{x}_0 + \cos\theta\sin\theta(\mathbf{x}_0'\mathbf{M}\mathbf{y} + \mathbf{y}'\mathbf{M}\mathbf{x}_0) + \sin^2\theta\mathbf{y}'\mathbf{M}\mathbf{y} \\
&= \cos^2\theta\lambda + 2\cos\theta\sin\theta\mu + \sin^2\theta\mathbf{y}'\mathbf{M}\mathbf{y}
\end{aligned}$$

where we substituted $\mathbf{M}\mathbf{x}_0 = \lambda\mathbf{x}_0 + \mu\mathbf{y}$, $\mathbf{x}_0'\mathbf{M} = (\mathbf{M}\mathbf{x}_0)' = \lambda\mathbf{x}_0' + \mu\mathbf{y}'$, using the symmetry of $\mathbf{M}$ to write $\mathbf{M} = \mathbf{M}'$. Therefore

$$\frac{d\varphi(\mathbf{x}_\theta)}{d\theta}\bigg|_{\theta=0} = 2\mu$$

and so $\mu = 0$. Thus $\mathbf{M}\mathbf{x}_0 = \lambda\mathbf{x}_0$, and $\mathbf{x}_0$ is an eigenvector.

---

[2] See Sutherland [40]. S^{n-1} is closed and bounded, hence compact by the Heine–Borel theorem. A general result about compact spaces says that any continuous real-valued function on such a space is bounded and attains its bounds.

In the remainder of the proof we will need to make use of two easily proved facts about orthogonal and symmetric matrices:

Fact 1. If M is symmetric and U is orthogonal, then $U'MU$ is symmetric. For since M is symmetric, $M' = M$, and so

$$(U'MU)' = U'M'U'' = U'MU.$$

Fact 2. If V and W are orthogonal, then so is VW. For

$$(VW)^{-1} = W^{-1}V^{-1} = W'V' = (VW)'.$$

We have shown that any symmetric matrix M has a unit eigenvector x_0. We can form an orthonormal basis $x_0, \ldots, x_{n-1}$ whose first element is this unit eigenvector. Let $V = (x_0 | \ldots | x_{n-1})$ be the orthogonal matrix whose columns are $x_0, \ldots, x_{n-1}$. Because the first column of V is the eigenvector x_0 with eigenvalue λ, the first column of MV is λx_0, and so the first column of $V'MV$ is

$$\lambda \begin{pmatrix} x_0^t x_0 \\ x_1^t x_0 \\ \vdots \\ x_{n-1}^t x_0 \end{pmatrix} = \begin{pmatrix} \lambda \\ 0 \\ \vdots \\ 0 \end{pmatrix}.$$

But the matrix $V'MV$ is symmetric (by Fact 1) and so the zeros in its first column must correspond to zeros in its first row. Thus we find that

$$V'MV = \begin{pmatrix} \lambda & 0 & \cdots & 0 \\ \hline 0 & & & \\ \vdots & & M' & \\ 0 & & & \end{pmatrix}$$

where M' is an $(n-1) \times (n-1)$ symmetric matrix.

Now we are in a position to apply our inductive hypothesis. This states that for any $(n-1) \times (n-1)$ symmetric matrix such as M', there is an $(n-1) \times (n-1)$ orthogonal matrix U' such that $U''M'U'$ is equal to a diagonal matrix D'. So if we let W be the $n \times n$ matrix

$$W = \begin{pmatrix} 1 & 0 & \cdots & 0 \\ \hline 0 & & & \\ \vdots & & U' & \\ 0 & & & \end{pmatrix}$$

then W is orthogonal and $W^t(V^tMV)W$ is a diagonal matrix

$$D = \begin{pmatrix} \lambda & 0 & \cdots & 0 \\ \hline 0 & & & \\ \vdots & & D' & \\ 0 & & & \end{pmatrix}.$$

Finally, therefore, we can put $U = VW$, which is an orthogonal matrix by Fact 2, and then

$$U^tMU = W^tV^tMVW = D$$

so that U^tMU is diagonal, as required. $\square$

Remark: The first part of this proof in fact gave us a way to find an eigenvector: namely, as a unit vector x for which x^tMx is as great as possible. The corresponding eigenvalue is then the maximum value of x^tMx. This way of obtaining eigenvalues and eigenvectors is sometimes known as *Rayleigh's principle*. In computer calculations with large matrices, it is hopelessly inefficient to try to find the eigenvectors by first finding the characteristic polynomial, then solving that, and then solving linear equations to find the eigenvectors. Methods based on Rayleigh's principle are sometimes far more successful.

You may have encountered the method of *Lagrange multipliers* for maximizing or minimizing a function of several variables subject to a constraint. The argument leading to Rayleigh's principle can also be phrased in terms of Lagrange multipliers. Recall the technique: to maximize a function $\varphi(x)$ subject to a constraint $\psi(x) = $ constant, one looks for solutions of

$$\nabla\varphi(x) = \lambda\nabla\psi(x)$$

where the number λ (which is also to be found) is called a *Lagrange multiplier*. Now in the case of Rayleigh's principle, we're trying to maximize $\varphi(x) = x^tMx$ subject to $\psi(x) = x^tx = 1$. In the next section we will find a formula (10.4.8) for the gradients of φ and ψ. Using this formula, and cancelling a factor of 2, we see that the Lagrange multiplier equation is just the eigenvalue equation $Mx = \lambda x$.

Remark: We used the symmetry of the matrix M several times in the proof of theorem 10.2.1. It is *not true* that a general matrix M has n independent eigenvectors (still less that it has n orthonormal ones). A simple example of this is given by the matrix $M = \begin{pmatrix} 1 & 1 \\ 0 & 1 \end{pmatrix}$. It is easily checked that the only eigenvalue is 1, and that the only eigenvectors for this eigenvalue are multiples of $\begin{pmatrix} 1 \\ 0 \end{pmatrix}$.

(10.3.1)   **Example:** An important example to which the general form of theorem 10.2.1 can be applied occurs in mechanics. We mentioned in 5.2.5 the idea that the behaviour of a complex mechanical system can be interpreted by the motion of a point in some high-dimensional *configuration space*. In simple cases, the motion is completely described by a potential energy function $V(q_i)$ of the coordinates q_i on configuration space. The system will be in stable equilibrium at a point where the potential energy attains its minimum.

Now suppose that the system is slightly disturbed from rest. Because the first derivatives of V are all zero at the minimum, the behaviour of the system will be governed by the matrix $\partial^2 V/\partial q_i \partial q_j$ of second derivatives. This is a symmetric matrix which can therefore be diagonalized. It turns out that the eigenvectors correspond to so-called *normal modes* of oscillation of the system, and that any oscillation can be regarded as a superposition of independent normal modes.

10.4 Intersection of a quadric with a line or plane

(10.4.1)   **Proposition:** *Let Q be a quadric and let L be a line. Then either $L \cap Q$ has at most two points, or else L lies entirely in Q.*

Proof: This is similar to 7.4.1. A general point on L has coordinates $x = x_0 + kt$, $y = y_0 + lt$, $z = z_0 + mt$, where t is a parameter. Substituting these into the equation of Q we obtain a quadratic for t. Either this quadratic has at most two roots (giving at most two points of intersection) or it is identically zero (in which case L lies entirely in Q). $\square$

The second alternative can occur; quadrics, despite the fact that they are curved surfaces, can contain straight lines. One obvious example is a cylinder, which contains an infinite family of parallel straight lines. There are also some more surprising possibilities:

(10.4.2)   **Example:** Consider the hyperboloid of one sheet whose equation is $x^2/a^2 + y^2/b^2 - z^2/c^2 = 1$. We may write this equation in the form

(10.4.3)
$$\left(\frac{x}{a} - \frac{z}{c}\right)\left(\frac{x}{a} + \frac{z}{c}\right) = \left(1 - \frac{y}{b}\right)\left(1 + \frac{y}{b}\right).$$

Let λ and μ be two real numbers, not both zero. We can define two planes P_1 and P_2 by

$$P_1 : \quad \lambda\left(\frac{x}{a} - \frac{z}{c}\right) = \mu\left(1 - \frac{y}{b}\right)$$

$$P_2 : \quad \mu\left(\frac{x}{a} + \frac{z}{c}\right) = \lambda\left(1 + \frac{y}{b}\right).$$

These planes have normal vectors $\frac{\lambda}{a}\mathbf{i} + \frac{\mu}{b}\mathbf{j} - \frac{\lambda}{c}\mathbf{k}$ and $\frac{\mu}{a}\mathbf{i} - \frac{\lambda}{b}\mathbf{j} + \frac{\mu}{c}\mathbf{k}$. These two vectors are never parallel, so that the planes intersect in a straight line L

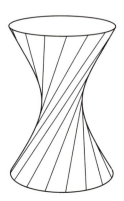

Fig. 10.6. A one-parameter family of straight lines on a hyperboloid.

(see proposition 8.2.6). Any point on $\mathcal{L}$ satisfies the equations both of $\mathcal{P}_1$ and $\mathcal{P}_2$, so substituting into equation 10.4.3 we find that $\mathcal{L}$ is completely contained within the hyperboloid.

As λ and μ vary we obtain a whole family of straight lines on the hyperboloid, illustrated in Figure 10.6. It is a one-parameter family of lines, because only the ratio of λ to μ really counts.

We can obtain a second one-parameter family of straight lines on the hyperboloid by intersecting the planes

$$\mathcal{P}_1' : \quad \lambda\left(\frac{x}{a} + \frac{z}{c}\right) = \mu\left(1 - \frac{y}{b}\right)$$
$$\mathcal{P}_2' : \quad \mu\left(\frac{x}{a} - \frac{z}{c}\right) = \lambda\left(1 + \frac{y}{b}\right).$$

These are lines 'sloping the other way'. It can be shown that in general each line of the second family meets each line of the first family in a point. The two families of lines are said to be *transversal* to each other. Because of the existence of these two families of lines on it, the hyperboloid is said to be a *doubly ruled surface*. It can be proved that the only nontrivial doubly ruled surfaces are quadrics; see Exercise 10.5.8.

(10.4.4)   **Example:** The hyperbolic paraboloid $z = x^2/a^2 - y^2/b^2$ also contains two families of straight lines. You can easily check that the two planes $\lambda z = \mu(x/a \pm y/b)$ and $\mu = \lambda(x/a \mp y/b)$ intersect in a straight line that lies in the paraboloid.

(10.4.5)   **Tangent lines:** We say that the line $\mathcal{L}$ is *tangent* to the quadric $\mathcal{Q}$ at a point P if the quadratic equation describing the intersection of $\mathcal{L}$ and $\mathcal{Q}$ has a double root at P. (We allow the possibility that this quadratic is identically zero, so that $\mathcal{L}$ is contained in $\mathcal{Q}$.) As in Section 7.1, it is possible to reformulate this definition using calculus. Thus, let the quadric be defined by an equation $f(x, y, z) = 0$. The *gradient* of such a function of three variables is defined by

$$\nabla f(x, y, z) = \frac{\partial f}{\partial x}\mathbf{i} + \frac{\partial f}{\partial y}\mathbf{j} + \frac{\partial f}{\partial z}\mathbf{k}.$$

The arguments leading to proposition 7.1.5 can be repeated with an extra variable z, so as to prove

(10.4.6) **Proposition:** *A line $\mathcal{L}$ is tangent to a quadric with equation $f(x, y, z) = 0$ at a point (x_0, y_0, z_0) on both of them if and only if $\mathcal{L}$ is perpendicular to the gradient vector $\nabla f(x_0, y_0, z_0)$.*

(10.4.7) **Corollary:** *At a point on the quadric $f(x, y, z) = 0$ where $\nabla f \neq \mathbf{0}$, the tangent lines form a plane, the tangent plane at the given point.*

The condition that the gradient be nonzero is a regularity condition for surfaces, analogous to the regularity condition for curves studied in Section 7.1. One can introduce the same terminology of regular and singular points. For example, every point of the cone $x^2 + y^2 = z^2$ is regular except the origin $(0, 0, 0)$ which is singular.

When calculating tangents to a quadric, the following lemma is often useful.

(10.4.8) **Lemma:** *Let*
$$f(x, y, z) = \mathbf{x}^t \mathbf{M} \mathbf{x} + \mathbf{N} \mathbf{x} + k$$

where $\mathbf{x}$ denotes the column vector $\begin{pmatrix} x \\ y \\ z \end{pmatrix}$. Then $\nabla f(x, y, z)$ is represented by the column vector $2\mathbf{M}\mathbf{x} + \mathbf{N}^t$.

Proof: Write

$$f(x, y, z) = Ax^2 + By^2 + Cz^2 + 2Dxy + 2Eyz + 2Fzx + Px + Qy + Rz + k.$$

Then

$$\frac{\partial f}{\partial x} = 2(Ax + Dy + Fz) + P$$

$$\frac{\partial f}{\partial y} = 2(Dx + By + Ez) + Q$$

$$\frac{\partial f}{\partial z} = 2(Fx + Ey + Cz) + R$$

and this can be written in matrix form as

$$\nabla f = 2\mathbf{M}\mathbf{x} + \mathbf{N}^t. \qquad \square$$

Therefore if the matrix $\mathbf{M}$ is invertible there is at most one singular point, because there is only one root $\mathbf{x}$ of the equation $2\mathbf{M}\mathbf{x} + \mathbf{N}^t = 0$. Usually this root will not lie on the surface, so there will be no singular points at all. Moreover for a central quadric we have:

(10.4.9) **Proposition:** *Let f be as above. If P = Q = R = 0, so that the quadric defined by f(x, y, z) = 0 is central, and k ≠ 0, then the quadric has no singular points.*

Proof: The equation of the quadric is $x^t Mx + k = 0$. But by lemma 10.4.8, ∇f is represented by the column vector $2Mx$. Thus, at a singular point $Mx = 0$ and so $x^t Mx + k = k$. Since $k \neq 0$, this is a contradiction; so there can be no singular points. □

It is also interesting to consider the intersection of a quadric with a *plane*. The quadric will meet the plane in a curve, which will in general be a conic. When the quadric is a cone, we will recover in this way the original Greek definition of conics.

(10.4.10) **Proposition:** *Let $\mathcal{Q}$ be a quadric and $\mathcal{P}$ a plane in a three-dimensional Euclidean space. Suppose that in the equation*

$$x^t Mx + Nx + k = 0$$

the determinant of M is nonzero. Then $\mathcal{Q} \cap \mathcal{P}$ is a conic in $\mathcal{P}$.

Proof: Let **n** be a unit normal vector to $\mathcal{P}$. Change to new Cartesian coordinates with origin somewhere on $\mathcal{P}$ and such that **n** is the third basis vector. Then, in these new coordinates, $\mathcal{P}$ has equation $z = 0$. If the quadric has equation

$$Ax^2 + By^2 + Cz^2 + 2Dxy + 2Eyz + 2Fzx + Px + Qy + Rz + k = 0$$

in the new coordinates, then its intersection with $\mathcal{P}$ is clearly given by the equation

$$Ax^2 + By^2 + 2Dxy + Px + Qy + k = 0.$$

This certainly looks like the equation of a conic; it could only fail to be one if A, B, and D were all zero. But then the matrix M relative to the new coordinates would be

$$\begin{pmatrix} 0 & 0 & F \\ 0 & 0 & E \\ F & E & C \end{pmatrix}$$

and this matrix has determinant zero, because its first two columns are linearly dependent. □

Remark: Under a coordinate change, M gets replaced by $U^t MU$, which has the same determinant as M. Thus the condition that det M ≠ 0 is independent of the choice of coordinate system.

(10.4.11) Example: A cone with equation $x^2 + y^2 = c^2z^2$ is represented by a matrix with nonzero determinant $-c$. So its intersection with any plane is a conic. This reconciles our definition of conics with the original Greek one discussed at the beginning of section 7.3.

10.5 Exercises

1. Find out what kinds of quadrics are represented by each of the following equations:

(i) $3x^2 + 6y^2 - 2z^2 + 4xy - 12zx + 6yz = 1$;

(ii) $3x^2 - 2y^2 - z^2 - 4xy - 12yz - 8zx = 1$;

(iii) $x^2 + y^2 + z^2 = \frac{3}{4}(x + y + z)^2$;

(iv) $9x^2 + 5y^2 + 5z^2 + 12xy + 6xz + 5x - 6y - 3z = 2$.

2. A plane $\mathcal{P}$ in $\mathbf{R}^3$ is given by the equation $z = \frac{1}{2}y + 1$. Find a coordinate change to new coordinates x', y', z' for which $\mathcal{P}$ has equation $z' =$constant. Hence find the eccentricity of the conic obtained by intersecting $\mathcal{P}$ with the cone $x^2 + y^2 = z^2$.

Why do you not get the right answer if you just substitute $z = \frac{1}{2}y + 1$ into the equation of the cone and think of the resulting equation as a conic in x and y?

3. An ellipsoid with centre at the origin meets the plane $z = 0$ in a circle with radius r. Show that the equation of the ellipsoid may be written

$$\frac{x^2}{r^2} + \frac{y^2}{r^2} + \frac{z^2}{s^2} + 2pxz + 2qyz = 1.$$

Show that one of the principal axes of the ellipsoid has length r, and that if the lengths of the principal axes are $a > b > c > 0$, then $r = b$. Show also that the principal axis which has length b lies in the plane $z = 0$.

4. The equation of a quadric $\mathcal{Q}$ is given in matrix form as $\mathsf{x}'\mathsf{M}\mathsf{x} + \mathsf{N}\mathsf{x} + k = 0$, where M is a 3×3 symmetric matrix, N is a 3×1 row matrix, and x is the coordinate vector.

Let u be some unit vector. Suppose that a is the midpoint of a chord of $\mathcal{Q}$ parallel to u; in other words, that there is some nonzero constant λ for which the two points $\mathsf{a} \pm \lambda\mathsf{u}$ both lie on $\mathcal{Q}$. Prove that a lies on the plane with equation

$$2\mathsf{a}'\mathsf{M}\mathsf{u} + \mathsf{N}\mathsf{u} = 0.$$

(This is called the *diametral plane* corresponding to u.) Show further that if the matrix M is non-singular, then there is a point that lies on every diametral plane. What point is it?

5. An ellipsoid has equation

$$\frac{x^2}{a^2} + \frac{y^2}{b^2} + \frac{z^2}{c^2} = 1.$$

The normal is drawn at a point P on the section of the ellipsoid by the plane with equation $z = k$, where $-c < k < c$. Let Q be the point in which this normal meets the plane with equation $x = 0$. Show that, as P varies, the locus of Q is a line segment parallel to the y-axis; and find the distance between the extreme positions of Q.

6. Complete the following outline to prove that any continuous function φ on S^{n-1} is bounded and attains its bounds.

(i) Let M be the least upper bound of φ on S^{n-1} (for all we know at this stage, M may be $+\infty$). Let C_0 be the (hyper)cube with vertices $(\pm 1, \ldots, \pm 1)$, so that $S^{n-1} \subseteq C_0$. By dividing C_0 into 2^n cubes of side 1, show that there is such a cube C_1 with $\sup\{\varphi(x) : x \in C_1 \cap S^{n-1}\} = M$.

(ii) By subdividing C_1, show that there is a cube $C_2 \subseteq C_1$ of side $\frac{1}{2}$ with $\sup\{\varphi(x) : x \in C_2 \cap S^{n-1}\} = M$. By induction show that there are cubes $C_n \subseteq C_{n-1}$ of side 2^{-n+1} with $\sup\{\varphi(x) : x \in C_n \cap S^{n-1}\} = M$.

(iii) Let $x_n \in C_n \cap S^{n-1}$. Prove that the sequence x_n is convergent, say to x_∞.

(iv) Using the continuity of φ, show that $M = \varphi(x_\infty)$.

7. Let M be a symmetric matrix and let u and v be two eigenvectors with distinct eigenvalues λ and μ. By evaluating the expression

$$u'Mv = v'Mu$$

in two different ways, prove that the two eigenvectors u and v are perpendicular.

8. Prove that it is possible to find a quadric passing through nine given noncoplanar points in three-dimensional space. Prove also that a straight line that meets a quadric in three distinct points lies entirely in this quadric.

Hence show that given three skew lines it is possible to find a quadric passing through them, and that this quadric is in fact the surface swept out by all the transversals to the three given lines (a *transversal* is a line that meets all three of the given lines).

Deduce that the only doubly ruled surfaces are quadrics.

9. Show that the equation

$$(x^2 + y^2 + z^2)^2 + (a^2 - b^2)^2 - 2(x^2 + y^2)(a^2 + b^2) + 2z^2(a^2 - b^2) = 0$$

(where $a > b$) defines a *torus* in three-dimensional space, obtained by rotating the circle in the xz-plane with equation

$$(x - a)^2 + z^2 = b^2$$

about the z-axis.

Show that the torus meets the plane $bx - (\sqrt{a^2 - b^2})z = 0$ in a pair of circles. Attempt a sketch.

(By rotation we can find two families of circles lying in the torus, analogous to the two families of lines on a hyperbolic paraboloid. They are called the *Villarceau circles* of the torus. According to Berger [4], one can see the Villarceau circles represented in sculpture at the top of the spiral staircase of the l'Oeuvre Notre-Dame museum in Strasbourg.)

11

Differential geometry of curves

11.1 Differentiation of vector functions

Most of the geometry that we have been studying so far has been what is called *algebraic geometry*. Algebraic geometry studies curves and surfaces that are represented by polynomial equations — such as the equation $x^2 + y^2 = 1$ of the circle — using only algebraic facts about polynomials and their roots. For instance, we discussed the tangent lines to a conic in terms of double roots of the equation for the intersection of a line and a conic. This is a typical kind of algebraic-geometry argument.

We also gave a different, though equivalent, definition of tangent lines in terms of calculus. This involves the nonalgebraic concepts of limit, differentiation, and so on. It can be extended to more general curves than the algebraic definition, but on the other hand it relies more heavily on analytic ideas which are specially associated with the fields of real or complex numbers. Applying the ideas of differential calculus to geometry gives rise to the subject called *differential geometry*. One of the key ideas in differential geometry is that of *curvature*, which is some combination of derivatives that measures the deviation of a figure from being straight or flat.

In this chapter we will study curves in three-dimensional space from the point of view of differential geometry. Such curves were first studied by Clairault (1731) who called them 'curves of double curvature'— the idea being that the curve has as it were two degrees of freedom ('bending' and 'twisting') in its motion at any point. In making this idea precise we will need to use differential calculus for *vector-valued* functions of a real variable, as well as ordinary or *scalar-valued* functions. We therefore begin with a short introduction to vector differentiation. We do not give proofs, though, as they are entirely analogous to the familiar proofs that one gives in the scalar case.

(11.1.1) **Definition:** *We will say that a vector-valued function* $\mathbf{r}(t)$ *tends to the limit*

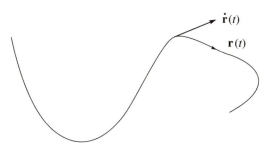

Fig. 11.1. The velocity vector of a particle.

a *as* $t \to t_0$ *if the ordinary real-valued function* $|\mathbf{r}(t) - \mathbf{a}|$ *tends to zero as* $t \to t_0$. *We will say that a vector-valued function* $\mathbf{r}(t)$ *is* differentiable *at* t_0 *if the vector*

$$\frac{\mathbf{r}(t) - \mathbf{r}(t_0)}{t - t_0}$$

tends to a limit as $t \to t_0$. *The limit will be called the* derivative *of the function at* $t = t_0$, *and will be denoted by* $\dot{\mathbf{r}}(t_0)$ *or* $\left.\dfrac{\mathrm{d}\mathbf{r}}{\mathrm{d}t}\right|_{t=t_0}$.

Notice that the derivative of a vector-valued function is another vector-valued function. Differentiation can therefore be repeated as often as one likes. If $\mathbf{r}(t)$ is thought of as giving the position at time t of a particle moving in space, then $\dot{\mathbf{r}}(t)$ gives the *velocity vector* of the particle at time t — see Figure 11.1. Similarly, the second derivative $\ddot{\mathbf{r}}(t)$ gives the acceleration vector at time t.

Vector differentiation obeys the following rules, analogous to rules for ordinary differentiation:

- If $\mathbf{r}$ is a constant vector (independent of t), then $\dot{\mathbf{r}} = \mathbf{0}$.

- Differentiation is linear: $\dfrac{\mathrm{d}(\mathbf{r}_1 + \mathbf{r}_2)}{\mathrm{d}t} = \dfrac{\mathrm{d}\mathbf{r}_1}{\mathrm{d}t} + \dfrac{\mathrm{d}\mathbf{r}_2}{\mathrm{d}t};$

- There is a product formula for multiplication by a scalar function of t: if $f(t)$ is a scalar function of t then

$$\frac{\mathrm{d}(f\mathbf{r})}{\mathrm{d}t} = \frac{\mathrm{d}f}{\mathrm{d}t}\mathbf{r} + f\frac{\mathrm{d}\mathbf{r}}{\mathrm{d}t}.$$

- There are also product formulae involving the scalar and vector products: if $\mathbf{r}_1$ and $\mathbf{r}_2$ are vector functions of t, then

$$\frac{\mathrm{d}(\mathbf{r}_1 \cdot \mathbf{r}_2)}{\mathrm{d}t} = \frac{\mathrm{d}\mathbf{r}_1}{\mathrm{d}t} \cdot \mathbf{r}_2 + \mathbf{r}_1 \cdot \frac{\mathrm{d}\mathbf{r}_2}{\mathrm{d}t}$$

$$\frac{d(\mathbf{r}_1 \wedge \mathbf{r}_2)}{dt} = \frac{d\mathbf{r}_1}{dt} \wedge \mathbf{r}_2 + \mathbf{r}_1 \wedge \frac{d\mathbf{r}_2}{dt}.$$

These rules imply that if $\mathbf{r}(t) = x(t)\mathbf{i} + y(t)\mathbf{j} + z(t)\mathbf{k}$ relative to some orthonormal basis $\mathbf{i}, \mathbf{j}, \mathbf{k}$, then $\dot{\mathbf{r}}(t) = \dot{x}(t)\mathbf{i} + \dot{y}(t)\mathbf{j} + \dot{z}(t)\mathbf{k}$. Thus you can calculate the derivative of a vector function simply by differentiating its components relative to some orthonormal basis.

Remark: If $\mathbf{r} = \mathbf{r}(u, v)$ is a vector function of two real variables then we can define vector partial derivatives $\dfrac{\partial \mathbf{r}}{\partial u}$ and $\dfrac{\partial \mathbf{r}}{\partial v}$ in the usual way, by holding one variable fixed and differentiating with respect to the other. Vector partial derivatives will be needed when we come to study the differential geometry of surfaces in the next chapter.

We will often need the following useful fact:

(11.1.2)   **Lemma:** *Let* $\mathbf{r}(t)$ *be a vector function of* t, *and suppose that the modulus* $|\mathbf{r}(t)|$ *is constant in* t. *Then* $\mathbf{r}(t) \cdot \dot{\mathbf{r}}(t) = 0$ *for all* t; *in other words,* $\mathbf{r}$ *is perpendicular to* $\dot{\mathbf{r}}$.

What this is saying geometrically is that any vector tangent to a circle or sphere at some point must be perpendicular to the radius vector from the centre to that point.

Proof: Since $|\mathbf{r}|$ is constant,

$$0 = \frac{d}{dt}\left(|\mathbf{r}|^2\right) = \frac{d}{dt}\left(\mathbf{r} \cdot \mathbf{r}\right) = \dot{\mathbf{r}} \cdot \mathbf{r} + \mathbf{r} \cdot \dot{\mathbf{r}} = 2\mathbf{r} \cdot \dot{\mathbf{r}}. \qquad \square$$

We can now define the curves that we will be studying in this chapter. Let S be a three-dimensional oriented Euclidean space. We assume that an origin O has been chosen in S, so that points of S can be represented by their position vectors. A *smooth curve* in S is then given by a smooth (that is, infinitely differentiable) vector-valued function $t \mapsto \gamma(t)$ from an interval $[a, b]$ of real numbers to S. The curve is *regular* if the derivative $\dot{\gamma}(t)$ is nonzero for every $t \in [a, b]$. This definition is the same as that given for plane parameterized curves in Section 7.1. As in that section, we can also define *piecewise regular* curves if we wish; but in fact we will not have any use for piecewise regular space curves.

(11.1.3)   **Example:** A *straight line* is of course the simplest kind of curve. It can be given in coordinates by

$$\gamma(t) = (a_1 + b_1 t, a_2 + b_2 t, a_3 + b_3 t)$$

where a_1, a_2, a_3, b_1, b_2, and b_3 are constants.

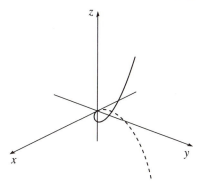

Fig. 11.2. A twisted cubic.

(11.1.4) **Example:** The curve with equation $\gamma(t) = (a\cos t, a\sin t, bt)$ is called a *helix*. It represents the path of a point on an aeroplane's propellor, moving forward and turning around at the same time.

(11.1.5) **Example:** The curve with equation $\gamma(t) = (at, bt^2, ct^3)$ is called a *twisted cubic*. It is illustrated in Figure 11.2.

Remark: If you are careful you will have noticed that our definition of a smooth curve in S depends on a choice of origin. Would the definition have been the same if we had chosen a different origin? The answer, of course, is that the choice of origin makes no difference. To see this, observe that a change of origin would change $\gamma(t)$ by a *constant* vector, and the derivative of a constant vector is zero.

11.2 Planetary orbits

Kepler stated in 1609 that the planets move in ellipses around the sun, having the sun at one focus. His laws were based on the astronomical observations of Tycho Brahe. One of the triumphs of seventeenth-century theoretical physics was the derivation, by Newton and others, of Kepler's laws from a unifying theoretical principle: Newton's *inverse square law* of gravitation, which states that two small massive bodies attract one another with a force that is inversely proportional to the distance between them and is directed along the line joining them. It is not known exactly when Newton found this result, but he was stimulated to publish it by a visit from his friend the astronomer Edmund Halley in 1684. Here is an account (written years later by Newton's son-in-law) of that famous occasion:

In 1684 Dr. Halley came to visit him [Newton] at Cambridge, after they had been some time together, the Dr. asked him what he thought the Curve would be that would be described by the Planets supposing the force of attraction towards the Sun to be reciprocal to the square of their distance from it. Sir Isaac replied immediately that it would be an Ellipsis, the Doctor struck with joy and amazement asked him how he knew it. Why saith he I have calculated it, whereupon Dr. Halley asked him for his calculations without further delay, Sir Isaac looked among his papers but could not find it, but he promised him to renew it, and then to send it to him ...

Newton's calculation is given in his famous book *Principia Mathematica Naturalis Philosophiae* (Mathematical Principles of Natural Philosophy). It used the classical Apollonian geometry of conics. A more modern derivation of the elliptic orbit provides a nice example of vector differentiation at work.

To a first approximation we may think of the sun and a planet as being pointlike particles, with the sun fixed. We take a fixed origin O at the position of the sun, and let $\mathbf{r} = \mathbf{r}(t)$ be the position vector of the planet, considered as a function of the time t. Assuming that the planet does not actually fall into the sun, $\mathbf{r}(t)$ is nonzero; so we may write

$$\mathbf{r}(t) = R(t)\mathbf{w}(t)$$

where $R = |\mathbf{r}|$ and $\mathbf{w}$ is a unit vector. Newton's law of gravitation may then be stated in the form

$$\ddot{\mathbf{r}} = -\frac{K}{R^2}\mathbf{w},$$

where K is some constant.

From the equation of motion we can find the orbit, as follows:

First step: We notice first that $\mathbf{r} \wedge \dot{\mathbf{r}} = \mathbf{H}$ is a constant vector. For $\dot{\mathbf{H}} = \mathbf{r} \wedge \ddot{\mathbf{r}} = \mathbf{0}$ because $\mathbf{r}$ and $\ddot{\mathbf{r}}$ are both multiples of $\mathbf{w}$. In physical terms, $\mathbf{H}$ is the *vector angular momentum* of the planet and what we have just proved is the law of conservation of vector angular momentum. Since $\mathbf{H}$ is perpendicular to $\mathbf{r}$, $\mathbf{r} \cdot \mathbf{H} = 0$. But since $\mathbf{H}$ is constant this is simply the equation of a plane (see 8.2.3), in which the planet always remains.

Second step: Now we can calculate $\dot{\mathbf{r}} = \dot{R}\mathbf{w} + R\dot{\mathbf{w}}$, and so $\mathbf{H} = \mathbf{r} \wedge \dot{\mathbf{r}} = R^2\mathbf{w} \wedge \dot{\mathbf{w}}$. Now $\mathbf{w}$ is of constant magnitude, and so $\dot{\mathbf{w}}$ is perpendicular to $\mathbf{w}$ (see 11.1.2). Therefore the magnitude of the vector $\mathbf{H}$ is equal to $R^2|\dot{\mathbf{w}}|$. We may now introduce polar coordinates in the plane of the motion (in other words, we may write the unit vector $\mathbf{w}$ as $(\cos\theta, \sin\theta)$); if we do this, then $|\dot{\mathbf{w}}| = \dot{\theta}$, so

$$R^2\dot{\theta} = h$$

where $h = |\mathbf{H}|$ is a constant.

The quantity $\frac{1}{2}R^2\dot{\theta}$ is the rate at which the line joining the planet to the sun sweeps out area[1] — the so-called *areal velocity* of the planet. It was one

---

[1]This follows from the formula for the area enclosed by a curve as a contour integral (9.6.5).

of Kepler's observations that this areal velocity is a constant. Here, then, is a theoretical explanation.

Third step: We can differentiate again to find

$$-\frac{K}{R^2}\mathbf{w} = \ddot{\mathbf{r}} = \ddot{R}\mathbf{w} + 2\dot{R}\dot{\mathbf{w}} + R\ddot{\mathbf{w}}.$$

Take the dot product with $\mathbf{w}$ to obtain

$$\ddot{R} + R\ddot{\mathbf{w}} \cdot \mathbf{w} = -\frac{K}{R^2}.$$

But $\ddot{\mathbf{w}} \cdot \mathbf{w} + \dot{\mathbf{w}} \cdot \dot{\mathbf{w}}$ is the derivative of $\dot{\mathbf{w}} \cdot \mathbf{w} = 0$, and so is zero; therefore

$$\ddot{\mathbf{w}} \cdot \mathbf{w} = -\dot{\mathbf{w}} \cdot \dot{\mathbf{w}} = -\frac{h^2}{R^4}.$$

Using this, we get the equation

(11.2.1) $$\ddot{R} - \frac{h^2}{R^3} + \frac{K}{R^2} = 0.$$

Fourth step: Equation 11.2.1 describes the distance R as a function of time t. However, we are primarily interested in the *shape* of the orbit rather than the rate at which it is traced out; so we want to know R as a function of the polar coordinate θ, rather than of t. We therefore calculate the θ-derivatives of R in terms of the t-derivatives.

By the chain rule

$$\frac{dR}{dt} = \frac{dR}{d\theta}\dot{\theta} = \frac{h}{R^2}\frac{dR}{d\theta}.$$

The form of this equation suggests that we should take $u = 1/R$ as independent variable in place of R. Then

$$\frac{du}{d\theta} = \frac{du/dt}{d\theta/dt} = -\frac{\dot{R}}{h}.$$

Differentiating again, we get

$$\frac{d^2u}{d\theta^2} = -\frac{1}{\dot{\theta}}\frac{\ddot{R}}{h} = -\frac{R^2\ddot{R}}{h^2},$$

and so $\ddot{R} = -\frac{h^2}{R^2}\frac{d^2u}{d\theta^2}.$

Fifth step: Substitute this last formula into 11.2.1 and cancel $1/R^2$. You get

$$-h^2 \left(\frac{d^2 u}{d\theta^2} + u \right) + K = 0.$$

This is just a simple harmonic motion equation with the origin offset by K/h^2. The general solution is

$$u = \frac{K}{h^2} - a \cos(\theta + \alpha)$$

where a and α are arbitrary constants. By choosing the initial line for our polar coordinate system we can make α equal to anything we want; let us choose things so that $\alpha = 0$. Then

$$\frac{1}{R} = u = \frac{K}{h^2} + a \cos \theta.$$

So

$$R = \frac{el}{1 - e \cos \theta}$$

where $e = ah^2/K$ and $l = 1/a$. We recognize this as the polar equation of a conic (7.2.1).

Thus we have seen that Newton's gravitational law implies that the planets move in conics around the sun. The only possible periodic orbits are therefore ellipses.

11.3 The length of a curve

Let γ be a curve in some Euclidean space S. We would like to define the *length* of the curve. To see what the length should be, consider a short piece of the curve between parameter values t and $t + \Delta t$. Then

$$\gamma(t + \alpha \Delta t) \approx \gamma(t) + \alpha \dot{\gamma}(t)$$

for $0 \le \alpha \le 1$, and so the piece of curve is approximately a straight line segment of length $|\dot{\gamma}(t)| \Delta t$. If we think of the curve as being put together from such segments, this suggests the following definition:

(11.3.1)   **Definition:** *The length of the curve γ between the points $\gamma(a)$ and $\gamma(b)$ (or, as one sometimes says, between the parameter values a and b) is defined to be*

$$\int_a^b |\dot{\gamma}(t)| \, dt.$$

If the end points a and b are understood, the length will be denoted by $\ell(\gamma)$. One can also think of this definition in mechanical terms; $\dot{\gamma}(t)$ is the velocity vector, so $|\dot{\gamma}(t)|$ is the speed and its integral is the distance covered.

In the notation of 9.6.3, the length of a curve can be expressed by

$$\int_\gamma \sqrt{\mathrm{d}x^2 + \mathrm{d}y^2 + \mathrm{d}z^2}.$$

As usual, this should be thought of as an abbreviation for

$$\int_a^b \sqrt{\left(\frac{\mathrm{d}x}{\mathrm{d}t}\right)^2 + \left(\frac{\mathrm{d}y}{\mathrm{d}t}\right)^2 + \left(\frac{\mathrm{d}z}{\mathrm{d}t}\right)^2} \, \mathrm{d}t.$$

One can use the chain rule to show that this expression does not depend on the choice of parameterization t for the curve γ.

(11.3.2) **Example:** Let us work out the length of the helix $t \mapsto (\cos t, \sin t, t)$ between the points $(1,0,0)$ (corresponding to $t = 0$) and $(1,0,2\pi)$ (corresponding to $t = 2\pi$). We find

$$\dot{\gamma}(t) = -\sin t\mathbf{i} + \cos t\mathbf{j} + \mathbf{k}, \quad |\dot{\gamma}(t)| = \sqrt{2}.$$

Therefore the length of the curve is $\int_0^{2\pi} \sqrt{2} \, \mathrm{d}t = 2\sqrt{2}\pi$.

Notice that the length of the helix in this example is greater than the distance 2π between the two end points of the curve — the particle has not travelled in a straight line, and so it has had to go further. In fact we can prove from our definition of the length that a straight line is always the shortest curve between two points. This is not a surprise, of course; but it does show that our definition of length has some of the properties we would intuitively expect.

(11.3.3) **Proposition:** *Let A and B be points in S, and let* $\gamma: [a, b] \rightarrow S$ *be a curve with* $\gamma(a) = A$ *and* $\gamma(b) = B$. *Then the length of* γ *is greater than or equal to* $|AB|$, *and equality occurs only if* γ *is a straight line.*

Proof: Take an origin at A and write $\mathbf{r}(t)$ for the position vector of $\gamma(t)$, in other words $\mathbf{r}(t) = \overrightarrow{A\gamma(t)}$. We may assume without loss of generality that $\mathbf{r}(t) \neq \mathbf{0}$ for $t > a$ — if you are looking for the shortest path from A to B, it would be pretty stupid to spend some time going round in a loop and then coming back to A again. Therefore we may write

$$\mathbf{r}(t) = R(t)\mathbf{w}(t)$$

where $R(t) = |\mathbf{r}(t)|$ and $\mathbf{w}(t)$ is a unit vector.

Differentiating this identity, we get

$$\dot{\mathbf{r}}(t) = \dot{R}(t)\mathbf{w}(t) + R(t)\dot{\mathbf{w}}(t).$$

But $\mathbf{w}$ and $\dot{\mathbf{w}}$ are perpendicular because $\mathbf{w}$ has constant magnitude, and so by Pythagoras' theorem

$$|\dot{\mathbf{r}}(t)|^2 = \dot{R}(t)^2 + R(t)^2|\dot{\mathbf{w}}(t)|^2 \geq \dot{R}(t)^2$$

and it follows that $|\dot{\mathbf{r}}(t)| \geq \dot{R}(t)$. Therefore

$$\ell(\gamma) = \int_a^b |\dot{\mathbf{r}}(t)|\, dt \geq \int_a^b \dot{R}(t)\, dt = R(b) - R(a) = |AB| - 0 = |AB|.$$

Equality can occur only if $\dot{\mathbf{w}}(t) \equiv 0$, which means that $\mathbf{w}$ is a constant vector and so the path is straight. $\square$

In 12.1.7 we will use an analogous argument to find the shortest path between two points on a sphere.

Remember that the 'same' curve may have many different parameterizations, related by parameter change maps (7.1.10). However, we can define a standard parameterization of any regular curve in terms of its length. Remember that we call a curve γ *regular* if it has no singular points, that is if $\dot{\gamma}(t) \neq \mathbf{0}$ at any point in the parameter range.

(11.3.4) **Proposition:** *Let* $\gamma: [a, b] \to S$ *be a regular curve. Let* ℓ *be the length of* γ. *Then the* arc length *function* $\varphi: [a, b] \to [0, \ell]$ *defined by*

$$\varphi(x) = \int_a^x |\dot{\gamma}(t)|\, dt$$

is a parameter change map. If $\tilde{\gamma} = \gamma \circ \varphi^{-1}$ *is the reparameterized curve, then* $|\dot{\tilde{\gamma}}(s)| = 1$ *for all allowable parameter values* s.

Proof: The proof is shorter than the proposition. Clearly $\dot{\varphi}(x) = |\dot{\gamma}(x)| > 0$ by the fundamental theorem of calculus, so φ is a parameter change map. By the chain rule, if $s = \varphi(t)$, then the derivative $|\dot{\gamma}(t)| = |\dot{\tilde{\gamma}}(s)\dot{\varphi}(t)| = |\dot{\tilde{\gamma}}(s)||\dot{\varphi}(t)|$. Cancelling $|\dot{\gamma}(t)| = \dot{\varphi}(t)$, we get $|\dot{\tilde{\gamma}}(s)| = 1$. $\square$

The curve $\tilde{\gamma}$ defined by this proposition is called a *unit speed parameterization* of γ. Because the parameter s is equal to the arc length measured along the curve, it is also often called an *arc length parameterization*. What the proposition shows is that every regular curve has an arc length parameterization. This means that for theoretical work we can make the assumption that our curves are parameterized by arc length, which often simplifies the calculations. Of course, curves that one encounters in practice are often not given in such a convenient form.

As a matter of notation, we will usually denote differentiation with respect to an arc length parameter s by a dash — like in γ' — rather than by a dot — like in $\dot{\gamma}$. This is particularly convenient when we have to work simultaneously with an arc length parameterization and a general parameterization of the same curve — the dashes and dots help you keep track of the different derivatives that are involved.

11.4 The Serret–Frenet formulae

When we were discussing conics, one of our aims was to find a coordinate system in which the equation of the curve took a particularly simple form. We looked for a coordinate system in which the xy terms — and if possible the x and y terms too — were eliminated from the equation. We will now try to do something similar for a general space curve γ.

Notice, however, that there is a big difference between a conic and a general curve — a difference that exemplifies a fundamental contrast between algebraic and differential geometry. If you are given just a small piece of a conic, you can reconstruct the whole of the rest of the conic from it. In fact we have seen that just five points are usually enough to reconstruct the whole conic (7.4.4). This kind of 'rigidity' is characteristic of algebraic geometry. By contrast, the behaviour of a general curve γ in one small piece imposes no constraint on its behaviour elsewhere. This means that we can't hope in general to find a frame in which the *whole curve* γ is simplified; the best we can hope for is to find a frame in which γ is simplified *in a small neighbourhood of some point*, and the frame will depend on the point we choose. One often refers to such a frame as a 'moving frame' attached to the curve — the idea being that the basis vectors of the frame change as they 'move along the curve'. In other words, the frame is a function of the curve parameter.

Suppose then that $s \mapsto \gamma(s)$ is a regular curve in a three-dimensional Euclidean space. As we have just seen, we may then assume (by reparameterization if necessary) that the parameter s is equal to arc length along the curve, so that $\mathbf{t} = \gamma'(s)$ is a unit vector. $\mathbf{t}$ is going to be the first basis vector of our moving frame; it is called the *unit tangent vector* to the curve γ at parameter value s.

Since $\mathbf{t}$ has constant magnitude, the vector $\mathbf{t}'$ is perpendicular to $\mathbf{t}$. Its magnitude $|\mathbf{t}'|$ is called the *curvature* of our space curve, and is denoted by the Greek letter κ. In general, $\mathbf{t}' \neq \mathbf{0}$; we will assume that this is the case. (Otherwise, our frame cannot be defined.) We may therefore write

(11.4.1) $$\mathbf{t}' = \kappa \mathbf{n}$$

where $\kappa = |\mathbf{t}'| > 0$ and $\mathbf{n}$ is a unit vector, called the *principal normal* to the curve γ at the parameter value s.

Here is a simple example. Suppose that the curve is a circle in the xy-plane, of radius a and centre the origin. Thus

$$\gamma(s) = \left(a\cos\frac{s}{a}, a\sin\frac{s}{a}, 0\right).$$

By differentiation we find that

$$\mathbf{t} = -\sin\frac{s}{a}\mathbf{i} + \cos\frac{s}{a}\mathbf{j}$$

(notice that this is a unit vector) and

$$\mathbf{t}' = -\frac{1}{a}\cos\frac{s}{a}\mathbf{i} - \frac{1}{a}\sin\frac{s}{a}\mathbf{j}.$$

This is a vector of magnitude $1/a$, so the curvature $\kappa = 1/a$, and

$$\mathbf{n} = -\cos\frac{s}{a}\mathbf{i} - \sin\frac{s}{a}\mathbf{j}.$$

Notice that the curvature is the reciprocal of the radius of the circle. The smaller the circle, the 'more tightly' it is curved, and so the greater the curvature. If $\kappa \equiv 0$, the curve is a straight line; the curvature measures how much the curve deviates from a straight line.

But we are not only interested in plane curves. Just as we introduced the curvature to measure the extent to which the curve deviates from being a straight line, we can introduce a new quantity called the *torsion* which measures the extent to which the curve deviates from lying in a plane. To do this we differentiate again.

Let $\mathbf{b} = \mathbf{t} \wedge \mathbf{n}$. $\mathbf{b}$ is called the *binormal* vector to the curve. Because $\mathbf{t}$ and $\mathbf{n}$ are unit vectors at right angles to one another, $\mathbf{b}$ is again a unit vector, at right angles to both of them; so $\{\mathbf{t}, \mathbf{n}, \mathbf{b}\}$ is an orthonormal basis (depending on the parameter s) for the space of vectors. It is called the *Frenet frame* for the curve γ.

Now we will calculate the derivative of $\mathbf{n}$. By the usual argument (11.1.2), $\mathbf{n}'$ is perpendicular to $\mathbf{n}$, so $\mathbf{n}'$ must be a linear combination of $\mathbf{t}$ and $\mathbf{b}$. Moreover

$$\mathbf{n}' \cdot \mathbf{t} + \mathbf{n} \cdot \mathbf{t}' = \frac{\mathrm{d}}{\mathrm{d}s}(\mathbf{n} \cdot \mathbf{t}) = 0.$$

Therefore $\mathbf{n}' \cdot \mathbf{t} = -\mathbf{n} \cdot \mathbf{t}' = -\kappa$ (by 11.4.1). We define the *torsion* τ by $\tau = \mathbf{n}' \cdot \mathbf{b}$; we may therefore write

(11.4.2)
$$\mathbf{n}' = -\kappa\mathbf{t} + \tau\mathbf{b}.$$

To complete the picture, we find the derivative of **b**. Using 11.4.1 and 11.4.2, we find

(11.4.3) $\mathbf{b}' = (\mathbf{t} \wedge \mathbf{n})' = \mathbf{t}' \wedge \mathbf{n} + \mathbf{t} \wedge \mathbf{n}' = (\kappa \mathbf{n}) \wedge \mathbf{n} + (-\kappa \mathbf{t} + \tau \mathbf{b}) \wedge \mathbf{t} = -\tau \mathbf{n}.$

The equations 11.4.1, 11.4.2, and 11.4.3 are collectively known as the *Serret–Frenet formulae:*

$$
\begin{aligned}
\mathbf{t}' &= \kappa \mathbf{n} \\
\mathbf{n}' &= -\kappa \mathbf{t} + \tau \mathbf{b} \\
\mathbf{b}' &= -\tau \mathbf{n}
\end{aligned}
$$

Remember (once again!) that everything is a function of the curve parameter s.

(11.4.4)   **Example:** Let us find the Frenet frame and the curvature and torsion for the helix whose equation is $\gamma(s) = \frac{1}{\sqrt{2}}(\cos s, \sin s, s)$. We calculate the derivative

$$\gamma'(s) = \frac{1}{\sqrt{2}}(-\sin s \mathbf{i} + \cos s \mathbf{j} + \mathbf{k}).$$

Thus $\gamma'(s)$ is a unit vector **t**, and this shows that the curve was already parameterized by arc length; we don't need to change the parameterization. Differentiating,

$$\mathbf{t}' = \frac{1}{\sqrt{2}}(-\cos s \mathbf{i} - \sin s \mathbf{j}) = \kappa \mathbf{n}.$$

Therefore the curvature $\kappa = 1/\sqrt{2}$, and $\mathbf{n} = -\cos s \mathbf{i} - \sin s \mathbf{j}$.
    We can now calculate the binormal vector $\mathbf{b} = \mathbf{t} \wedge \mathbf{n}$. We get

$$\mathbf{b} = \frac{1}{\sqrt{2}} \begin{vmatrix} \mathbf{i} & \mathbf{j} & \mathbf{k} \\ -\sin s & \cos s & 1 \\ -\cos s & -\sin s & 0 \end{vmatrix} = \frac{1}{\sqrt{2}}(-\sin s \mathbf{i} + \cos s \mathbf{j} + \mathbf{k}).$$

The torsion τ is equal to $\mathbf{n}' \cdot \mathbf{b}$, which is $-1/\sqrt{2}$.
    So for this particular curve, the curvature and torsion are constants. Conversely, it can be proved that any curve whose curvature and torsion are constants is part of a helix.

(11.4.5)   **Example:** Now we will try to find the Frenet frame for the twisted cubic $\gamma(t) = (2t, t^2, t^3/3)$ at $t = 0$. Differentiating, we calculate

$$\dot{\gamma}(t) = 2\mathbf{i} + 2t\mathbf{j} + t^2\mathbf{k}.$$

This is *not* a unit vector; its magnitude is $2 + t^2$. So we need to make a parameter change to a new arc length parameter s which satisfies $ds/dt = 2 + t^2$. Then $\mathbf{t} = \gamma'(s)$ will be a unit vector in the direction of $\dot{\gamma}$; in other words, $\mathbf{t} = \dfrac{1}{t^2 + 2}(2\mathbf{i} + 2t\mathbf{j} + t^2\mathbf{k})$. At $t = 0$ this gives $\mathbf{t} = \mathbf{i}$.

Now we must differentiate again. To relate differentiation with respect to s and differentiation with respect to t we use the chain rule:

$$\frac{d\mathbf{t}}{ds} = \frac{d\mathbf{t}/dt}{ds/dt} = \frac{1}{(t^2 + 2)^3}(-4t\mathbf{i} + (2 - 2t^2)\mathbf{j} + 4t\mathbf{k}).$$

The curvature κ is the magnitude of this vector, and the principal normal $\mathbf{n}$ is a unit vector in its direction. At $t = 0$, therefore, it is clear that $\mathbf{n} = \mathbf{j}$. Thus at $t = 0$, $\mathbf{b} = \mathbf{i} \wedge \mathbf{j} = \mathbf{k}$.

Therefore the Frenet frame for the twisted cubic at $t = 0$ is the standard coordinate frame. Conversely, it can be shown that *any* curve in its Frenet frame looks locally like a twisted cubic in the standard frame — where 'locally' means that we neglect terms of higher than third order in the Taylor expansion of the curve as a function of the parameter.

We have already mentioned that the curvature measures the deviation of the curve from being a straight line, and that the torsion measures the deviation of a curve from lying in a plane. The next result reinforces this message.

(11.4.6)   **Proposition:** *Let γ be a regular curve in a three-dimensional Euclidean space. Let κ and τ be its curvature and torsion (if defined). Then*

(i) *If $\kappa \equiv 0$, the curve is part of a straight line.*

(ii) *If $\tau \equiv 0$, the curve lies in a plane.*

(iii) *If $\tau \equiv 0$ and $\kappa \neq 0$ is constant, the curve is part of a circle.*

Proof: We may assume that the curve is parameterized by arc length s. Choose an origin O and let $\mathbf{r}(s)$ be the position vector of $\gamma(s)$ relative to O.

If $\kappa \equiv 0$, then $\mathbf{t}' \equiv \mathbf{0}$ and so $\mathbf{t}$ is a constant vector. Thus

$$\frac{d}{ds}\left(\mathbf{r} \wedge \mathbf{t}\right) = \mathbf{t} \wedge \mathbf{t} = \mathbf{0}.$$

So $\mathbf{r} \wedge \mathbf{t} = \mathbf{a}$ is a constant vector. But this is just the equation of a straight line (8.2.2) so the curve is part of a straight line.

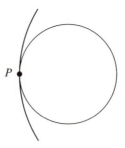

Fig. 11.3. The circle of curvature.

If $\tau \equiv 0$, then **b** is a constant vector, by a similar argument. Therefore

$$\frac{d}{ds}\left(\mathbf{r}\cdot\mathbf{b}\right) = \mathbf{t}\cdot\mathbf{b} = \mathbf{0}.$$

So $\mathbf{r}\cdot\mathbf{b}$ is a constant, and this is the equation of a plane (8.2.3).

If $\tau \equiv 0$ and κ is constant, then by the previous result the curve lies in a plane. Consider

$$\frac{d}{ds}\left(\mathbf{r} + \frac{1}{\kappa}\mathbf{n}\right) = \mathbf{t} + \frac{-\kappa\mathbf{t} + \tau\mathbf{b}}{\kappa} - \frac{\kappa'}{\kappa^2}\mathbf{n} = \mathbf{0}.$$

Thus $\mathbf{r} + \mathbf{n}/\kappa$ is a constant vector, say **p**. Clearly, then, $|\mathbf{r} - \mathbf{p}| = 1/\kappa$, so **r** lies at a fixed distance from the fixed point **p**. Since the curve is in a plane, it must be part of a circle. □

Remark: Let γ be a regular curve, which we assume is parameterized by arc length s, and let $\mathbf{r}(s)$ be the position vector of $\gamma(s)$ relative to some origin. For any point $P = \gamma(s)$ on the curve for which $\kappa \neq 0$, we can consider the circle in the **t, n** plane with centre $\mathbf{r} + \mathbf{n}/\kappa$ and radius $1/\kappa$, as in the previous proof. We notice that the circle touches the curve at P, and that at P the circle has the same curvature and the same tangent and normal vectors as the original curve. (See Figure 11.3.) This circle is the circle which 'best approximates' the curve at P; it's called the *circle of curvature* or *osculating circle* at P, and its radius $1/\kappa$ is called the *radius of curvature*.

It is sometimes helpful to know how to calculate the curvature and torsion of a curve which is not parameterized by arc length, without having to go through the calculation of an appropriate parameter change map.

(11.4.7)   **Proposition:** *Let* $t \mapsto \gamma(t)$ *be a regular curve in some Euclidean space, not necessarily parameterized by arc length. Then one has the following formulae for the curvature and torsion:*

$$\kappa = \frac{|\dot{\gamma} \wedge \ddot{\gamma}|}{|\dot{\gamma}|^3}, \qquad \tau = \frac{[\dot{\gamma}, \ddot{\gamma}, \dddot{\gamma}]}{|\dot{\gamma} \wedge \ddot{\gamma}|^2}.$$

The first formula is always valid, and the second is valid whenever the torsion is defined, i.e. *whenever* $\kappa \neq 0$.

Proof: Let s be arc length along the curve, so that $ds/dt = |\dot{\gamma}|$. Thus for any vector function $\mathbf{v}(t)$,

$$\dot{\mathbf{v}} = \mathbf{v}'|\dot{\gamma}|$$

from the chain rule, $\dfrac{d\mathbf{v}}{dt} = \dfrac{d\mathbf{v}}{ds}\dfrac{ds}{dt}$.

We use this to differentiate the identity $\dot{\gamma} = |\dot{\gamma}|\mathbf{t}$. This gives

$$\ddot{\gamma} = \frac{d|\dot{\gamma}|}{dt}\mathbf{t} + |\dot{\gamma}|\dot{\mathbf{t}} = \frac{d|\dot{\gamma}|}{dt}\mathbf{t} + |\dot{\gamma}|^2\mathbf{t}' = \frac{d|\dot{\gamma}|}{dt}\mathbf{t} + |\dot{\gamma}|^2\kappa\mathbf{n}$$

by the first Serret–Frenet formula. Therefore

$$\dot{\gamma} \wedge \ddot{\gamma} = \kappa|\dot{\gamma}|^3\mathbf{b}.$$

Taking the magnitude of both sides, we get the formula for the curvature.

To get the formula for the torsion, differentiate again. This gives

$$\frac{d}{dt}\left(\dot{\gamma} \wedge \ddot{\gamma}\right) = \frac{d(\kappa|\dot{\gamma}|^3)}{dt}\mathbf{b} + \kappa|\dot{\gamma}|^3\dot{\mathbf{b}} = \frac{d(\kappa|\dot{\gamma}|^3)}{dt}\mathbf{b} - \tau\kappa|\dot{\gamma}|^4\mathbf{n}$$

by the third Serret–Frenet formula. But

$$\frac{d}{dt}\left(\dot{\gamma} \wedge \ddot{\gamma}\right) = \ddot{\gamma} \wedge \ddot{\gamma} + \dot{\gamma} \wedge \dddot{\gamma} = \dot{\gamma} \wedge \dddot{\gamma}.$$

Now calculate in the orthonormal basis $\mathbf{t}, \mathbf{n}, \mathbf{b}$ and use the formula already obtained for $\ddot{\gamma}$ to find

$$[\dot{\gamma}, \ddot{\gamma}, \dddot{\gamma}] = -(\dot{\gamma} \wedge \ddot{\gamma}) \cdot \ddot{\gamma} = \tau\kappa^2|\dot{\gamma}|^6.$$

This gives the formula for the torsion. $\square$

One can refine the concept of curvature slightly if it is known from the start that the curve γ lies in an oriented *plane*, rather than in three-dimensional space. Our original definition of κ ensures that it is always positive; this is because there is no prescribed choice of one unit normal direction rather than the other. But in an oriented plane, there is such a choice: the vector $\mathbf{t}^{\perp}$. We can therefore define the *signed curvature* $\hat{\kappa}$ by

(11.4.8) $$\mathbf{t}' = \hat{\kappa}\mathbf{t}^{\perp}.$$

$\hat{\kappa}$ can have either sign, and its absolute value is equal to the curvature as previously defined.

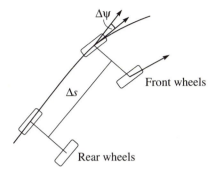

Fig. 11.4. Path of a car.

Suppose that **t** makes an oriented angle ψ with some fixed vector, say **i**. Then $\hat{\kappa} = \dfrac{d\psi}{ds}$; the curvature measures the rate of change of the angle of the tangent vector. This is easy to prove; just write

$$\mathbf{t} = \cos\psi\mathbf{i} + \sin\psi\mathbf{j}$$

and differentiate with respect to s.

To a first approximation, the position of a car's steering wheel controls the curvature of its path. Figure 11.4 illustrates why; the deviation $\Delta\psi$ of the front wheels from straight ahead is controlled by the steering wheel. If the car has length Δs, and if both front and rear wheels follow the same path, then $\Delta\psi/\Delta s$ is an approximation to the curvature of the path.

If you know where a car starts from, and you know the position of the steering wheel at each point of the journey, you expect to be able to work out where it ended up. This illustrates a theorem which says that a plane curve is determined by its starting point and its *intrinsic equation*, giving ψ as a function of s.

11.5 The general isoperimetric inequality

Results in differential geometry can be divided into 'local' and 'global'. Local results, like the Serret–Frenet formulae, tell you about the geometry of a curve or surface in a small neighbourhood of some point; global results tell you about the whole thing. Global results are often more interesting, as well as being harder to obtain.

One of the oldest global questions in differential geometry is the *isoperimetric problem*. Recall that in 9.2.10, we showed that of all triangles with a given perimeter the 'most symmetrical' one, an equilateral triangle, has the greatest area. The isoperimetric problem is the more general question; which (simple closed) *curve* of given perimeter encloses the greatest area? The answer is again the 'most symmetrical' curve, which in this case is a circle. This fact was known to the Greeks and is no doubt familiar to the reader, but rigorous proofs are hard to find. The one that we will give in this section is now part of mathematical folklore, and seems to originate from the work of Hurwitz in the first decade of the twentieth century.

The difficulty in the proof is quite subtle. There are several methods available which show that *if* there is a figure which has the greatest possible area for a given perimeter, *then* that figure is a circle. One such method with which the reader may be familiar is the classical *calculus of variations*, which is based on the principle that *if* a function has a maximum or minimum at some point, *then* its derivative must vanish at that point. Unfortunately these methods leave open the question of whether there *is* any figure which has the greatest possible area for a given perimeter. That this is not a trivial matter was shown by Weierstrass in the nineteenth century; he gave several examples of similar-sounding 'extremal problems' for which the desired maximum is not attained. In fact, the methods developed for the isoperimetric problem are now applied much more widely to show that solutions exist to other calculus of variations problems also.

Our solution to the isoperimetric problem will be based on an interesting inequality which relates the size of a function to the size of its first derivative.

(11.5.1)   **Lemma:** (WIRTINGER INEQUALITY) *Let* $f\colon [0, \pi] \to \mathbf{R}$ *be a smooth function with* $f(0) = f(\pi) = 0$. *Then*

$$\int_0^\pi f'(t)^2 \, dt \geq \int_0^\pi f(t)^2 \, dt.$$

Equality is achieved if and only if $f(t)$ *is a multiple of* $\sin t$.

Proof: Consider the function

$$g(t) = \frac{f(t)}{\sin t}.$$

This function is smooth on $[0, \pi]$; even at the end points 0 and π, where it takes the indeterminate form $\frac{0}{0}$, it has a smooth limiting value by l'Hôpital's rule[2]. Notice that

$$f'(t) = g(t) \cos t + g'(t) \sin t.$$

Therefore

---

[2] See Binmore [5].

$$\int_0^\pi f'(t)^2 \, dt$$

$$= \int_0^\pi g(t)^2 \cos^2 t \, dt + 2 \int_0^\pi g(t)g'(t) \cos t \sin t \, dt + \int_0^\pi g'(t)^2 \sin^2 t \, dt.$$

But on integration by parts,

$$2 \int_0^\pi g(t)g'(t) \cos t \sin t \, dt = -\int_0^\pi g(t)^2 (\cos^2 t - \sin^2 t) \, dt.$$

Therefore

$$
\begin{aligned}
\int_0^\pi f'(t)^2 \, dt &= \int_0^\pi \left(g(t)^2 + g'(t)^2 \right) \sin^2 t \, dt \\
&= \int_0^\pi f(t)^2 \, dt + \int_0^\pi g'(t)^2 \sin^2 t \, dt \\
&\geq \int_0^\pi f(t)^2 \, dt
\end{aligned}
$$

with equality if and only if $g' = 0$, so that g is constant and f is a multiple of $\sin t$. $\square$

To prove the isoperimetric inequality, we will need to know an expression for the length of a simple closed curve in a plane γ in terms of *polar* coordinates (r, θ). In fact

(11.5.2) $$\ell(\gamma) = \int_\gamma \sqrt{dx^2 + dy^2} = \int_\gamma \sqrt{dr^2 + r^2 \, d\theta^2}$$

where as usual these integrals should be interpreted in terms of a parameterization as

$$\int_a^b \sqrt{\left(\frac{dx}{dt}\right)^2 + \left(\frac{dy}{dt}\right)^2} \, dt = \int_a^b \sqrt{\left(\frac{dr}{dt}\right)^2 + r^2 \left(\frac{d\theta}{dt}\right)^2} \, dt.$$

To check this we begin from the identities

$$x = r \cos \theta, \quad y = r \sin \theta$$

relating Cartesian and polar coordinates. Differentiating and using the chain rule, we find

$$\frac{dx}{dt} = \frac{dr}{dt} \cos \theta - r \sin \theta \frac{d\theta}{dt}, \quad \frac{dy}{dt} = \frac{dr}{dt} \sin \theta + r \cos \theta \frac{d\theta}{dt}.$$

Therefore

$$\left(\frac{dx}{dt}\right)^2 + \left(\frac{dy}{dt}\right)^2 = \left(\frac{dr}{dt}\right)^2 + r^2 \left(\frac{d\theta}{dt}\right)^2$$

as required.

It turns out to be useful also to introduce another quantity, the *action* of the curve γ, defined in terms of a parameterization by

(11.5.3) $$\mathcal{E}(\gamma) = \frac{1}{2}\int_a^b \left(\frac{dx}{dt}\right)^2 + \left(\frac{dy}{dt}\right)^2 dt = \frac{1}{2}\int_a^b \left(\frac{dr}{dt}\right)^2 + r^2 \left(\frac{d\theta}{dt}\right)^2 dt.$$

Unlike the length and the area enclosed, the action of a curve depends on the way that it is parameterized; in particular, if we rescale the parameterization by a factor of k, the action gets divided by k. We will therefore restrict attention to *standard parameterizations*, which we define as those for which the parameter interval $[a, b]$ is $[0, \pi]$.

(11.5.4)   **Proposition:** *Let γ be a curve with a standard parameterization. Then*

$$\ell(\gamma)^2 \le 2\pi\mathcal{E}(\gamma),$$

and equality is attained if and only if the parameter t is a multiple of arc length along γ.

Proof: Recall the Cauchy–Schwarz inequality for integrals (Exercise 4.6.11): for two functions f and g,

$$\left(\int_0^\pi f(t)g(t)\,dt\right)^2 \le \int_0^\pi f(t)^2\,dt \int_0^\pi g(t)^2\,dt$$

and equality is attained if and only if f and g are multiples of one another. Apply this inequality with

$$f(t) = \sqrt{\left(\frac{dx}{dt}\right)^2 + \left(\frac{dy}{dt}\right)^2}, \quad g(t) = 1.$$

This gives the inequality we want. The condition for equality is that

$$\sqrt{\left(\frac{dx}{dt}\right)^2 + \left(\frac{dy}{dt}\right)^2} = \text{constant};$$

this says that the point $(x(t), y(t))$ moves along the curve with constant speed, which implies that t is a multiple of arc length. □

Remark: It follows from this proposition, together with the fact that a straight line gives the shortest distance between its end points (11.3.3), that the minimum action path between two given points is a straight line traversed at constant speed. This may remind you of mechanics; Newton's first law of motion says that a particle not acted on by external forces moves in a straight line at constant speed. In other words, a particle acted on by no forces moves in a way that will minimize the action.

This idea was taken up and generalized by Lagrange and Hamilton. They showed that the motion of any kind of mechanical system (perhaps consisting of many particles with forces between them) can be thought of as determined by a single 'action functional' or *Lagrangian*. The system moves in such a way as to make the action stationary; in other words, the actual motion is distinguished from other conceivable motions by the fact that a small change in it would not alter the action. This approach (nowadays called *Lagrangian mechanics*) leads to both theoretical and practical simplifications.

There is an interesting 'explanation' of the principle of stationary action in terms of quantum mechanics. According to this explanation, a mechanical system in fact 'tries out' all the possible routes from its initial to its final state; but the routes that are far from stationary action cancel each other out, and only those very close to stationary action are left. For a beautiful account, see Feynman [16].

We will use action as a tool to prove the isoperimetric inequality. We have already shown how to relate action and length; now we will use Wirtinger's inequality (11.5.1) to relate action and area.

(11.5.5)   **Proposition:** *Let* γ *be a simple closed curve with a standard parameterization. Then*

$$\mathcal{E}(\gamma) \geq 2\mathcal{A}(\mathrm{Int}(\gamma))$$

and equality holds if and only if γ *is a circle.*

Proof: We calculate using the integral formula 9.6.5 for $\mathcal{A}(\mathrm{Int}(\gamma))$. We take polar coordinates whose origin is the point on the curve with parameter value $t = 0$. Then

$$\mathcal{E}(\gamma) - 2\mathcal{A}(\mathrm{Int}(\gamma)) = \frac{1}{2}\int_0^\pi \left(\left(\frac{dr}{dt}\right)^2 + r^2\left(\frac{d\theta}{dt}\right)^2 - 2r^2\frac{d\theta}{dt}\right)\,dt.$$

The right-hand side may be rewritten as

$$\frac{1}{2}\int_0^\pi r^2\left(\frac{d\theta}{dt} - 1\right)^2\,dt + \frac{1}{2}\int_0^\pi \left(\left(\frac{dr}{dt}\right)^2 - r^2\right)\,dt,$$

and here the first term is non-negative because the integrand is a square and the second term is non-negative by Wirtinger's inequality, which can be applied because $r(0) = r(\pi) = 0$. For equality it is necessary both that $\frac{d\theta}{dt} - 1 = 0$ and that there is equality in Wirtinger's inequality, the condition for which is that $r = a\sin t$ for some constant a. Putting these two facts together, we get

$$r = a\sin(\theta + \varphi)$$

for constants a and φ, and this is the polar equation of a circle (see 6.3.3). $\square$

By combining these two results, we can solve the isoperimetric problem.

(11.5.6)   Theorem: (ISOPERIMETRIC INEQUALITY) *Let γ be a simple closed curve; then*

$$A(\text{Int}(\gamma)) \le \frac{1}{4\pi} \ell(\gamma)^2$$

and equality is attained only if γ is a circle.

Proof: Choose a standard parameterization by a multiple of arc length, so that equality is attained in 11.5.4. Then from 11.5.4 and 11.5.5,

$$A(\text{Int}(\gamma)) \le \frac{1}{2}\mathcal{E}(\gamma) = \frac{1}{4\pi} \ell(\gamma)^2$$

and equality is attained if and only if γ is a circle. □

There is a huge literature on the isoperimetric problem and many different methods have been used to solve it. We will sketch one of these, which is quite different from the one just given and is based on the idea of 'mixing' two subsets of the plane.

Suppose that we agree to fix an origin O, so that points of the plane are represented by their position vectors. Then for two subsets A and B of the plane, and two numbers α and β, we can define the 'mixing'

$$\alpha A + \beta B = \{\alpha \mathbf{a} + \beta \mathbf{b} : \mathbf{a} \in A,\ \mathbf{b} \in B\}.$$

For instance, if A and B are rectangles with sides parallel to the axes, A having sides x and y and B having sides r and s, then $\alpha A + \beta B$ is also a rectangle with sides parallel to the axes, and the lengths of its sides are $\alpha x + \beta r$ and $\alpha y + \beta s$.

The *Brunn–Minkowski inequality* says that for any two convex subsets A and B,

$$\alpha A(A)^{\frac{1}{2}} + \beta A(B)^{\frac{1}{2}} \le A(\alpha A + \beta B)^{\frac{1}{2}}.$$

We can check this 'by hand' in the case of rectangles, where it says

$$\alpha\sqrt{xy} + \beta\sqrt{rs} \le \sqrt{(\alpha x + \beta r)(\alpha y + \beta s)}.$$

To prove this, square both sides, cancel terms, and use the arithmetic–geometric mean inequality. The general result can be proved by exhausting the sets A and B by disjoint rectangles.

Now suppose that A is the interior of some curve γ and let B be the disc of centre 0 and radius 1. Then $A + \beta B$ consists of A together with a thin strip

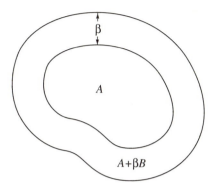

Fig. 11.5. $A + \beta B$ where B is a disc.

of width β around the outside (see Figure 11.5). It is intuitively clear that if β is small, then the area of this strip is approximately $\beta \ell(\gamma)$, and so

$$\mathcal{A}(A + \beta B) = \mathcal{A}(A) + \beta \ell(\gamma) + o(\beta),$$

where $o(\beta)$ denotes an error term that is negligible compared to β. Using the binomial theorem we find that

$$\mathcal{A}(A + \beta B)^{\frac{1}{2}} = \mathcal{A}(A)^{\frac{1}{2}} + \frac{\beta \ell(\gamma)}{2\mathcal{A}(A)^{\frac{1}{2}}} + o(\beta).$$

But by the Brunn–Minkowski inequality

$$\mathcal{A}(A + \beta B)^{\frac{1}{2}} \geq \mathcal{A}(A)^{\frac{1}{2}} + \beta \pi^{\frac{1}{2}}.$$

Combine these two facts and cancel to obtain

$$\frac{\beta \ell(\gamma)}{2\mathcal{A}(A)^{\frac{1}{2}}} \geq \beta \pi^{\frac{1}{2}} + o(\beta).$$

Divide by β and then let $\beta \to 0$ to obtain

$$\ell(\gamma)^2 \geq 4\pi \mathcal{A}(A)$$

which is the isoperimetric inequality.

An advantage of this proof is that it generalizes easily to n dimensions (the exponent $\frac{1}{2}$ in the Brunn–Minkowski inequality gets replaced by $\frac{1}{n}$). For full details, see Berger [4].

11.6 Exercises

1. A vector function of time $\mathbf{r}(t)$ has the property that $\ddot{\mathbf{r}}(t) = f(t)\mathbf{r}(t) + g(t)\mathbf{n}$, where f and g are functions of time and $\mathbf{n}$ is a constant vector. Prove that the scalar triple product $[\mathbf{r}, \dot{\mathbf{r}}, \mathbf{n}]$ is a constant. Can you give a mechanical interpretation?

2. Suppose that a particle P moves in orbit under the action of a force directed towards a centre O, in such a way that its acceleration is $K/R^2 + a/R^3$ towards O, where R denotes the distance $|OP|$ and a is a small quantity. Show that the *perihelion* (the point where the particle is nearest to O) will advance by approximately $a\pi/h^2$ radians for each revolution of the particle about O, where h is equal to the constant value of $R^2\dot{\theta}$.

(One of the triumphs of Einstein's theory of gravity was to explain the anomalous perihelion advance of the planet Mercury, of about 2×10^{-6} radians per year. However, Einstein's theory was not simply a matter of making an *ad hoc* modification to the force law, as we have done with our a/R^3. It was based on entirely new geometric principles. See Misner *et al.* [30, chapter 40].)

3. A particle P moves in orbit under the action of a central force directed towards a fixed point O. The force depends only on the distance from O to P, and is proportional to a certain power of that distance. If the orbit of the particle is an arc of a circle passing through O, find the force law.

4. Find the length of the segment of the curve

$$x = a\cosh t, \quad y = a\sinh t, \quad z = at$$

between the parameter values $t = 0$ and $t = k$. Find also its curvature and torsion at the point with parameter value t.

5. Find the Frenet frame at a general point of the curve given parametrically by

$$x = \frac{1}{3}(1+t)^{\frac{3}{2}}, \quad y = \frac{1}{3}(1-t)^{\frac{3}{2}}, \quad z = \frac{t}{\sqrt{2}}.$$

6. Find the curvature of the curve

$$x = t - \sin t, \quad y = 1 - \cos t, \quad z = 4\sin t/2.$$

7. A plane curve parameterized by arc length has equation $\mathbf{r} = \mathbf{r}(s)$. A new curve is given by $\mathbf{r}(s) = \mathbf{r}(s) + s\tan\theta\mathbf{p}$, where θ is constant and $\mathbf{p}$ is a unit vector normal to the plane of the original curve. Show that the curvature and torsion of the new curve are, respectively, $\kappa\cos^2\theta$ and $\frac{1}{2}\kappa\sin 2\theta$, where κ is the curvature of the original curve.

8. Prove that the following four properties of a space curve are equivalent:

(i) the tangents make a constant angle with a certain fixed direction;

(ii) the principal normals are parallel to a certain fixed plane;

(iii) the binormals make a constant angle with a certain fixed direction;

(iv) the ratio of curvature to torsion is constant.

(You may assume that the curvature and torsion are nonzero.)

9. Let γ be a space curve parameterized by arc length. Suppose that for some parameter value $s = s_0$, the distance from $\gamma(s)$ to some fixed point O attains its maximum value r_0. Prove that $\kappa(s_0) \geq 1/r_0$.

10. Let γ be a space curve parameterized by arc length s, with non-constant curvature and non-zero torsion. Let $K = 1/\kappa$ and $T = 1/\tau$. Suppose that $K^2 + \left(T\dfrac{dK}{ds}\right)^2$ is a constant. Prove that the image of γ lies on a sphere.

11. Show that a curve whose curvature and torsion are constant and nonzero is part of a helix.

12. What is the greatest area that can be enclosed between a fixed straight line and a curve of length l whose end points lie on the straight line? What is the shape of the curve when this greatest area is enclosed?

12

Differential geometry of surfaces

12.1 Parameterized surfaces

In this chapter we will study the differential geometry of curved surfaces in three-dimensional space. This subject owes its origin to Euler, but it came to maturity at the hands of Carl Friedrich Gauss (1777–1855), one of the greatest mathematicians of all time. In a memoir entitled *General Investigations of Curved Surfaces*, published in 1827, Gauss showed that a curved surface could be considered, not just as sitting in some extrinsic three-dimensional space, but as having its own intrinsic geometry; a geometry which, moreover, might be non-Euclidean.

To appreciate what is meant by this distinction between intrinsic and extrinsic geometry, consider first a curve in space, which we can imagine is made out of wire. Looking down on the curve from outside, we can see whether or not it is straight, and we can define the curvature and the torsion which measure how far it is from straightness or flatness. These invariants are part of extrinsic geometry. But now imagine[1] a race of antlike beings whose whole universe is that one-dimensional curve. They can measure distances along the curve, but they have no perception of any 'outside'. Then they cannot tell any difference between their 'curved' universe and a straight one. The mathematical expression of this is 11.3.4, which says that any curve can be parameterized by arc length. Thus curvature in one dimension cannot be detected by measurements internal to the curve, and this is what is meant by saying that it is not part of intrinsic geometry.

Things are different, Gauss found, in two dimensions. For example, we live on the surface of the earth, a near-perfect sphere. One way to realize its curvature is to go up in a spacecraft and take a photograph. This is the extrinsic approach; looking down from outside. But the curvature of the earth

---

[1]A good reference for such imaginings is Abbott's classic [1].

could also be perceived by intrinsic measurements, which do not leave the surface at all. For instance, the sum of the angles of a large triangle on the earth's surface will be slightly greater than π.

One piece of modern science which arises from Gauss' insight is Einstein's theory of gravity, *general relativity*. According to this theory, the four-dimensional spacetime world in which we live is in fact intrinsically curved. (The natural question is 'curved in what?'; but there is no need to answer this, because we're talking about *intrinsic* geometry.) The curvature means that the shortest route for a body to take through spacetime may not be what we (with our Euclidean perception) see as a straight line. According to Einstein, this is the explanation for gravity; a body which appears to be deviated from its path by gravity is in fact taking the shortest path available to it.

(12.1.1)   **Parametric equations for surfaces:** In Section 10.1 we defined certain surfaces (quadrics) in terms of *locus equations* of the form $f(x, y, z) = 0$. For differential geometry, however, it is more convenient to study a surface by means of parametric equations. A well-known example of a parameterization for a surface is that of the unit sphere $\{(x, y, z) : x^2 + y^2 + z^2 = 1\}$ by means of latitude and longitude; a general point on the sphere is given by

$$x = \cos\theta\cos\varphi, \quad y = \cos\theta\sin\varphi, \quad z = \sin\theta$$

where θ is the *latitude* (positive in the 'northern' hemisphere and negative in the 'southern' hemisphere) and φ is the *longitude*.

This illustrates the fact that we will need *two* parameters to describe a surface, because a point on a surface can move in two independent directions. A surface will therefore be described by a smooth map

$$\sigma : U \to S, \quad (u, v) \mapsto \sigma(u, v)$$

where U is a region in $\mathbf{R}^2$. As in the case of curves, we will need a 'regularity' condition on the derivatives of σ to ensure that the image of the map σ conforms to our intuition of a surface.

What condition should we impose? We want to ensure that our surface has a well-defined tangent plane at each point. Now the tangent plane should be made up of the tangent vectors to all curves lying in the surface. Since the surface is supposed to be parameterized by the map σ, any curve in the surface must be representable by a curve in (u, v)-space: $u = u(t)$, $v = v(t)$. But then the tangent vector will be

$$\frac{d\sigma}{dt} = \frac{\partial\sigma}{\partial u}\frac{du}{dt} + \frac{\partial\sigma}{\partial v}\frac{dv}{dt}$$

by the chain rule. We see therefore that the tangent 'plane' will be made up of linear combinations of the vectors $\dfrac{\partial\sigma}{\partial u}$ and $\dfrac{\partial\sigma}{\partial v}$. In order that it should actually be a plane, we need to require that these two vectors be linearly independent. To simplify the notation, we will denote partial derivatives by subscripts; thus σ_u will mean $\dfrac{\partial \sigma}{\partial u}$, σ_{uv} will mean $\dfrac{\partial^2 \sigma}{\partial u \partial v}$, and so on.

(12.1.2)   **Definition:** *Let U be an open[2] subset of* $\mathbf{R}^2$. *Let* σ *be a smooth map from U to a three-dimensional Euclidean space S. Then* σ *defines a* regular patch of surface *in S if the partial derivatives* σ_u *and* σ_v *are linearly independent at all points of U.*

(12.1.3)   **Example:** Let us check the definition in the case of the sphere parameterized by latitude and longitude. Here we have

$$\sigma(u, v) = (\cos u \cos v, \cos u \sin v, \sin u),$$

so $\sigma_u = -\sin u \cos v\mathbf{i} - \sin u \sin v\mathbf{j} + \cos u\mathbf{k}$ and $\sigma_v = -\cos u \sin v\mathbf{i} + \cos u \cos v\mathbf{j}$. According to 8.2.1, the vector product $\sigma_u \wedge \sigma_v$ will be nonzero if and only if σ_u and σ_v are independent. We calculate

$$\sigma_u \wedge \sigma_v = \cos u(\cos u \cos v\mathbf{i} + \cos u \sin v\mathbf{j} + \sin u\mathbf{k}).$$

So σ_u and σ_v are independent except when $u = \pm\frac{\pi}{2}$, that is at the north and south poles. At these points the meridians of longitude come together and our parameterization becomes singular.

Of course there is nothing intrinsically singular about the sphere itself at these two points, any more than there is something intrinsically singular about the origin of polar coordinates in a plane. The problem is with our parameterization. But it can be proved that *any* parameterization of the whole sphere must suffer a similar problem. This is why we referred to σ as a 'patch' of surface; in general it will take more than one such patch to cover a complete surface. The different patches are called *charts* for the surface, and a complete set of patches covering the whole surface is called an *atlas*. When proving local theorems, we can confine our attention to one patch. For global theorems it is necessary to think about how the patches fit together.

We will mainly be looking at local theorems in this chapter. However, we will need to consider one global issue: orientation. The vector $\dfrac{\sigma_u \wedge \sigma_v}{|\sigma_u \wedge \sigma_v|}$ is a unit normal to the tangent plane, and so by 8.2.4 defines an orientation on this plane. We will assume that whenever two charts overlap they define the same orientation on the tangent planes of the overlap. This hypothesis of global orientability rules out surfaces such as the Möbius band (see 6.1.2).

---

[2]This means that U contains a neighbourhood of each of its points; it has no 'boundary points'. See Sutherland [40]. The hypothesis is necessary in order that the partial derivatives be well defined.

(12.1.4) **The first fundamental form:** Let σ be a regular patch of surface. The parameter values u and v give 'curvilinear coordinates' on the patch, and any point of the patch can be described in terms of its u and v values. In particular, a *curve* γ in the surface can be described in terms of u and v values by $u = u(t), v = v(t)$. As we calculated before, the tangent vector to such a curve is given by

$$\dot{\gamma} = \sigma_u \dot{u} + \sigma_v \dot{v}$$

where we use the dot to denote differentiation with respect to t.

How can we calculate the length of the curve γ? By definition 11.3.1, the length of γ is obtained by integrating $|\dot{\gamma}|$ with respect to t. We can calculate

$$|\dot{\gamma}|^2 = |\sigma_u|^2 \dot{u}^2 + 2\sigma_u \cdot \sigma_v \dot{u}\dot{v} + |\sigma_v|^2 \dot{v}^2.$$

Therefore, with our usual notation for curve integrals,

$$\ell(\gamma) = \int_\gamma \sqrt{E\, du^2 + 2F\, du\, dv + G\, dv^2}$$

where

$$E = |\sigma_u|^2, \quad F = \sigma_u \cdot \sigma_v, \quad G = |\sigma_v|^2.$$

The quantities E, F, and G make up the *first fundamental form* or *metric tensor* of σ. Notice that they do not depend on the curve γ, but only on the geometry of the patch of surface that we are considering. In fact, they completely determine the intrinsic geometry of this patch of surface, because they determine the lengths of all curves in it. A race of surface-dwelling ants could perceive no geometrical difference between two surfaces having the same first fundamental form, even though the two surfaces might appear very different from our extrinsic viewpoint in three-dimensional space.

(12.1.5) **Example:** For the sphere parameterized by latitude and longitude, as above, we have

$$E = 1, \quad F = 0, \quad G = \cos^2 u.$$

(12.1.6) **Example:** The surface of a circular cylinder can be parameterized by

$$\sigma(u, v) = (\cos u, \sin u, v).$$

We can work out that

$$\sigma_u = -\sin u \mathbf{i} + \cos u \mathbf{j}, \quad \sigma_v = \mathbf{k}$$

and so that

$$E = 1, \quad F = 0, \quad G = 1.$$

Notice that this first fundamental form is the same as that of a plane

$$\sigma(u, v) = (u, v, 0).$$

The cylinder and the plane appear quite different when we look at them as surfaces in S. But because their first fundamental forms are equal, there is no way that one can distinguish a patch of cylinder from a patch of plane simply by examining the intrinsic geometry of curves in the patch. We can see why if we notice that a cylinder can be cut along a generating line and then 'unrolled' onto a plane without any stretching or tearing; the lengths of all curves that do not cross the cut are preserved by this 'unrolling' procedure.

A *great circle* on a sphere is a circle obtained by intersecting the sphere with a plane through its centre. On a sphere parameterized by latitude and longitude, the *meridians of longitude* (the lines $v = $ constant) are examples of great circles, whereas the *parallels of latitude* (the lines $u = $ constant) are not great circles in general.

We can use our formula for arc length to prove

(12.1.7) **Proposition:** *The shortest path between two points on a sphere is an arc of a great circle.*

Proof: Because of the symmetry of the sphere, we may as well assume that one of the points is the north pole. Parameterize the sphere by latitude and longitude, and let (u_1, v_1) be the parameter values of the other point. If γ is a curve in the surface joining the two points, then

$$\ell(\gamma) = \int_\gamma \sqrt{du^2 + \cos^2 u \, dv^2} \geq \int_\gamma |du| \geq \tfrac{\pi}{2} - u_1.$$

Equality is attained only when v is constant along the path, in other words when it is a meridian of longitude. □

Thus great circles will play the same rôle in the geometry of the sphere as straight lines do in the geometry of the plane.

Remark: The same patch of surface may have more than one parameterization, just as in the case of curves. For example, we could equally well parameterize the northern hemisphere by

$$\tilde{\sigma}(\tilde{u}, \tilde{v}) = (\tilde{u}, \tilde{v}, \sqrt{1 - \tilde{u}^2 - \tilde{v}^2}).$$

The new parameters $\tilde{u}$ and $\tilde{v}$ will be related to the old parameters u and v by means of a two-variable 'parameter change map', analogous to the one-variable parameter change maps that we used for curves. Whenever two patches of surface overlap, the two parameterizations on the overlap should be related by such a map. For a full treatment of differential geometry it is therefore important to develop a systematic theory of parameter change maps, and this is done in textbooks such as do Carmo [12]. We will not do so here. However, the reader may be interested to know that the condition in two variables analogous to our single-variable condition that a parameter change must have positive derivative is that the *Jacobian*

$$\begin{vmatrix} \partial\tilde{u}/\partial u & \partial\tilde{v}/\partial u \\ \partial\tilde{u}/\partial v & \partial\tilde{v}/\partial v \end{vmatrix}$$

should be strictly positive.

The first fundamental form depends on the parameterization. (Exercise 12.5.2 works out exactly how.)

(12.1.8)   **Definition:** *Let S_1 and S_2 be two surfaces in S. Then an* isometry *between them is a bijective map which takes any curve in S_1 to a curve of the same length in S_2. If there is an isometry between them, the surfaces are called* isometric.

This fits with our previous definition of isometry for Euclidean space in 4.5.1.

If the two surfaces can be parameterized so as to have the same first fundamental form, then they are isometric; an isometry between them can be given by mapping the point with parameter values (u, v) in S_1 to the point with the same parameter values in S_2. For example, a patch of cylinder and a patch of plane are isometric.

(12.1.9)   **Surface area:** We would like to define not just the lengths of curves in a surface, but also the areas of pieces of the surface. We will see that these areas are also determined by the first fundamental form.

How should we define the area? Let σ be a patch of surface, and let Ω be a contented subset of the (u, v)-plane. Then $\sigma(\Omega)$ is a piece of the surface, and we would like to express its area in terms of the first fundamental form. Suppose first of all that Ω is a small rectangle, of sides Δu and Δv parallel to the coordinate axes. Then

$$\sigma(u + \alpha\Delta u, v + \beta\Delta v) \approx \sigma(u, v) + \alpha\sigma_u\Delta u + \beta\sigma_v\Delta v$$

for $0 \le \alpha, \beta \le 1$, and so $\sigma(\Omega)$ is approximately a parallelogram with sides given by the vectors $\sigma_u\Delta u$ and $\sigma_v\Delta v$. We therefore expect that its area should be approximately equal to the area of such a parallelogram. By the vector formula for the area of a triangle (9.2.7), this area is

$$|\sigma_u\Delta u \wedge \sigma_v\Delta v| = |\sigma_u \wedge \sigma_v|\mathcal{A}(\Omega).$$

By the scalar quadruple product expansion (8.1.7),

$$|\sigma_u \wedge \sigma_v|^2 = |\sigma_u|^2|\sigma_v|^2 - (\sigma_u \cdot \sigma_v)^2 = EG - F^2.$$

So we expect that the area of the small piece of surface $\sigma(\Omega)$ will be roughly $\sqrt{EG - F^2}\mathcal{A}(\Omega)$.

So far our arguments have been heuristic, but they motivate the following definition.

(12.1.10) Definition: *Let* $\sigma: U \rightarrow S$ *be a regular patch of surface, and let* $\Omega \subseteq U$ *be a contented set. Then the* surface area $A_s(X)$ *of the piece of surface* $X = \sigma(\Omega)$ *is defined to be*

$$A_s(X) = \int_\Omega \sqrt{EG - F^2}\, dA,$$

where E, F, and G are the coefficients of the first fundamental form.

(12.1.11) Example: For the unit sphere parameterized by latitude and longitude we have

$$\sqrt{EG - F^2} = \cos u.$$

The area of a region $\sigma(\Omega)$ on the sphere is therefore given by

$$\int_\Omega \cos u\, dA(u, v).$$

In particular we can calculate the area of the whole sphere as

$$\int_{u=-\frac{\pi}{2}}^{\frac{\pi}{2}} \int_{v=0}^{2\pi} \cos u\, dv\, du = 4\pi.$$

For a sphere of radius r, the area would be $4\pi r^2$.

Let S be the sphere, and let C be a circular cylinder surrounding it as shown in Figure 12.1. We would like to relate the area of a region on the sphere to the area of the corresponding region on the cylinder obtained by projecting outwards from the axis. For this purpose we parameterize the cylinder C by

$$\tilde\sigma(u, v) = (\cos v, \sin v, \sin u).$$

The projection map then takes the point $\sigma(u, v)$ on the sphere to the corresponding point $\tilde\sigma(u, v)$ on the cylinder.

If we calculate the first fundamental form of $\tilde\sigma$ we get

$$\tilde E = \cos^2 u, \quad \tilde F = 0, \quad \tilde G = 1.$$

The first fundamental forms are different, so the projection is not an isometry. Nevertheless, $\tilde E \tilde G - \tilde F^2 = EG - F^2$. This proves

(12.1.12) Proposition: (ARCHIMEDES' TOMBSTONE THEOREM) *The area of any region on the sphere is the same as the area of the corresponding projected region on the cylinder.*

This result was proved by Archimedes, who was so pleased with it that he requested that it be engraved on his tombstone. After his death in 212 BC this

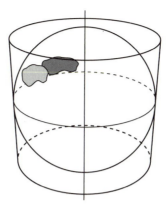

Fig. 12.1. Archimedes' tombstone theorem.

was duly done. Many years later (around 75 BC) the Roman author Cicero visited the tomb and restored the inscription.

Cartographers call the process of projecting a sphere onto a cylinder and then unrolling onto a plane the *cylindrical equal-area projection*. Archimedes' theorem shows that this projection gives an accurate representation of area, even though it distorts shape. It is not possible to find a map projection which gives an accurate representation both of size and shape; we will prove this later (12.2.5).

As an example let us work out the area of a *lune*, that is a region enclosed between two great circular arcs meeting at some angle θ (see Figure 12.2). By the symmetry of the sphere, we may assume that two vertices of the lune are at the north and south poles. Projection then maps the lune to a rectangle of sides 2 and θ, so the area of the lune is 2θ.

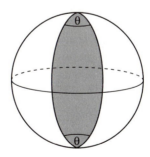

Fig. 12.2. A lune of angle θ.

(12.1.13) Definition: *If $X = \sigma(\Omega)$ is a piece of surface as above, and f is a function defined on X, then we define the* surface integral *of f by*

$$\int_X f \, dA_s = \int_\Omega (f \circ \sigma) \sqrt{EG - F^2} \, dA.$$

Thus, by definition, $A_s(X) = \int_X 1 \, dA_s$. If X is a piece of surface too large to be contained in a single patch σ, we can still define the surface integral by dividing X into pieces each of which lies in a patch, and then adding up the integrals over the pieces.

12.2 Spherical geometry

Before studying the geometry of a general surface we will look in more detail at the geometry of the surface of a sphere. There are several reasons to do this. A practical one is that we all live on the surface of a sphere, at least to a good approximation.

Let A, B, and C be three points on the surface of the unit sphere S with centre O, no two of which are opposite one another. The three points, together with the three shortest great circle arcs joining them, are said to make up a *spherical triangle*. The *sides* a, b, and c of the spherical triangle are the lengths of the arcs BC, CA, and AB, which are equal to the angles $\widehat{BOC}$, $\widehat{COA}$, and $\widehat{AOB}$. The *angles* α, β, and γ of the spherical triangle are the angles between the tangent vectors to the arcs AB, BC, and CA at the vertices where they meet. (See Figure 12.3.)

In working out the relationships between the sides and angles of a spherical triangle it is helpful to use vector algebra. Let $\mathbf{a}$, $\mathbf{b}$, and $\mathbf{c}$ be the position vectors of A, B, and C relative to origin O. Then for the sides of the triangle we have the formulae

$$\cos a = \mathbf{b} \cdot \mathbf{c}, \quad \sin a = |\mathbf{b} \wedge \mathbf{c}|$$

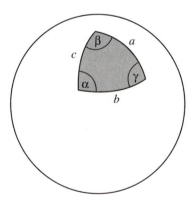

Fig. 12.3. A spherical triangle.

(with corresponding ones for the sides b and c), whereas for the angles we have

$$\cos\alpha = \frac{(\mathbf{a}\wedge\mathbf{c})\cdot(\mathbf{a}\wedge\mathbf{b})}{\sin b\,\sin c}$$

(together with corresponding formulae for β and γ), because $\mathbf{a}\wedge\mathbf{c}$ is normal to the plane AOC and $\mathbf{a}\wedge\mathbf{b}$ is normal to the plane AOB.

The scalar quadruple product appearing above can be expanded using 8.1.7:

$$(\mathbf{a}\wedge\mathbf{c})\cdot(\mathbf{a}\wedge\mathbf{b}) = \mathbf{b}\cdot\mathbf{c} - (\mathbf{a}\cdot\mathbf{b})(\mathbf{a}\cdot\mathbf{c}) = \cos a - \cos b\cos c.$$

So we obtain the *cosine rule for spherical triangles*:

(12.2.1)
$$\cos a = \cos\alpha\,\sin b\,\sin c + \cos b\cos c.$$

We could also work out $\sin\alpha$ instead of $\cos\alpha$. This gives

$$\sin\alpha = \frac{|(\mathbf{a}\wedge\mathbf{c})\wedge(\mathbf{b}\wedge\mathbf{c})|}{\sin b\,\sin c} = \frac{|[\mathbf{a},\mathbf{b},\mathbf{c}]|}{\sin b\,\sin c}.$$

Therefore
$$\frac{\sin\alpha}{\sin a} = \frac{|[\mathbf{a},\mathbf{b},\mathbf{c}]|}{\sin a\,\sin b\,\sin c}.$$

The symmetry of this expression gives us the *sine rule for spherical triangles*:

(12.2.2)
$$\frac{\sin\alpha}{\sin a} = \frac{\sin\beta}{\sin b} = \frac{\sin\gamma}{\sin c}.$$

Although the exact formulae are different, these rules are obviously analogous to the Euclidean ones (4.4.1 and 4.4.3). In fact the Euclidean ones can be recovered as limiting cases when a, b, and c become small; see Exercise 12.5.6. A more startling deviation from Euclidean geometry becomes apparent when we consider *area*. Is there an analogue of Héron's formula (9.2.9) for the area of a triangle?

(12.2.3) **Proposition:** *The area of a spherical triangle with angles α, β, and γ is* $\alpha+\beta+\gamma-\pi$.

Proof: Let A', B', and C' be the points opposite to A, B, and C. Then (see Figure 12.4) triangles ABC and $A'BC$ together make up a lune with angle α, so

$$A_s(ABC) + A_s(A'BC) = 2\alpha.$$

Similarly

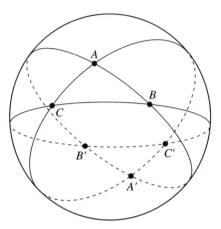

Fig. 12.4. Finding the area of a spherical triangle.

$$\mathcal{A}_s(ABC) + \mathcal{A}_s(AB'C) = 2\beta, \quad \mathcal{A}_s(ABC) + \mathcal{A}_s(ABC') = 2\gamma.$$

Adding these three equations

(12.2.4) $$3\mathcal{A}_s(ABC) + \mathcal{A}_s(A'BC) + \mathcal{A}_s(AB'C) + \mathcal{A}_s(ABC') = 2(\alpha + \beta + \gamma).$$

The triangles ABC, ACB', $AB'C'$, and $AC'B$ together make up a hemisphere of area 2π. Moreover $AB'C'$ is opposite to $A'BC$, and so has the same area. This gives

$$\mathcal{A}_s(ABC) + \mathcal{A}_s(A'BC) + \mathcal{A}_s(AB'C) + \mathcal{A}_s(ABC') = 2\pi.$$

Subtract this from equation 12.2.4 to obtain the result. □

This result has many consequences. Here are some of them.

(12.2.5) **Example:** There is *no isometry* between the sphere and the plane (or even between any patch of the sphere and a patch of the plane). In cartographic terms, it is impossible to make a (flat) map of any portion of the earth's surface that will represent all distances with complete accuracy. Why? Because such an isometry would have to preserve distances and angles, and would have to map great circles (which are shortest paths on the sphere) onto straight lines (which are shortest paths in the plane). So the angle sum of a spherical triangle would have to be the same as the angle sum of the corresponding Euclidean triangle, that is π radians. And this would imply that the triangle had zero area.

(12.2.6) **Example:** There is no concept of *similarity* in spherical geometry. Two similar triangles in Euclidean geometry have the same angles but are of different size. But in spherical geometry, the angles of a triangle determine its area and so its size as well as its shape. This is because spherical geometry has an absolute standard of length: the radius of the sphere.

(12.2.7) **Example:** Another application reveals a surprising link with topology. We need to know that the area formula 12.2.3 for a triangle can be extended to an area formula for any *convex spherical polygon* (given by an intersection of n hemispheres); if $\alpha_1, \ldots, \alpha_n$ are the interior angles of such a polygon, then its area is equal to

$$\sum_{i=1}^{n} \alpha_i - (n-2)\pi.$$

This formula can easily be proved by dividing the polygon into triangles and using 12.2.3. Now suppose we divide up the whole surface of the sphere into convex spherical polygons. (The division can be quite general; it need not be done in any kind of symmetrical way.) Let V be the number of vertices, E the number of edges, and F the number of 'faces' (polygons). What is the sum of all the angles of all the polygons? On the one hand it is clear that each vertex contributes 2π to the total, so that the sum is $2\pi V$. On the other hand, if we sum up the formula

$$\sum_{i=1}^{n} \alpha_i = (n-2)\pi + \text{Area of polygon}$$

over all the polygons, and remember that each edge is an edge of two polygons and that the total area is 4π, we will find that the total angle is

$$2\pi E - 2\pi F + 4\pi.$$

Equating this to $2\pi V$ and dividing by 2π, we get Euler's famous formula

$$V - E + F = 2.$$

12.3 Holonomy

We have seen that a sphere is an example of an 'intrinsically curved' surface. Its intrinsic geometry, given by the first fundamental form, is essentially different from that of a plane. In this section and the next we will discuss a concept of intrinsic curvature that can be applied to all surfaces. It will turn out that the curvature of a sphere is the same all over, whereas a general surface can be more curved in some places and less curved in others. Nevertheless, we will find that there is an analogue to the 'excess formula' 12.2.3 which revealed the curvature of the sphere to us.

Throughout the next two sections, we will work with a simplifying assumption. Most of the examples of patches of surface that we have given so far have had the coefficient F of the first fundamental form equal to zero. Such a parameterization is called *orthogonal* because the coordinate lines $u =$ constant and $v =$ constant are perpendicular to one another. It can be proved that any patch of surface has an orthogonal parameterization, and we will therefore assume in what follows that all the parameterizations we use are orthogonal. While this assumption is not a necessary one, its omission would complicate still further some algebra which you may well think is already messy enough.

Let σ be a regular patch of surface. Let γ be a regular curve in the surface, so that γ is given in terms of the parameterization by

$$\gamma(t) = \sigma(u(t), v(t)).$$

A *tangent vector field* $\mathbf{w}$ along γ is a smooth map which to each t associates a vector $\mathbf{w}(t)$ belonging to the tangent plane to the surface at $\gamma(t)$. This tangent plane is spanned by the two vectors σ_u and σ_v; so what we require is that $\mathbf{w}(t)$ can be expressed as a linear combination of these two vectors. If $|\mathbf{w}(t)| = 1$ for all t, we call $\mathbf{w}$ a *unit tangent vector field*.

Let $\mathbf{p} = \sigma_u/E$ and $\mathbf{q} = \sigma_v/G$. Because of our assumption that $F = 0$, the vectors $\mathbf{p}$ and $\mathbf{q}$ form an orthonormal basis for the tangent plane. Using A.1.8, we find that a unit tangent vector field can always be written as

(12.3.1)
$$\mathbf{w}(t) = \cos\theta(t)\mathbf{p} + \sin\theta(t)\mathbf{q}$$

where θ is a smooth function of t.

(12.3.2) **Definition:** *A vector field* $\mathbf{w}(t)$ *along a curve* $\gamma(t)$ *in a surface is* parallel *if the derivative* $\dot{\mathbf{w}}(t)$ *is perpendicular to the tangent plane at each point.*

A parallel vector field is as constant as it can be while staying tangent to the surface at each point. A tangent vector field has to vary if the tangent plane varies, but we can require that the variation is not 'detectable' in the tangent plane, and this is the condition of parallelism. For instance, the unit tangent vectors to a great circle on the sphere form a parallel vector field. This suggests another definition.

(12.3.3) **Definition:** *A regular curve in a surface is called a* geodesic *if its own unit tangent vectors form a parallel vector field along it.*

Since a parallel vector field is as constant as it can be while staying tangent to the surface, a geodesic is a curve which is as straight as it can be while remaining in the surface. Great circles on the sphere, and straight lines in the plane, are examples of geodesics. These examples suggest that geodesics are as short as they can be, as well as as straight as they can be, and this is indeed true in a certain sense. But we won't go into the details.

The concept of a parallel vector field is *a priori* part of extrinsic geometry, because it makes use of the extrinsic concept of the normal to the surface. Remarkably, though, one can prove that the concept is in fact part of *intrinsic* geometry; it depends only on the first fundamental form.

To prove this we will first need to calculate the derivatives of the unit vectors **p** and **q** in terms of the first fundamental form.

(12.3.4) **Lemma:** *With notation as above, the partial derivatives of the unit vectors* **p** *and* **q** *satisfy the following relations:*

$$\mathbf{p}_u \cdot \mathbf{p} = \mathbf{p}_v \cdot \mathbf{p} = \mathbf{q}_u \cdot \mathbf{q} = \mathbf{q}_v \cdot \mathbf{q} = 0,$$

$$\mathbf{p}_u \cdot \mathbf{q} = -\mathbf{p} \cdot \mathbf{q}_u = \frac{-E_v}{2\sqrt{EG}},$$

$$\mathbf{p}_v \cdot \mathbf{q} = -\mathbf{p} \cdot \mathbf{q}_v = \frac{G_u}{2\sqrt{EG}}.$$

Proof: The first set of identities is easy to obtain, because the derivative of any vector of constant magnitude is perpendicular to that vector (11.1.2).

For the second set, consider the definition $E = \sigma_u \cdot \sigma_u$. Differentiating with respect to v, we find that

$$E_v = 2\sigma_{uv} \cdot \sigma_u.$$

Now $\mathbf{q} = \sigma_v / \sqrt{G}$, and so by differentiation

$$\mathbf{q}_u = \frac{\sigma_{vu}}{\sqrt{G}} - \frac{\sigma_v}{2G^{\frac{3}{2}}} G_u.$$

Take the dot product with $\mathbf{p} = \sigma_u / \sqrt{E}$ to get

$$\mathbf{p} \cdot \mathbf{q}_u = \frac{\sigma_u \cdot \sigma_{vu}}{\sqrt{EG}}.$$

Now a theorem in calculus of several variables says[3] that 'mixed partial derivatives commute', in other words that $\sigma_{vu} = \sigma_{uv}$. Using this fact and substituting in the calculation of E_v we made earlier, we see that

$$\mathbf{p} \cdot \mathbf{q}_u = \frac{E_v}{2\sqrt{EG}}.$$

Finally, we see that $\mathbf{p}_u \cdot \mathbf{q} = -\mathbf{p} \cdot \mathbf{q}_u$ by differentiating the identity $\mathbf{p} \cdot \mathbf{q} = 0$ with respect to u. This completes the proof of the second set of identities, and the proof of the third set is similar. $\square$

---

[3]Under suitable hypotheses, for instance that all the functions involved are smooth. See Apostol [3] or Kaplan [25] for a precise statement and proof.

(12.3.5) **Proposition:** *Let σ be an orthogonal parameterization of a patch of surface, and let γ be a regular curve in the surface. Then a unit tangent vector field along γ given by*

$$\mathbf{w}(t) = \cos \theta(t)\mathbf{p} + \sin \theta(t)\mathbf{q}$$

is parallel if and only if

$$\frac{d\theta}{dt} = \frac{1}{2\sqrt{EG}}\left(\frac{\partial E}{\partial v}\frac{du}{dt} - \frac{\partial G}{\partial u}\frac{dv}{dt}\right).$$

Proof: Let $\mathbf{n} = \mathbf{p} \wedge \mathbf{q}$ be the unit normal vector field to the surface. Since $\mathbf{w}$ is of constant magnitude, the derivative $\dot{\mathbf{w}}$ is perpendicular to $\mathbf{w}$. To prove that $\dot{\mathbf{w}}$ is perpendicular to the entire tangent plane, therefore, it will be enough to prove that it is also perpendicular to $\mathbf{n} \wedge \mathbf{w}$. So the field will be parallel if and only if the scalar triple product $[\mathbf{n}, \mathbf{w}, \dot{\mathbf{w}}]$ is equal to zero.

We can calculate that

$$\mathbf{n} \wedge \mathbf{w} = -\sin \theta\mathbf{p} + \cos \theta\mathbf{q},$$

$$\dot{\mathbf{w}} = \dot{\theta}(-\sin \theta\mathbf{p} + \cos \theta\mathbf{q}) + \cos \theta(\dot{u}\mathbf{p}_u + \dot{v}\mathbf{p}_v) + \sin \theta(\dot{u}\mathbf{q}_u + \dot{v}\mathbf{q}_v).$$

Now take the dot product of these two expressions and use lemma 12.3.4 to find that

$$[\mathbf{n}, \mathbf{w}, \dot{\mathbf{w}}] = \dot{\theta} + \frac{1}{2\sqrt{EG}}\left(G_u\dot{v} - E_v\dot{u}\right).$$

Set this equal to zero to obtain the result. □

(12.3.6) **Corollary:** *If $\mathbf{w}_1$ and $\mathbf{w}_2$ are two parallel vector fields along the same curve γ, then the oriented angle between $\mathbf{w}_1$ and $\mathbf{w}_2$ is constant.*

Proof: If the vector fields are described by angles θ_1 and θ_2, then the differential equation 12.3.5 shows that $\dot{\theta}_1 - \dot{\theta}_2 = 0$. □

Remark: Here and in what follows, we will frequently need to speak of oriented angles, orientation, and so on, on a patch of surface. The orientation that is assumed is always that given by the unit normal vector field $\mathbf{n} = \mathbf{p} \wedge \mathbf{q}$ described above; that is, $\mathbf{p}^\perp = \mathbf{q}$. Remember that we have assumed that the charts have been chosen so that their orientations are consistent.

Suppose now that the curve $\gamma: [a, b] \to S$ is a loop, which begins and ends at the same point $\gamma(a) = \gamma(b)$. Let $\mathbf{w}$ be a unit vector field that is parallel along γ. It need not be the case that $\mathbf{w}(a) = \mathbf{w}(b)$; when we take the vector $\mathbf{w}(a)$ and 'parallel transport' it along the curve it may not return to its original position. We define the *holonomy* Hol(γ) of the curve γ to be the oriented angle between $\mathbf{w}(b)$ and $\mathbf{w}(a)$. By 12.3.6, this angle does not depend on the choice of parallel vector field $\mathbf{w}$.

Suppose that the path γ is contained in a single coordinate patch. Then we may write $\mathbf{w}(s) = \cos \theta(s)\mathbf{p} + \sin \theta(s)\mathbf{q}$, and the holonomy is then equal to $\theta(b) - \theta(a)$. From proposition 12.3.5, therefore, we find that

(12.3.7)
$$\mathrm{Hol}(\gamma) \equiv \int_\gamma \frac{1}{2\sqrt{EG}}\left(E_v\,du - G_u\,dv\right).$$

Because the holonomy is defined as an oriented angle, we have used the special symbol $\equiv$ which denotes equality of oriented angles in this equation. Remember (6.2.2) that an oriented angle is defined 'only up to a multiple of 2π', so that $a \equiv b$ means $a = b + 2\pi k$, where k is some integer.

(12.3.8)   **Definition:** *Let σ be an (orthogonal) parameterization of a patch of surface. The* Gaussian curvature *is the function K on the surface defined in terms of the first fundamental form by*

$$K = \frac{-1}{2\sqrt{EG}}\left\{\frac{\partial}{\partial v}\left(\frac{E_v}{\sqrt{EG}}\right) + \frac{\partial}{\partial u}\left(\frac{G_u}{\sqrt{EG}}\right)\right\}.$$

(12.3.9)   **Theorem:** (Holonomy Theorem) *Let γ be a positively oriented regular simple closed curve in a patch of surface. Then the holonomy of γ is related to the Gaussian curvature by the formula*

$$\mathrm{Hol}(\gamma) \equiv \int_{\mathrm{Int}(\gamma)} K\,d\mathcal{A}_s.$$

Proof: Simply apply Green's theorem (9.6.4) to equation 12.3.7. This gives

$$\mathrm{Hol}(\gamma) \equiv -\frac{1}{2}\int_{\mathrm{Int}(\gamma)}\left\{\frac{\partial}{\partial v}\left(\frac{E_v}{\sqrt{EG}}\right) + \frac{\partial}{\partial u}\left(\frac{G_u}{\sqrt{EG}}\right)\right\}d\mathcal{A}.$$

The result now follows from the definitions of Gaussian curvature and of surface integral (12.1.13). $\square$

If our surface were part of a plane, then a parallel vector field would actually be constant, and the holonomy of every curve would therefore be zero. By the holonomy theorem, this could happen only if the Gaussian curvature were zero. The Gaussian curvature is therefore a measure of how far the intrinsic geometry of our surface deviates from that of a plane.

(12.3.10)   **Example:** Let us work out the Gaussian curvature of a sphere of radius r. We will use the parameterization by latitude and longitude, for which the first fundamental form is

$$E = r^2, \quad F = 0, \quad G = r^2 \cos^2 u.$$

Then $E_v = 0$, $G_u / \sqrt{EG} = -2 \sin u$, and so the Gaussian curvature K is

$$K = \frac{-1}{2r^2 \cos u} \frac{\partial}{\partial u} \left(-2 \sin u \right) = + \frac{1}{r^2}.$$

The bigger the sphere, the less the curvature. In this example the curvature is constant.

(12.3.11)   **Example:** Now we will work out the Gaussian curvature of a *surface of revolution*. Suppose that $u \mapsto (x(u), y(u))$ is the equation of a curve parameterized by arc length in the xy-plane, with $y(u) > 0$ for all u. The surface obtained by revolving this curve around the x-axis is called a *surface of revolution*, and it can be parameterized by

$$\sigma(u, v) = (x(u), y(u) \cos v, y(u) \sin v).$$

Computing the first fundamental form, we find that

$$E = x'(u)^2 + y'(u)^2 = 1, \quad F = 0, \quad G = y(u)^2.$$

Therefore

$$K = \frac{-1}{2y(u)} \frac{\partial}{\partial u} \left(\frac{2y(u) y'(u)}{y(u)} \right) = -\frac{y''(u)}{y(u)}.$$

If $y(u) = \cos u$ we have a sphere of curvature $+1$. If $y(u) = a + bu$ then the curvature is zero. In this case the curve is a straight line, so the surface is a cone or cylinder, which can be 'unwrapped' onto a plane. In general the curvature can be either positive or negative. As you can see in Figure 12.5, the curvature is positive at a point where the surface is convex or concave, staying on one side of its tangent plane; it is negative at a point where the surface is saddle shaped and passes through its own tangent plane.

12.4 The Gauss–Bonnet theorem

In 12.2.5, we considered the angle sum of a spherical triangle in order to show that the geometry of the sphere is intrinsically non-Euclidean. Now we have

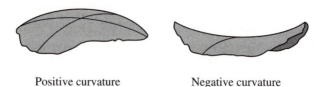

Positive curvature   Negative curvature

Fig. 12.5. Positive and negative Gaussian curvature.

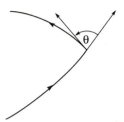

Fig. 12.6. A vertex of a piecewise regular curve.

introduced the Gaussian curvature, which shows the same thing. The link between the two concepts is provided by the *Gauss–Bonnet theorem*.

This theorem uses the concept of the *geodesic curvature* of a curve in a surface. The geodesic curvature measures how far the curve is from being a geodesic, just as the ordinary curvature of a curve in a plane measures how far it is from being a straight line.

(12.4.1)   **Definition:** *Let γ be a regular curve in a patch of surface σ, and assume that γ is parameterized by arc length s. Let $\mathbf{w}$ be a unit vector field that is parallel along γ, and write[4] the unit tangent vector $\gamma'(s)$ as $\gamma'(s) = \cos \psi \mathbf{w} + \sin \psi \mathbf{w}^{\perp}$, so that ψ is the oriented angle between $\mathbf{w}$ and $\gamma'(s)$. Then the geodesic curvature $\kappa_g(s)$ is defined by*

$$\kappa_g(s) = \frac{d\psi}{ds}.$$

This definition is analogous to the description of the signed curvature in the plane given in 11.4.8. Notice that if γ is a loop, then

$$\mathrm{Hol}(\gamma) \equiv - \int_{\gamma} \kappa_g(s)\, ds$$

since the tangent vector $\gamma'(s)$ returns to its original position after passing once round the loop, whereas the parallel vector field $\mathbf{w}$ is turned through an angle of $\mathrm{Hol}(\gamma)$.

It is not hard to extend this relationship to piecewise regular curves. Let γ be a curve with a vertex V, as shown in Figure 12.6, and let $\mathbf{w}$ be a parallel vector field along γ. As we pass from a parameter value just before the vertex to a parameter value just after the vertex, the parallel vector field $\mathbf{w}$ remains almost constant, but the tangent vector $\gamma'(s)$ jumps through an oriented angle

---

[4]We are using A.1.8 here.

θ, the *external angle* at the vertex. Arguing as before, we find that for a piecewise regular curve

$$\text{Hol}(\gamma) \equiv -\int_\gamma \kappa_g(s)\,ds - \sum_{i=1}^n \theta_i$$

where $\theta_1, \ldots, \theta_n$ are the oriented angles at the vertices.

A vertex where $\theta \equiv \pi$ is called a *cusp*. We will assume that our piecewise regular curves do not have cusps. This allows us to choose numerical values for the external angles; we will always choose the determination θ of the oriented angle for which $-\pi < \theta < \pi$.

(12.4.2)   **Theorem:** (GAUSS–BONNET THEOREM — LOCAL VERSION) *Let* γ *be a positively oriented piecewise regular simple closed curve without cusps in a patch of surface. Let* $\theta_1, \ldots, \theta_n$ *be the external angles. Then*

$$\int_\gamma \kappa_g(s)\,ds + \sum_{i=1}^n \theta_i = 2\pi - \int_{\text{Int}(\gamma)} K\,dA_s$$

where K is the Gaussian curvature.

Proof: From the holonomy theorem (12.3.9),

$$\text{Hol}(\gamma) \equiv \int_{\text{Int}(\gamma)} K\,dA_s \equiv -\int_\gamma \kappa_g(s)\,ds - \sum_{i=1}^n \theta_i.$$

Therefore

(12.4.3) $$\int_\gamma \kappa_g(s)\,ds + \sum_{i=1}^n \theta_i = 2\pi k(\gamma) - \int_{\text{Int}(\gamma)} K\,dA_s,$$

and we must prove that the integer $k(\gamma)$ is always $+1$.

A priori, the integer $k(\gamma)$ might depend upon the curve γ. However, the following is in fact true:

(12.4.4)   **Claim:** *The integer* $k(\gamma)$ *is the same for all positively oriented piecewise regular simple closed curves without cusps.*

If we grant this claim, we can go on to work out what $k = k(\gamma)$ is. Let ζ be a regular curve, divided into two curves ζ_1 and ζ_2 by a cross-cut (see Figure 12.7). Let θ_1 and φ_1 be the exterior angles of ζ_1 at its two vertices, θ_2 and φ_2 the corresponding exterior angles of ζ_2. Notice that

$$\theta_1 + \theta_2 = \varphi_1 + \varphi_2 = \pi.$$

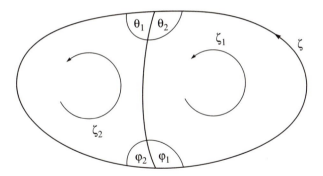

Fig. 12.7. Calculation of k.

By 12.4.3,

$$\int_{\zeta_1} k_g(s)\,\mathrm{d}s + \theta_1 + \varphi_1 = 2\pi k - \int_{\mathrm{Int}(\zeta_1)} K\,\mathrm{d}A_s$$

$$\int_{\zeta_2} k_g(s)\,\mathrm{d}s + \theta_2 + \varphi_2 = 2\pi k - \int_{\mathrm{Int}(\zeta_2)} K.\,\mathrm{d}A_s$$

Add these two equations, and notice that the two contributions from the integral of the geodesic curvature over the cross-cut cancel because they are taken in opposite directions. This gives

$$\int_\zeta k_g(s)\,\mathrm{d}s + 2\pi = 4\pi k - \int_{\mathrm{Int}(\zeta)} K\,\mathrm{d}A_s.$$

Comparing this with 12.4.3 for ζ, we find that $k = 1$.

It remains to check Claim 12.4.4. It rests ultimately on certain facts about plane topology that are related to the Jordan curve theorem, and whose proofs I think are too difficult to be included in this book. Heuristically, though, we can justify the claim as follows. It is 'clear' that the curve γ can be *smoothly deformed* into any other simple closed curve lying in its interior, $\mathrm{Int}(\gamma)$. This means that one can find a family γ_τ of simple closed curves (without cusps) depending smoothly on the parameter τ with $\gamma_0 = \gamma$ and γ_1 equal to any prescribed simple closed curve in $\mathrm{Int}(\gamma)$. A deformation of this kind is illustrated in Figure 12.8.

Since the deformation is smooth, the integer $k(\gamma_\tau)$ must depend continuously on the parameter τ (because this is true of all the other quantities appearing in equation 12.4.3). But a continuous integer-valued function is a constant, by the intermediate value theorem of real analysis. Therefore, $k(\gamma_1) = k(\gamma_0)$. But γ_1 could be any curve in the interior of γ. Hence the constant k is the same for γ and for all curves in its interior. □

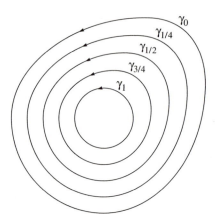

Fig. 12.8. Deforming a curve into its interior.

As a particular case of the Gauss–Bonnet theorem, consider a *geodesic polygon P*, that is the interior of a piecewise regular simple closed curve whose sides are geodesics. The *internal angles* α_i are defined by $\alpha_i = \pi - \theta_i$. Since a geodesic has zero geodesic curvature, we find

(12.4.5) **Corollary:** *For a geodesic polygon P with internal angles* $\alpha_1, \ldots, \alpha_n$,

$$\sum_{i=1}^{n} \alpha_i - (n-2)\pi = \int_P K \, \mathrm{d}\mathcal{A}_s.$$

This is analogous to the formula that we found for spherical polygons in 12.2.7, except that the area is replaced by the integral of the Gaussian curvature. (The Gaussian curvature of the unit sphere is $+1$, so the formulae are consistent.)

A surface S is said to be *closed* if it is bounded and has no 'edge'. Examples are spheres and tori. It can be proved that any closed surface can be subdivided into geodesic polygons. If V, E, and F are the numbers of vertices, edges, and faces in such a decomposition, then the same argument as in 12.2.7 can be used to prove

(12.4.6) **Theorem:** (GAUSS–BONNET THEOREM — GLOBAL VERSION) *For any closed surface S,*

$$V - E + F = \frac{1}{2\pi} \int_S K \, \mathrm{d}\mathcal{A}_s.$$

In fact it is not even necessary to assume that the sides of the polygons are geodesics. Without this assumption there are some extra geodesic curvature terms; but they all cancel out because each side is traversed twice in opposite directions.

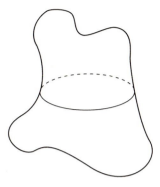

Fig. 12.9. A deformed sphere. Total curvature is still 4π.

This is a wonderful theorem. On the left hand of the equation we have a topological invariant, which takes no account of the local geometry; it does not change at all when the surface is deformed. In fact $V - E + F = 2 - 2g$, where g is the *genus* or number of holes, mentioned earlier (7.5.6). On the right hand of the equation we have something which depends strongly on the local geometry, but which has no explicit dependence on the overall topology; you can calculate the curvature at a point without knowing anything about the shape of the surface far away from that point. One has only to think about the very different shapes into which a sphere can be deformed (see Figure 12.9), for example, to feel that it is remarkable that in each case the total amount of curvature manages to balance out at 4π. The theorem is surprising, profound, full of consequences and leading to deep generalizations, and yet in retrospect inevitable. It also provides a suitably high point on which to end this book.

12.5 Exercises

1. Let $\gamma: [a, b] \rightarrow \mathbf{R}^3$ be a curve parameterized by arc length. For each $u \in [a, b]$, let Π_u denote the plane through $\gamma(u)$ normal to the curve, and let S denote the surface swept out as u varies by the circle in Π_u of centre $\gamma(u)$ and constant radius r.

(i) Explain why the surface S can be parameterized by

$$\sigma(u, v) = \gamma(u) + r(\mathbf{n}(u) \cos v + \mathbf{b}(u) \sin v),$$

where $\mathbf{n}$ and $\mathbf{b}$ are the unit normal and binormal vectors to γ.

(ii) Show that the unit normal vector to S at the point parameterized by (u, v) is $\mathbf{n}(u, v) = -(\mathbf{n}(u) \cos v + \mathbf{b}(u) \sin v)$.

(iii) Why is this question not sensible unless r is smaller than the radius of curvature of γ?

2. Suppose that a patch of surface has two parameterizations (u, v) and $(\tilde{u}, \tilde{v})$. Let E, F, G and $\tilde{E}$, $\tilde{F}$, $\tilde{G}$ be the coefficients of the first fundamental form relative to the two parameterizations. Show that

$$\begin{pmatrix} E & F \\ F & G \end{pmatrix} = J \begin{pmatrix} \tilde{E} & \tilde{F} \\ \tilde{F} & \tilde{G} \end{pmatrix} J^t,$$

where J is the Jacobian matrix

$$J = \begin{pmatrix} \partial\tilde{u}/\partial u & \partial\tilde{v}/\partial u \\ \partial\tilde{u}/\partial v & \partial\tilde{v}/\partial v \end{pmatrix}.$$

3. Compute the first fundamental forms of the following parameterized surfaces:

(i) the *catenoid*: $\sigma(u, v) = (u, \cosh u \cos v, \cosh u \sin v)$;

(ii) the *helicoid*: $\sigma(u, v) = (u, v \cos u, v \sin u)$.

A *meridian* on the catenoid is a curve $v = $ constant. Prove that if one meridian is removed from the catenoid, the resulting surface is isometric to part of the helicoid, in such a way that meridians on the catenoid map to straight lines $u = $ constant on the helicoid.

4. Let S be the surface described in question 12.5.1. Prove that the first fundamental form of S is

$$\left((1 - \kappa r \cos v)^2 + r^2 \tau^2 \right) du^2 + 2r^2 \tau \, du \, dv + r^2 \, dv^2,$$

where κ and τ denote the curvature and torsion of the original curve γ. Hence prove that the area of S is equal to $2\pi r$ times the length of γ.

5. A sailor circumnavigates Australia by a route consisting of a triangle of great circular arcs. Prove that one angle of the triangle is at least $\frac{\pi}{3} + \frac{1}{16}$ radians. (Assume that the earth is a sphere of radius 4,000 miles, and that the area of Australia is 3 million square miles.)

6. Show that in the limit as the sides a, b, and c become small the spherical sine rule (12.2.2) becomes the ordinary sine rule. Carry out a similar investigation for the spherical cosine rule (12.2.1); you will need to consider terms up to second order in the small quantities a, b, and c.

7. The *tractoid* is the surface of revolution obtained by rotating the tractrix (Exercise 7.6.3) with equation $x = \log \cot \frac{1}{2}u - \cos u$, $y = \sin u$ about the x-axis. Show that the total area of the tractoid is 2π, and that its Gaussian curvature is -1 everywhere.

(The tractoid is sometimes called a *pseudosphere* because it provides a model for Lobatchewsky and Bolyai's hyperbolic geometry of constant negative curvature, just as the ordinary sphere provides a model for the geometry of constant positive curvature. But there is an important difference in that the tractoid has an 'edge' — it models only a part of the hyperbolic plane. It can be shown that *no* curved surface in three-dimensional space can model the entire hyperbolic plane.)

8. Find all the geodesics on a right circular cylinder, and show that there are usually an infinite number of geodesics between two given points.

9. Compute the first fundamental form and the Gaussian curvature for the torus in $\mathbf{R}^3$ obtained by rotating the circle $(x - a)^2 + y^2 = b^2$ about the y-axis. Verify that the integral of the Gaussian curvature over the surface is zero, in accordance with the Gauss–Bonnet theorem.

10. The unit sphere in $\mathbf{R}^3$ is covered by m nonoverlapping triangles with geodesic sides. The internal angles of each triangle are π/p, π/q, and π/r, where p, q, r are integers greater than unity.

   (i) Show that $1/p + 1/q + 1/r > 1$.

  (ii) Determine all integers p, q, r $(1 < p \leq q \leq r)$ which satisfy this inequality.

 (iii) In each case find the number m of triangles needed to provide a non-overlapping covering of the sphere.

11. Let γ be a positively oriented simple closed curve in the plane $\mathbf{R}^2$. Use the Gauss–Bonnet theorem to show that the tangent vector $\dot{\gamma}$ turns through an angle of exactly 2π when the curve is traversed once in the positive direction.

    (This topological fact, called the *theorem of turning tangents*, is in fact needed in a rigorous proof of Claim 12.4.4. For a proof independent of the Gauss–Bonnet theorem, see do Carmo [12].)

12. (For those with some knowledge of the calculus of variations.) In this exercise we will prove that a path which is the shortest possible one between its two endpoints must necessarily be a geodesic. Let $\gamma(t) = \sigma(u(t), v(t))$ give a path in a patch of surface equipped with an orthogonal parameterization σ, and suppose that the path is the shortest possible between parameter values $t = a$ and $t = b$.

   (i) Show that the length of the path is $L = \displaystyle\int_a^b \sqrt{E\dot{u}^2 + G\dot{v}^2}\, dv$.

  (ii) Suppose that a small variation $\delta u(t)$, vanishing at the endpoints, is made to $u(t)$. Show that the corresponding variation in the length of the path is

$$\delta L = \int_a^b \frac{1}{2\sqrt{E\dot{u}^2 + G\dot{v}^2}} (E_u \dot{u}^2 \delta u + G_u \dot{v}^2 \delta u + 2E\dot{u}\delta\dot{u})\, dt$$

to first order.

 (iii) Assume now that the original path was parameterized by arc length. By integration by parts, show that

$$\delta L = \int_a^b \left(\frac{E_u \dot{u}^2 + G_u \dot{v}^2}{2} - \frac{d(E\dot{u})}{dt} \right) \delta u\, dt$$

to first order.

(iv) Explain why δL must be zero for all possible variations δu. Hence obtain the differential equation

$$\frac{d(E\dot{u})}{dt} = \frac{E_u\dot{u}^2 + G_u\dot{v}^2}{2}.$$

(v) Rewrite this differential equation as

$$\sqrt{E}\frac{d(\dot{u}\sqrt{E})}{dt} = \tfrac{1}{2}\dot{v}(G_u\dot{v} - E_v\dot{u}).$$

(vi) If the tangent vector field to the curve is written in terms of the standard basis vectors $\mathbf{p}$ and $\mathbf{q}$ as $\cos\theta\mathbf{p} + \sin\theta\mathbf{q}$, explain why $\cos\theta = \dot{u}\sqrt{E}$ and $\sin\theta = \dot{v}\sqrt{G}$.

(vii) Prove that

$$\frac{d\theta}{dt} = \frac{1}{2\sqrt{EG}}(E_v\dot{u} - G_u\dot{v}),$$

and deduce that the curve γ is a geodesic.

(viii) We have shown that a length-minimizing curve must be a geodesic. But must a geodesic be a length-minimizing curve? (Consider great circles on the sphere.)

Appendix A

The trigonometric functions

The trigonometric functions may be defined by the power series

$$\sin\theta = \sum_{n=0}^{\infty} \frac{(-1)^n \theta^{2n+1}}{(2n+1)!}, \quad \cos\theta = \sum_{n=0}^{\infty} \frac{(-1)^n \theta^{2n}}{(2n)!}.$$

From these one can deduce directly

(A.1.1) Trigonometric identities:

$$
\begin{aligned}
\cos(\theta + \varphi) &= \cos\theta\cos\varphi - \sin\theta\sin\varphi \\
\sin(\theta + \varphi) &= \cos\theta\sin\varphi + \sin\theta\cos\varphi \\
\cos^2\theta + \sin^2\theta &= 1 \\
\cos 2\theta &= 2\cos^2\theta - 1 = 1 - 2\sin^2\theta \\
\sin 2\theta &= 2\sin\theta\cos\theta \\
\cos 3\theta &= 4\cos^3\theta - 3\cos\theta \\
\sin 3\theta &= 3\sin\theta - 4\sin^3\theta
\end{aligned}
$$

(A.1.2) Limiting values:

$$\lim_{\theta \to 0} \cos\theta = 1, \quad \lim_{\theta \to 0} \frac{\sin\theta}{\theta} = 1.$$

(A.1.3) Differentiation:

$$\frac{d}{d\theta}\cos\theta = -\sin\theta, \quad \frac{d}{d\theta}\sin\theta = \cos\theta.$$

(A.1.4) The number π: Let p be the smallest positive root of the equation $\cos p = 0$. One can compute that $\cos\theta > 0$ for $0 \le \theta \le 1$, whereas $\cos 2 < 0$, so that $1 < p < 2$. The number π is defined to be $2p$.

(A.1.5) Periodicity relations:

$$\cos(\theta + \pi/2) = -\sin\theta, \quad \sin(\theta + \pi/2) = \cos\theta,$$

$$\cos(-\theta) = \cos\theta, \quad \sin(-\theta) = -\sin\theta,$$

$$\cos(\theta + 2\pi) = \cos\theta, \quad \sin(\theta + 2\pi) = \sin\theta.$$

(A.1.6) Special values:

θ	$\sin\theta$	$\cos\theta$
0	0	1
$\frac{\pi}{6}$	$\frac{1}{2}$	$\frac{\sqrt{3}}{2}$
$\frac{\pi}{4}$	$\frac{1}{\sqrt{2}}$	$\frac{1}{\sqrt{2}}$
$\frac{\pi}{3}$	$\frac{\sqrt{3}}{2}$	$\frac{1}{2}$

(A.1.7) Let x and y be real numbers with $x^2 + y^2 = 1$. Then one can find θ such that

$$\cos\theta = x, \quad \sin\theta = y;$$

and any two possible solutions θ differ by an integer multiple of 2π.

We will sometimes need a version of this result 'with parameters':

(A.1.8) **Proposition:** *Let $x(t)$ and $y(t)$ be two smooth functions of a real variable t with $x(t)^2 + y(t)^2 = 1$. Then there is a smooth function $\theta(t)$ of t such that*

$$\cos\theta(t) = x(t), \quad \sin\theta(t) = y(t)$$

for all t.

The point is that A.1.7 ensures that there is a θ for each t considered individually, but does not tell you anything about how these values of θ fit together.

Proof: Using A.1.7, choose ψ such that $\cos\psi = x(0)$ and $\sin\psi = y(0)$. Then define

$$\theta(t) = \int_0^t (x(\tau)\dot{y}(\tau) - y(\tau)\dot{x}(\tau))\, d\tau.$$

To prove that $\theta(t)$ has the required properties we will use several times the identities $x^2 + y^2 = 1$ and $x\dot{x} + y\dot{y} = 0$ (the second of which follows from the first by differentiation). Notice that

$$\frac{d}{dt}\left((x - \cos\theta)^2 + (y - \sin\theta)^2\right) = 2((x\sin\theta - y\cos\theta)\dot{\theta} - \cos\theta\dot{x} - \sin\theta\dot{y})$$

by expanding and using the second identity above. However, from the definition of θ,

$$\dot{\theta} = x\dot{y} - y\dot{x}.$$

Substituting this into the equation above, we obtain

$$2(x^2\dot{y} - xy\dot{x} - \dot{y})\sin\theta + 2(y^2\dot{x} - xy\dot{y} - \dot{x})\cos\theta$$

and this is zero by the identities above. Thus

$$(x - \cos\theta)^2 + (y - \sin\theta)^2$$

is a constant. This constant must be zero, because $\theta(0) = \psi$, so it follows that $x = \cos\theta$ and $y = \sin\theta$ for all t. $\square$

Hints and solutions to the exercises

Chapter 1. **1.** No; use the line–plane intersection axiom. **2.** What if A, B, and C are collinear? **3.** The plane axiom has something to do with it. **4.** Model 1 satisfies the parallel axiom, model 2 doesn't. **5.** The Nim-sum of three points gives the fourth point in their plane. To prove the plane–plane intersection axiom, use the fact that the Nim-sum of all the numbers from 0 to 7 is zero. If two planes intersected in one point only, then seven of the eight numbers would represent points in the two planes. What would the Nim-sum of these seven numbers be? **6.** Use the converse to the similarity axiom to see that WZ and XY are both parallel to BD, and argue in the same way for the other sides. **8.** Draw a line through X parallel to AB, meeting BC at Y. Apply the similarity axiom three times to show that $BY : YC = BX : XD = WX : XA = WY : YB$, and then use the fact that W is the midpoint of BC. **9.** Use the similarity axiom to show that both ratios are equal to $OP : PB'$. **10.** Consider the point of intersection of the two lines, and use the similarity axiom (twice) and its converse. **11.** Consider a line parallel to $\mathcal{L}$ through one of the vertices, say C, and work out in two different ways the ratio in which it divides the opposite side AB. **12.** Apply Menelaus' theorem to triangle ABX and to triangle AXC.

Chapter 2. **1.** C has position vector $5\mathbf{i} + 6\mathbf{j}$; D has position vector $9\mathbf{i} + 14\mathbf{j}$. **2.** See 5.2.1. **3.** AD is parallel to BC, and AB is parallel to CD. This implies that the points are coplanar; in fact they form a parallelogram. You could also use 2.2.4 to show that the points are coplanar. **4.** Yes; let Q have position vector $(\mathbf{p}_1 + \cdots + \mathbf{p}_n)/n$. **5.** With the obvious notation for position vectors, $(\mathbf{a} + \mathbf{c})/2 = (\mathbf{b} + \mathbf{d})/2$ if and only if $\mathbf{d} - \mathbf{a} = \mathbf{c} - \mathbf{b}$. For the last part, write the vector $\mathbf{c} - \mathbf{a}$ in two different ways and use the fact that $\mathbf{b} - \mathbf{a}$ and $\mathbf{d} - \mathbf{a}$ are linearly independent. **6.** $(\mathbf{a} + \mathbf{b} + \mathbf{c})/3$. **7.** The point with position vector $(\mathbf{a} + \mathbf{b} + \mathbf{c} + \mathbf{d})/4$ lies on all four lines. **8.** By Ceva's theorem, BX bisects AC, so X lies on the median through B. **9.** $\mathbf{c} = \mathbf{x} + \mathbf{y} - 2\mathbf{z}$, and so on, perhaps differently according to how you labelled the sides; XY is parallel to AB, and so on. **10.** The vectors $\vec{AB}$, $\vec{AC}$, and $\vec{CD}$ are linearly independent, so the four points are not coplanar. The points of intersection are A and D. **11.** Take position vectors relative to C, say, and work out when $\vec{XY}$ will be a multiple of $\vec{YZ}$. **12.** Represent points by their position vectors relative to A, expressed as linear combinations of $\mathbf{b}$ and $\mathbf{p}$. Find the position vectors of S and then of Q by simultaneous equations. A short alternative argument uses Pappus' harmonic-division theorem: C and a 'point at infinity' divide AB harmonically. **13.** If A, B, C are collinear and A', B', C' are collinear, then $AB' \cap BA'$, $BC' \cap CB'$, and $AC' \cap CA'$ are collinear. **14.** Take appropriate vectors $\mathbf{p}$ and $\mathbf{q}$ parallel to the two lines; they span $\vec{\mathcal{P}}$. Take position vectors relative

294

to O; X has position vector $(r + x)\mathbf{p} + s\mathbf{q}$ and Y has position vector $t\mathbf{p} + (u + y)\mathbf{q}$, where r, s, t, u are constants. Collinearity implies that these two vectors are linearly dependent. **15.** Use the coordinate formula for projection obtained in the previous exercise. **16.** Student B is right. Student A forgot that parallel lines need not look parallel in projection.

Chapter 3.   1. Use the uniqueness of perpendiculars. **2.** Use the uniqueness of perpendiculars, again. **3.** Each centre is equidistant from the two points of intersection. **4.** $ABCD$ need not be a parallelogram; consider what might happen if B and D were both on the same side of $\overleftrightarrow{AC}$. **5.** Either imitate the proof of the symmetry of the dot product, or just use the symmetry and linearity properties of the dot product.

Chapter 4.   1. By definition, $\mathbf{a} \cdot \mathbf{b} = \frac{1}{2}$. **2.** $\lambda = -2$; what if $\mathbf{b} \cdot \mathbf{c} = 0$? **3.** $\frac{\pi}{6}, \frac{\pi}{3}$, and $\frac{\pi}{2}$. **4.** $2\mathbf{i} - \mathbf{j}$ (or any positive multiple). **5.** Expand using the linearity of the dot product and notice that the terms in $\mathbf{u} \cdot \mathbf{v}$ cancel. **6.** The intersection of the bisectors of AB and BC is equidistant from A, B, and C. **7.** For the first method, take D to be the point of intersection of two altitudes and use the identity to show that the third altitude also passes through it. For the second, observe that the altitude from B divides AC in the ratio $\tan A : \tan C$. For the third, $A'B'C'$ is the triangle the midpoints of whose sides are A, B, C, as in Exercise 2.5.9. **8.** T is an isometry, in fact a reflection in the line or plane perpendicular to $\mathbf{n}$ through the origin. **9.** The centre has position vector $\mathbf{b}$. **10.** Let A, B, and C have position vectors $\mathbf{a}$, $\mathbf{b}$, and $\mathbf{c}$ relative to D as origin. Expand $|AC|^2 = |\mathbf{a} - \mathbf{c}|^2$ and $|BC|^2 = |\mathbf{b} - \mathbf{c}|^2$ using the dot product. Notice that since D is between A and B, $|AB| = |\mathbf{a}| + |\mathbf{b}|$ and $|\mathbf{a}|\mathbf{b} + |\mathbf{b}|\mathbf{a} = \mathbf{0}$. **11.** This is just the Cauchy–Schwarz inequality. **12.** BCD has the same angles as ABC; the ratios of the sides are therefore the same. For the last part, use the cosine rule to express $\cos A$ in terms of a/b. **13.** Show that a similarity preserves midpoints, and then imitate the proof of 4.5.5. **14.** Represent points by their position vectors relative to origin O, and suppose that the two planes are $U[X]$ and $U[X']$, U being a two-dimensional vector subspace and OXX' being a line perpendicular to the planes. Say $\mathbf{x}' = \rho\mathbf{x}$. If $T(A) = A'$ then $\mathbf{a}' = \lambda\mathbf{a}$ for some λ and $\mathbf{a} - \mathbf{x} \in U$, $\mathbf{a}' - \mathbf{x}' \in U$. Take dot product with $\mathbf{x}$ to find that $\lambda = \rho$.

Chapter 5.   1. D has coordinates $(2, 4, 6, 7)$. Two independent vectors perpendicular to the plane $ABCD$ are those with components $(1, -2, 1, 0)$ and $(-1, -1, 0, 1)$; you may have found other possibilities which are linear combinations of these. **2.** The points of intersection have coordinates $x = \frac{1}{2c}(a^2 + c^2 - b^2)$ and $y = \pm\frac{1}{2c}\sqrt{(a+b+c)(a+b-c)(b+c-a)(c+a-b)}$. For this square root to be real, the expression under the square root sign must be positive, so either 0, 2, or 4 of the factors must be positive. The possibilities that 0 or 2 are positive can be ruled out because they would imply that one of a, b, or c was negative. Geometrically, this is the triangle inequality. **3.** $(4c^2 - 16)x^2 + 4c^2y^2 = c^4 - 4c^2$. **4.** The equation is $(x^2 - 1)^2 + 2y^2(x^2 + 1) + y^4 = c^2$. To investigate the shape of the curve, try to find y^2 as a function of x. What is the condition for the quadratic equation for y^2 to have a positive root? **5.** $l\mathbf{i} + m\mathbf{j}$ is a unit vector perpendicular to the line. If $(x_0 + tl, y_0 + tm)$ lies on the line, what is t? **6.** If P has coordinates (x, y) then A has coordinates $(x, 2a)$ and B has coordinates $(\sqrt{2ay - y^2}, y)$. Write down the condition for these two points to be collinear with the origin. **7.** The key is to use the formula for the product of the roots of a quadratic equation in terms of the coefficients. **9.** $(x, y) = (\frac{3+\sqrt{5}}{12}, \frac{3-\sqrt{5}}{12})$ or vice versa. **10.** The possibilities are rotation through an angle $- \cos^{-1}(4/5)$ and reflection in the line $y = 3x$. **11.** Apply the 'first isomorphism theorem' to the homomorphism from isometries to orthogonal matrices which sends T to the matrix of its vectorialization. **12.** The line becomes $5x' + y' + 4\sqrt{2} = 0$. For the second part, any coordinate system will do whose second basis vector is $\pm(2\mathbf{i} + 3\mathbf{j})/\sqrt{13}$. **13.** There are many possibilities for such a coordinate system; the third basis vector has to be $\pm(\mathbf{i} + \mathbf{j} - \mathbf{k})/\sqrt{3}$. The curve of intersection is a circle. **14.** Express everything in terms of a basis p, q, where $\mathsf{U}\mathsf{p} = \mathsf{p}$ and $\mathsf{U}\mathsf{q} = -\mathsf{q}$. **15.** What must the isometry do to a unit vector perpendicular to H?

Chapter 6. 1. Find the position vectors of the centres of the squares in terms of the orientation operation $\perp$. **2.** The formula is $\mathbf{v} \mapsto \cos\theta\mathbf{v} + \sin\theta\mathbf{v}^{\perp} + \mathbf{u}$. Equate both sides and remember that $(1 - \cos\theta)/\sin\theta = \tan(\theta/2)$. **3.** Use induction on n, splitting off one triangle at time. The tricky part is to show that if $A_1 \ldots A_n$ is convex, then so is $A_1 \ldots A_{n-1}$. **4.** Use a 'total turning' argument. **5.** The locus consists of two perpendicular lines passing through the point of intersection of $\mathcal{L}_1$ and $\mathcal{L}_2$ and bisecting the angles between them. **6.** Use the previous question; the (internal and external) bisectors of the angles meet by threes to give the four required points. The existence of the incircle comes from the fact that the perpendicular distances from the incentre to the three sides are equal. **7.** From the construction, $|AY| = |AZ|$ and so on; now apply Ceva's theorem. **8.** $r = 2/(1 - \sin\theta)$. **9.** $r = a(1 + \cos\theta)$, relative to an initial line in the direction from O to the centre of the circle. **10.** Use a Möbius transformation to show that it is enough to prove the result for the case of *concentric* circles. **11.** Let X be the point of intersection of AC and BD, and use Pythagoras' theorem. For the last part, use the Möbius transformation $z \mapsto 1/z$. **14.** If it were zero, then $\sqrt{s}$ would be in the field $\mathcal{K}$.

Chapter 7. 1. The tangent line to the cissoid has equation $y_0(y - y_0) + x_0^2(x_0 - 3a)(x - x_0)/(x_0 - 2a)^2 = 0$; the only singular point is $(0, 0)$. The lemniscate has tangent line $(2(x_0^2 + y_0^2) - a^2)x_0 x + (2(x_0^2 + y_0^2) + a^2)y_0 y = a^2(x_0^2 - y_0^2)$; the only singular point is $(0, 0)$. **2.** $(2 - t_0^3)t_0 x - (1 - 2t_0^3)y = 3t_0^2$. **3.** Trigonometry gives $y = a\sin\varphi$, and $dy/dx = -\tan\varphi$ because φ is the slope of the tangent line. Thus $dx/d\varphi = -a\cot\varphi\cos\varphi = a\sin\varphi - a\csc\varphi$, and now integrate and check the constant. **4.** $x^2 + y^2/5 = 1$; $x^2/5 - y^2/20 = 1$; $x^2/13 + y^2/49 = 4$. **5.** Use the polar equation. **6.** The first equation says that (x_0, y_0) lies on the line, the second that the line is perpendicular to $\nabla(x^2/a^2 + y^2/b^2)$. An alternative solution to the last part: look for the condition that the quadratic describing the intersections of line and ellipse has a repeated root. **7.** Ellipse, eccentricity $\sqrt{3}/2$; hyperbola, eccentricity $3/\sqrt{7}$; ellipse, eccentricity $1/2$; parabola; single line counted twice; ellipse, eccentricity $\sqrt{2/3}$; two intersecting straight lines. **8.** To find the orthocentre, solve simultaneously the equations for two of the altitudes; to find the circumcircle, write its equation as $x^2 + y^2 - 2ax - 2by + k = 0$ and solve three simultaneous equations to find a, b, and k. **9.** There is just one, and its equation is $a^{1/3}x + b^{1/3}y + (ab)^{2/3} = 0$. **10.** Represent the parabola by $y^2 = 4ax$, let $P = (at^2, 2at)$ be a point on it, and let $\mathbf{t} = (t\mathbf{i} + \mathbf{j})/\sqrt{1 + t^2}$ be the unit tangent vector at P. It is sufficient to show that $\mathbf{t} \cdot \mathbf{i} = \mathbf{t} \cdot \mathbf{u}$, where $\mathbf{u}$ is a unit vector in the direction from the focus to P. **11.** Represent the hyperbola by the equation $xy = c^2$ relative to the asymptotes as axes. Let $(ct_i, c/t_i)$, $i = 1, 2, 3$, be the three given points. Show that the three altitudes have equations $y + cT = (Tx + c)/t_i$, where $T = t_1 t_2 t_3$. These meet at $(-c/T, -cT)$ which is on the hyperbola. **12.** The trigonometric parameterization shows that an ellipse is a continuous image of a circle, so it's compact and connected. The parabola and hyperbola are unbounded, hence noncompact. A parabola is connected because it's a continuous image of the real line; a hyperbola isn't because the two branches partition it into disjoint nonempty relatively open subsets. **13.** $(-(D + Et)/(A + 2Bt + Ct^2), -t(D + Et)/(A + 2Bt + Ct^2))$. **14.** An obvious one is $(1, 1)$. The next simplest is the point where the tangent at $(1, 1)$ meets the cubic again, which is $(-5/4, -7/20)$. **15.** The conic in question is the hyperbola $X^2 - 3Y^2 + X + Y + 5 = 0$, where $X = 1/x$ and $Y = 1/y$.

Chapter 8.   1. s $-$ **r** and **t** $-$ **r** must be parallel. **2.** Use the product rule for determinants. **3.** Use the formula for the vector triple product. **4.** The proof is analogous to that of 6.2.7. **5.** It is easy to check that the given vector *is* a solution; it is the only solution because the simultaneous equations are non-singular. **6.** Work out $[\mathbf{a} \wedge \mathbf{b}, \mathbf{b} \wedge \mathbf{c}, \mathbf{c} \wedge \mathbf{a}]$. **7.** Expand out the left hand side by using the expansion for the vector quadruple product. You will also need to use the fact that the scalar triple product of three vectors is unchanged by cyclic permutation of the vectors. In the last part, use the fact that three vectors are linearly dependent if and only if their scalar triple product is zero. **8.** Either solve simultaneous equations, or construct the point geometrically as the intersection of three planes which are the perpendicular bisectors of three noncoplanar line segments between the points. Centre $(1, 2, 3)$; radius 5. **9.** Consider the family of planes $\mathbf{r} \cdot \mathbf{n} = k$, where $\mathbf{n} = \mathbf{a}_1 \wedge \mathbf{a}_2$. **10. a** is a scalar multiple of $(\mathbf{b}_1 - \mathbf{r}_0 \wedge \mathbf{a}_1) \wedge (\mathbf{b}_2 - \mathbf{r}_0 \wedge \mathbf{a}_2)$. **11.** Axis $\mathbf{j} + \mathbf{k}$; angle $- \cos^{-1}(1/3)$. **12.** Because every orthogonal matrix with determinant $+1$ represents a rotation. Axis $(\mathbf{i} + \mathbf{j} + \mathbf{k})/\sqrt{3}$; angle $2\pi/3$. **13.** Rotation, axis $(3\mathbf{i} + \mathbf{j} + 3\mathbf{k})/\sqrt{19}$, angle $- \cos^{-1}(31/50)$; reflection, in plane perpendicular to $(3\mathbf{i} + 4\mathbf{k})/5$. **14.** Use the vector definition of a rotation (8.3.2) to show that $[\mathbf{n}, \mathbf{v}, T\mathbf{v}] = |\mathbf{n} \wedge \mathbf{v}|^2 \sin \theta$, where θ is the angle of rotation. **15.** The matrix $-M$ represents a rotation. **16.** The identity, reflections, rotations through π, and the central inversion. **17.** Think about an eigenvector of the linear transformation given by multiplying by a fixed element of A. **19.** Use the formula $|q_1 q_2| = |q_1||q_2|$. **20.** Verify that $x_\alpha y_\beta x_\gamma = [\cos(\alpha + \gamma) + \sin(\alpha + \gamma)\mathbf{i}] \cos \beta + [\cos(\alpha - \gamma)\mathbf{j} + \sin(\alpha - \gamma)\mathbf{k}] \sin \beta$, then use A.1.7 three times.

Chapter 9.   1. $2 \sinh a$. **2.** $\pi h^3/2$; less than volume of hemispherical bowl. **3.** Show that the semicircle on AC as diameter has the same area as the quarter-circle with centre B and radius $|AB| = |BC|$. **4.** Let $z = \lambda x + (1 - \lambda)y$. Use the mean value theorem to compare $(f(z) - f(x))/(z - x)$ with $(f(y) - f(z))/(y - z)$. **5.** Use formula 9.2.8. **6.** If $\mathbf{u}^\perp \cdot \mathbf{v} > 0$, then **v** is in an anticlockwise direction from **u**. **7.** Let γ be the piecewise regular curve given by the three sides of the triangle. **8.** The area is $3/2$. Parameterize the folium as in 7.5.3, and use 9.6.5 in the form area $= (1/2) \int x\, dy - y\, dx$. **9.** Let l be the length of a diagonal of the cyclic quadrilateral, and let θ and ψ be the angles opposite the diagonal. By concyclicity $\theta + \varphi = \pi$. Use the cosine rule twice to get two formulae for $\cos \theta = - \cos \varphi$, and eliminate l^2 between the two equations. Now express the area in terms of the sides and $\sin \theta = \sin \varphi$. **10.** Use 9.2.8, together with the calculation of the intersection of the tangents in Exercise 7.6.8. **12.** Formula 9.2.8 shows that the area of ABC is half an integer. Moreover, ABC lies inside the region bounded by the line segment AC and the arc of the circle from

A to C; the area of this region is $r^2(\theta - \sin\theta)$. **15.** To prove that the circular cylinder is contented, approximate it from outside and from inside by cylinders on polygonal bases, whose volume can be worked out by Cavalieri's principle.

Chapter 10. **1.** Hyperboloid of revolution (one sheet); hyperboloid of one sheet; circular cone; elliptic paraboloid. **2.** The eccentricity is $\sqrt{2/5}$. The substitution $z = 1 + y/2$ gives the equation of the projection of the conic onto the xz plane; projection changes the eccentricity. **3.** To show that the principal axis of length r is the middle one, you must show that the corresponding eigenvalue is the middle one. Use the fact that if a quadratic expression $\lambda^2 + K\lambda + L$ is negative for $\lambda = \lambda_0$, then it has one root on either side of λ_0. **4.** Write down the equations for the points $a \pm u$ to lie on the quadric, and subtract one from the other. If the matrix is nonsingular, the quadric is central, and the centre lies on every diametral plane. **5.** The distance is $2b(1 - a^2/b^2)\sqrt{1 - k^2/c^2}$. **7.** The expression evaluates both to $\lambda u'v$ and to $\mu u'v$. **8.** One can always solve nine homogeneous linear equations in the ten unknowns A, B, C, D, E, F, P, Q, R, k. Given the three skew lines, choose three points on each and find a quadric going through the nine points thus described. **9.** Clearly the surface has rotational symmetry about the z-axis, and it meets the plane $y = 0$ in a curve with equation $((x - a)^2 + z^2 - b^2)((x + a)^2 + z^2 - b^2) = 0$, that is two circles. For the last part introduce coordinates y and t in the given plane, where $z = bt/a$ and $x = (\sqrt{a^2 - b^2})t/a$.

Chapter 11. **1.** To give the proof, differentiate and use the fact that the scalar triple product of three coplanar vectors is zero. Mechanically, this shows that the angular momentum about the axis $\mathbf{n}$ is conserved. **2.** Make the same substitutions as in Section 11.2 and find that u executes simple harmonic motion with period $2\pi/\sqrt{1 - a/h^2}$. **3.** The force is proportional to the inverse fifth power of the distance. **4.** Length $\sqrt{2}a\sinh k$; curvature $1/(2a\cosh^2 t)$; torsion the same as curvature. **5.** $\mathbf{t} = \frac{1}{2}(1 + t)^{1/2}\mathbf{i} - \frac{1}{2}(1 - t)^{1/2}\mathbf{j} + \frac{1}{\sqrt{2}}\mathbf{k}$; $\mathbf{n} = \frac{1}{\sqrt{2}}(1 - t)^{1/2}\mathbf{i} + \frac{1}{\sqrt{2}}(1 + t)^{1/2}\mathbf{j}$; and $\mathbf{b} = -\frac{1}{2}(1 + t)^{1/2}\mathbf{i} + \frac{1}{2}(1 - t)^{1/2}\mathbf{j} + \frac{1}{\sqrt{2}}\mathbf{k}$. **6.** The curvature is $\frac{1}{4}\sqrt{1 + \sin^2(t/2)}$. **7.** Watch out — s is not equal to arc length along the new curve. **8.** Use the first and third Serret–Frenet formulae to show that the first three conditions are equivalent. Show they imply the fourth by dotting the second Serret–Frenet formula with a unit vector in the fixed direction. If κ/τ is constant, show that the vector $\tau\mathbf{t} + \kappa\mathbf{b}$ has constant direction. **9.** Let $\mathbf{r}(s)$ be the position vector of $\gamma(s)$ relative to O. Look at the condition for $|\mathbf{r}(s)|^2$ to have a local maximum in terms of the second derivative. Apply the Cauchy–Schwarz inequality. **10.** Consider the derivative of the vector $\mathbf{r} + (1/\kappa)\mathbf{n} - (\kappa'/\tau\kappa^2)\mathbf{b}$. **11.** Calculate the second derivative $\mathbf{n}''$ and solve the resulting differential equation. **12.** A semicircle, area l^2/π. To prove it, apply the ordinary isoperimetric inequality to the region enclosed by the given curve and its mirror image.

Chapter 12. **1.** Work out $\sigma_u \wedge \sigma_v$ using the Serret–Frenet formulae. If r were greater than the radius of curvature, then the surface would intersect itself somewhere and would no longer be regular. **2.** Use the chain rule for partial derivatives. **3.** For the catenoid, $E = G = \cosh^2 u$ and $F = 0$; for the helicoid, $E = 1 + v^2$, $F = 0$, and $G = 1$. An isometry takes the point $(u, \cosh u \cos v, \cosh u \sin v)$ on the catenoid to the point $(v, \sinh u \cos v, \sinh u \sin v)$ on the helicoid. **4.** Use the Serret–Frenet formulae again; for the area, work out $EG - F^2$. **5.** Use 12.2.3. **7.** Parameterize the tractoid by $(\log \cot(u/2) - \cos u, \sin u \sin v, \sin u \sin v)$. The first fundamental form then has $E = \cot^2 u$, $F = 0$, and $G = \sin^2 u$. **8.** The geodesics are helices (which become straight lines when the cylinder is 'unwrapped'). **9.** Parameterize the torus by $((a + b \cos u) \cos v, b \sin u, (a + b \cos u) \sin v)$. Then the first fundamental form has $E = b^2$, $F = 0$, $G = (a + b \cos u)^2$, and the Gaussian curvature is $\cos u / (b(a + b \cos u))$. **10.** Consider areas and use 12.2.3. The possibilities are $p = q = 2$, r arbitrary ($4r$ triangles) or $p = 2$, $q = 3$, $r = 3, 4$, or 5 (24, 48, or 120 triangles respectively).

Bibliography

The fundamental ancient work on geometry is Euclid's *Elements*. Heath's translation [22] is not hard to obtain and provides a wealth of historical and mathematical commentary. You can discover more about the history of the subject in the books by Gray [19], Hollingdale [24], Kline [27], and Stillwell [38].

Modern axiomatic approaches to geometry may be found in Blumenthal [7], Cederberg [9], and Moise and Downs [31]. The first two of these discuss projective and non-Euclidean geometries. The third is a detailed textbook intended for US high schools; it is based on a system of axioms for geometry introduced by Birkhoff [6]. More advanced 'classical' treatments of geometry are Coxeter [11] and Eves [14].

We have based most of our study of geometry on linear algebra. Berger [4] gives a more comprehensive development of geometry from this perspective. A shorter book covering Euclidean, projective, and hyperbolic geometry is Rees [34]. The books by Cohn [10] and Pogorelov [33] discuss three-dimensional coordinate geometry, scalar and vector products, conics and quadrics. A beautiful account of projective geometry, requiring some algebraic sophistication, is Samuel [36].

Two recent textbooks will introduce the student to the delights of algebraic geometry, which we touched on in our short tour around some higher-degree curves: Kirwan [26] (for the classical case of curves over **C**) and Reid [35]. The historical discussion in Stillwell [38] is also recommended.

To go further with the study of area and volume one really needs Lebesgue measure theory. Two possible texts here are Halmos [21] and Weir [42]. The book by Wagon [41] will tell you about some of the paradoxes that afflict a naive approach to the theory of volume.

A comprehensive reference for differential geometry is do Carmo [12]; there is a brief treatment in Pogorelov [33]. Faux and Pratt [15] discuss the practical use of differential geometry in the design and manufacture of curved objects.

Many approaches to geometry overlap in the study of (abstract) surfaces of *constant* curvature. A modern account which emphasizes the topological and group-theoretic aspects is Stillwell [39].

Details on the construction of the trigonometric functions can be found in Ebbinghaus *et al.* [13]; this book also contains much more information on the quaternions and other 'generalized number systems'.

Finally, the dictionary by Wells [43] lives up to its title!

1.   E. A. Abbott. *Flatland: A Romance of Many Dimensions*. Seeley & Co., London, 1884.

2.   D. J. Acheson. *Elementary Fluid Dynamics*. Oxford University Press, 1990.

3.   T. M. Apostol. *Calculus (volume II)*. John Wiley, New York, second edition, 1969.

4.   M. Berger. *Geometry*. Springer-Verlag, New York, 1987. Two volumes.

5.   K. G. Binmore. *Mathematical Analysis: A Straightforward Approach*. Cambridge University Press, second edition, 1982.

6.   G. D. Birkhoff. A set of postulates for plane geometry, based on scale and protractor. *Annals of Mathematics*, 33:329–345, 1932.

7.   L. G. Blumenthal. *A Modern View of Geometry*. Dover, New York, 1961.

8.   D. E. Bourne and P. C. Kendall. *Vector Analysis and Cartesian Tensors*. Nelson, London, 1977.

9.   J. N. Cederberg. *A Course in Modern Geometries*. Springer-Verlag, New York, 1989.

10.   P. M. Cohn. *Solid Geometry*. Routledge and Kegan Paul, London, 1961.

11.   H. S. M. Coxeter. *Introduction to Geometry*. John Wiley and Sons, New York, 1961.

12.   M. P. do Carmo. *Differential Geometry of Curves and Surfaces*. Prentice-Hall, Englewood Cliffs, New Jersey, 1976.

13.   H.-D. Ebbinghaus *et al. Numbers*. Springer-Verlag, New York, 1990.

14.   H. Eves. *A Survey of Geometry*. Allyn and Bacon, Boston, revised edition, 1972.

15.   I. D. Faux and M. J. Pratt. *Computational Geometry for Design and Manufacture*. Ellis Horwood, Chichester, 1979.

16.   R. P. Feynman. *QED: The Strange Theory of Light and Matter*. Penguin, Harmondsworth, 1990.

17.   M. Gardner. *The Ambidextrous Universe*. Penguin, Harmondsworth, second edition, 1982.

18.   E. Gombrich. *The Story of Art*. Phaidon, London, fifteenth edition, 1989.

19.   J. Gray. *Ideas of Space: Euclidean, Noneuclidean, and Relativistic*. Oxford University Press, second edition, 1988.

20.   P. R. Halmos. *Finite-Dimensional Vector Spaces*. Springer-Verlag, New York, 1974.

21.   P. R. Halmos. *Measure Theory*. Springer-Verlag, New York, 1974.

22.   T. L. Heath. *The Thirteen Books of Euclid's Elements*. Dover, New York, 1956. Three volumes.

23. D. Hilbert and S. Cohn-Vossen. *Geometry and the Imagination*. Chelsea Publishing Company, New York, 1952.

24. S. Hollingdale. *Makers of Mathematics*. Penguin, Harmondsworth, 1989.

25. W. Kaplan. *Advanced Calculus*. Addison-Wesley, Reading, Massachussets, third edition, 1984.

26. F. C. Kirwan. *Complex Algebraic Curves*. Cambridge University Press, 1992.

27. M. Kline. *Mathematics in Western Culture*. Oxford University Press, 1953.

28. S. Lang. Old and new conjectured diophantine inequalities. *Bulletin of the American Mathematical Society*, 23:37–76, 1990.

29. S. Lipschutz. *Linear Algebra*. Schaum's Outline Series. McGraw-Hill, New York, 1974.

30. C. W. Misner, K. S. Thorne, and J. A. Wheeler. *Gravitation*. W. H. Freeman, San Francisco, 1973.

31. E. E. Moise and F. L. Downs. *Geometry*. Addison-Wesley, Menlo Park, California, 1967.

32. A. O. Morris. *Linear Algebra: An Introduction*. Van Nostrand Reinhold, Wokingham, England, second edition, 1982.

33. A. Pogorelov. *Geometry*. Mir, Moscow, 1987.

34. E. P. Rees. *Notes on Geometry*. Springer-Verlag, New York, 1983.

35. M. Reid. *Undergraduate Algebraic Geometry*. Cambridge University Press, 1988.

36. P. Samuel. *Projective Geometry*. Springer-Verlag, New York, 1988.

37. I. Stewart. *Galois Theory*. Chapman and Hall, London, second edition, 1989.

38. J. Stillwell. *Mathematics and its History*. Springer-Verlag, New York, 1989.

39. J. Stillwell. *Geometry of Surfaces*. Springer-Verlag, New York, 1992.

40. W. Sutherland. *Introduction to Metric and Topological Spaces*. Oxford University Press, Oxford, 1975.

41. S. Wagon. *The Banach–Tarski Paradox*. Cambridge University Press, 1985.

42. A. J. Weir. *Lebesgue Integration and Measure*. Cambridge University Press, 1973.

43. D. F. Wells. *The Penguin Dictionary of Curious and Interesting Geometry*. Penguin, Harmondsworth, 1991.

Index